ALLE ZEIT WACH
1842

Aleksey N. Khramov

Paleomagnetology

Technical Editor
D. H. Tarling

With 117 Figures

Springer-Verlag
Berlin Heidelberg New York
London Paris Tokyo

Professor Dr. ALEKSEY N. KHRAMOV
All-Union Oil Research
Geological Prospecting Institution
VNIGRI
Liteiny 39
Leningrad 191104, USSR

Translated from Russian by:

VALENTINA P. GRUDINA
All-Union Geological Research Institute
VSEGEI
Sredney Prospect 74
Leningrad 199026, USSR

Technical Editor

Professor Dr. D. H. TARLING
Plymouth Polytechnic
Faculty of Science
Department of Geological Sciences
Drake Circus
Plymouth, Devon, PL4 8AA, Great Britain

Title of the original Russian edition:
Paleomagnitologiya

ISBN-13:978-3-642-71752-9 e-ISBN-13:978-3-642-71750-5
DOI: 10.1007/978-3-642-71750-5

Library of Congress Cataloging-in-Publication Data. Paleomagnitologiîa. Englisch. Paleomagnetology. Translation of: Paleomagnitologiîa. 1. Paleomagnetism. I. Khramov, A. N. (Aleseĭ Nikitich) II. Tarling, D. H. (Donald Harvey) III. Title. QE501.4.P35P3613 1987 538'.7 87-9810

Softcover reprint of the hardcover 1st edition 1987

Typesetting: Overseas Typographers, Inc., Makati, Philippines.

2132/3130-543210

Preface

Paleomagnetology is a separate branch of science with its own particular methods of study, with definite objects and spheres of application for its data.

Paleomagnetology is the study of the phenomena of paleomagnetism, that is the geomagnetic field of the past as recorded in the characteristic traces of this field: vectors of natural remanent magnetization of rocks. The geomagnetic field of the past is examined in both space and time, its laws being recorded by becoming "frozen" in rocks, which preserve these traces. All these phenomena find application in studies of the Earth's structure, its evolution and in practical geology. Thus, paleomagnetology, like any other field in the geosciences, has both direct and indirect applications. The direct applications are related to studies of the geologic history of the geomagnetic field; the indirect ones help to solve geologic questions by establishing laws in the distribution of the geomagnetic field and examining changes in then over the course of time.

In any paleomagnetic study the tasks to be solved are complex and various, as paleomagnetology is quite a new scientific branch and lies at the meeting point of geology with physics and geophysics. Thus it is urgent that a book should be made available which encompasses the basic principles and methods of paleomagnetology with reference to the available results and general recommendations for paleomagnetic procedures.

Those few books that embrace all these aspects were published 10–20 years ago and have become out of date due to the rapidly developing paleomagnetic studies. Examples are the classical monography of Irving (1964) as well as other, later publications (Khramov and Sholpo 1967; Collinson et al. 1967; Krs 1969; Tarling 1971; McElhinny 1973; Belokon' et al. 1973). The works that appeared still later (Khramov et al. 1974; Vander Voo and Channell 1980; Tarling 1983; Tretyak 1983) and others deal with important problems of paleomagnetology and related branches of science, but do not meet this need. However, in the book of Yanovsky (1978) *The Earth's Magnetism*, there is a short chapter on paleomagnetology and its applications, which is quite useful for workers in this field, although still inadequate.

While writing this book, the authors took into account that the problems of rock magnetism have been examined in detail in several monographies (Stacey and Banerjee 1974; Pechersky et al. 1975; Sholpo 1977; Kudryavtseva et al. 1982); methods and procedures of paleomagnetic studies are also the subject of the book by Collinson (1983).

A brief description of techniques and instrumentation can also be found with regard to problems of archeomagnetism; the readers are referred to Burlatskaya 1965 and Creer et al. 1983.

The following authors contributed to this book: Preface and Introduction – A.N. Khramov, Chapter I – S.A. Pisarevsky and A.N. Khramov, Chapter 2 – R.A. Komissarova, I.A. Pogarskaya, Yu.S. Rzhevsky and A.N. Khramov, Chapter 3 – R.A. Komissarova, S.A. Pisarevsky, I.P. Slautsitays, V.P. Rodionov and A.N. Khramov, Chapter 4 – A.N. Khramov, Chapter 5 – Yu.S. Rzhevsky, A.N. Khramov and V.P. Rodionov, Chapter 6 – G.I. Goncharov, Chapter 7 – R.A. Komissarova, I.A. Pogarskaya and A.N. Khramov; all authors cooperated in compiling the supplement. N.N. Forsh (VNIGRI) wrote Section 5.3 in this book, M.A. Pevzner (GIN Acad. Sci. USSR) – Sections 3.I.6 (in part) and 3.4.I, D.M. Pechersky (IFZ Acad. Sci. USSR) – Sections 3.2.3, 3.2.5 (in part) and 3.2.6. D.M. Pechersky, G.Z. Gurary, S.V. Shipunov and M.L. Bazhenov (GIN Acad. Sci. USSR) looked through the manuscript and made certain corrections and remarks. E.P. Osipova and M.T. Nemova (VNIGRI) also took part in preparation of the book. The authors extend their thanks to all workers who were envolved in this job.

While preparing this English version of the book, the authors have kept mainly to the text of the 1982 edition; certain corrections and supplements have, however been inserted with the aim of including the more important recent results in particular, the data presented at the 27th session of the IGC, Moscow, April 1987.

Leningrad, June 1987 ALEKSEY N. KHRAMOV

Contents

Symbols Used in Paleomagnetology

Magnetization of Rocks

J_n	Natural remanent, NRM
J_n^s	Initial
J_n^o	Primary (a part of the initial preserved up to the present)
J_n^h	Secondary
J_n^a	Ancient
J_n^m	Metachronous or ancient secondary
J_n^r	Relict
J_n^c	Synchronous
J_{nh}, J_{nt}, $J_{n\tau}$, J_{nc}	Characteristic, ChRM (component of J_n isolated after a corresponding cleaning)
J_{nv}	Viscous, VRM
J_i	Induced
J_r	Isothermal, IRM
J_{rt}	Thermoremanent, TRM
J_{rc}	Chemical, CRM
J_{ro}	Detrital, DRM
J_s	Saturation
J_{rs}	Saturation remanent
J_{ri}	Anhysteretic, ARM
$J_{T_1 H}^{T_2}$, J_{rpt}	Partial thermoremanent, PTRM
H_T	Total geomagnetic field vector
H	Steady magnetizing field
H_s	Saturation field
H'_{cs}	Coercivity of maximal remanence or saturation destructive field
H_c	Coercive force
H'_c	Coercivity of NRM or destructive field
h	Peak strength of alternating magnetic field
t	Time
τ	Relaxation time
T_b	Blocking temperature
d_b	Blocking diameter of particles
V_b	Blocking volume

Polarity of Field and of Rock Magnetization

N, n	Normal
R, r	Reversed
D	Declination of Geomagnetic field or magnetization
I	Inclination of geomagnetic field or magnetization
$Q = J_n/(\varkappa H_T)$	Q-factor, Koenigsberger ratio
$\varkappa$	Magnetic susceptibility
T_C	Curie point, temperature
T	Temperature
φ	Latitude of sampling locality
λ	Longitude of sampling locality
Φ	Latitude of paleomagnetic or geomagnetic pole
Λ	Longitude of paleomagnetic or geomagnetic pole
φ_m	Geomagnetic latitude
M	Earth's magnetic moment
N	Number of unit vectors included into statistics
R	length of their geometrical sum
K	Fishers's precision parameter for directions and poles
S	Measure of paleomagnetic stability
	Radius of circle of confidence at $p = 9{,}95$:
α_{95}	For a mean vector
A_{95}	For a mean pole
θ_1, θ_2	Semiaxes of an oval of confidence for a mean pole

Symbols Used in Figures

Projection of J vectors on the lower hemisphere

Projection of J vectors on the upper hemisphere

Projections of mean directions

Intersection points of remagnetization circles

Projections of remagnetization directions

Normal polarity zones

Reversed polarity zones

Unsampled strata

Mixed polarity zones

Transitional beds

Stratigraphic correlation lines

Paleomagnetic correlation lines

Paleomagnetic directions and paleomagnetic latitudes

Paleomagnetic parallels

Approximate boundaries of the epi-Hercynian platforms

Introduction

The prehistory of paleomagnetic studies, as we know, started in the middle of the 19th century when the thermoremanent magnetization of lavas became known (Melloni and Arago 1853). In the course of the following 100 years it became evident that many of the magmatic and sedimentary (clastic and chemogenic) rocks might carry traces of the ancient geomagnetic field (Koenigsberger 1930; Graham 1949), and some discoveries were interpreted as evidence of recurrent polarity reversals of the geomagnetic field and of continental drift (Mercanton 1932; Hospers 1953; Creer et al. 1957). All this has resulted in a growing interest in the phenomena of paleomagnetism, and systematic paleomagnetic studies have taken over the earlier irregular work in this field.

Lazarev (1924) first discussed the problem of the history of the geomagnetic field in the USSR, and some important preliminary work was done in the pre-war period. In the USSR, paleomagnetic studies proper started in 1953–1958, when the indirect applications of paleomagnetology – the paleomagnetic correlation of sedimentary sections – was solved for the first time (Khramov 1958). It must be noted that some years earlier Popov (1947) had shown that a similar problem in magmatic rocks could be solved by this means.

The importance of paleomagnetism is evident if we consider the fact that direct observations of the geomagnetic field cover only an extremely short time interval, from the viewpoint of geologic time. However, it is only possible to develop a physical theory of geomagnetism and to solve evolutionary problems of the Earth's interior when we understand the geologic history of the geomagnetic field. The relation between geomagnetism and the Earth's inner structure, with the processes acting in it, is quite evident, and studies of paleomagnetic phenomena have made it possible to obtain for the first time direct rather than extrapolated data on the Earth's interior of the past. The increased importance of the field of geomagnetism lies in its role in the studies of the Earth's inner structure and evolution (Petrova and Khramov 1970).

The development of geology today involves a new geotectonic hypothesis – the plate tectonic concept, that for the first time in the history of the Earth sciences combines certain fundamental geologic conceptions like rift genesis, continental drift, ocean origin and geosyncline development. The role of paleomagnetology in the development and substantiation of the plate tectonic hypothesis cannot be exaggerated (Le Pichon 1968; Dewey and Bird 1970). When the first results of systematic paleomagnetic studies became available in the 1950's, they stimulated a revival of the mobilistic concept (Irving 1956; Creer et al. 1957), and its promotion resulted in the development of the plate tectonic hypothesis. Both its components, the continental drift and the seafloor spreading hypotheses, however, were proposed without recourse to paleomagnetic data. However, only the results of paleomagnetic studies could substantiate these hypotheses, and quantify continental drift and seafloor spreading. It also became evident that paleomagnetic results might prove useful in the examination of horizontal movements on a different scale, that is of deformations associated with fold belts and island arcs and also with local fold structures (Hospers and Van Andel 1970).

Paleomagnetic studies are also useful in solving those tasks of geochronology and stratigraphy, paleogeography and structural geology that are related to practical questions of forecasting, surveying and exploration for mineral deposits and geologic mapping. In this case the emphasis must be on the elaboration of a new paleomagnetic-stratigraphic and geochronologic scale based on the history of the geomagnetic field and, above all, on the history of its reversals (Cox 1969; Khramov and Sholpo 1967; Molostovsky and Khramov 1984). Just as the fossils and organic traces provide a possibility for studying the evolution of organic life, correlating the fossil-bearing beds and determining the geologic age of these beds, "fossil geomagnetism", that is the natural remanent magnetization, J_n, of rocks makes it possible to study the history of the Earth's magnetic field, to correlate the rocks that bear these records and to determine their age. Just as any fossil or organic imprint is not an organism proper, so the direction of J_n is generally not the direction of the geomagnetic field at the time and site of rock formation, and the lengths of these vectors exhibit a complex pattern of relationships that depend on numerous factors.

The natural remanent magnetization of a rock generally comprises several components that appeared at different stages of rock formation and have also been destroyed, to varying extents up to the present. Only in the case of weakly magnetized rocks and in the absence of spontaneous reversals of their magnetization are these components consistent with the direction of the geomagnetic field at the time of rock formation. Meanwhile, at least two questions must be answered in order to solve the direct and indirect applications of paleomagnetology: the elements of the Earth's magnetism must be defined and they must be assigned to a certain geologic interval (or to a certain stage in rock formation). It is also necessary as a rule, to determine the paleogeographic coordinates of the actual site and the orientation of the paleogeographic meridian at this site. The task consists in understanding the physical and geochronologic aspects of those components of the natural remanent magnetization of the rock which can be recognized in laboratory studies and in different statistical procedures.

In interpreting the paleomagnetic data that represent a combination of vectors of natural remanent magnetization "cleaned" from the secondary components, the initial model is based on three fundamental assumptions, usually called the hypotheses of paleomagnetism (Khramov and Sholpo 1967):

1. Rocks become magnetized in the direction of the geomagnetic field at the time and site of their formation (the "freezing-in" hypothesis).

2. The acquired primary magnetization is preserved (if only partially) in a rock over the geologic time scale and can be separated from a resultant, multicomponent natural remanent magnetization of a rock (the preservation hypothesis).

3. The geomagnetic field, averaged for any time intervals of the order of 10^5 years (except the reversal epochs), is a field attributed to a dipole placed at the

centre of the Earth and oriented along its rotational axis (the hypothesis of the central axial dipole).

The validity of the first two hypotheses is not so evident as might appear at first sight. Numerous physical factors and processes might disturb the consistency between the ancient field and the component of rock magnetization that is commonly accepted as a primary magnetization. The sources of errors in paleomagnetic data appear to result from this magnetization having been acquired, not in the epoch of rock formation, but later (due to precipitation of iron oxides, for instance) under the effect of currents during sedimentation or movements of cooling lava, compaction of sediments, static or dynamic loads, and finally due to a complete substitution of primary magnetization by secondary components.

The first two hypotheses must be tested in every paleomagnetic study (for any rock type), and the extent of the disturbances that are produced under the effect of other factors established. The hypothesis of a central axial dipole represents a typical geophysical problem that can only be solved by a generalization of all paleomagnetic data available world wide and by specialized work on a specific epoch of the geologic past.

CHAPTER 1

Physical and Geophysical Aspects of Paleomagnetology

1.1 Magnetic Values and Magnetic Units

A magnetic field is characterized by two vector values: intensity H and magnetic induction B, that in the equation, given in the international SI system, are related in the form:

$$B = \mu_0 (H + J),$$

where μ_0 is the magnetic permeability of a vacuum and J is the magnetic moment of a volume unit, called medium magnetization ($J = m/V$).

These vectors are linearly relationed

$$J = \varkappa H; \qquad B = \mu_0 \mu H,$$

where $\varkappa$ is the magnetic susceptibility of a substance; μ is its magnetic permeability (relative) and $\mu = 1 + \varkappa$.

Magnetic field sources are closed electric currents and magnetized bodies. The latter are equivalent in their performance to a combination of molecular magnets or molecular current loops; in both cases they are called dipoles.

The relation between the strength of current i in a closed loop and a generated field with intensity H is given by the Biot-Savart's law which defines circular loop with a radius r as

$$H = i/(2r),$$

where H is the intensity of the field in the centre of the circular loop.

The product of the strength of current and the area covered by the loop creates a vector with a magnetic moment m so that the vectors m and H are directed along the normal to the loop area. For a circular loop

$$m = i\pi r^2.$$

The relationships between the magnetic values and the electric current thus determines the dimensions of magnetic values in SI and the relation between the SI and CGS units (Table 1.1).

The application of SI magnetic units values in paleomagnetology presents certain difficulties (McElhinny 1973), inherent in the peculiarities of the SI system, which, while highly appropriate for engineering calculations, are, strictly speaking, misleading or confusing from the viewpoint of electromagnetic units, the units of magnetization J in particular. In fact, besides the above-mentioned method of determination of the magnetic moment of the current loop, which gives the magnetization unit there is another which defines magnetic intensity in terms of the

Table 1.1. Relationships of magnetic units in SI (rationalized MKS) and CGS systems

Quantity	Symbol	Unit CGS	System SI	Conversion CGS/SI	Factors SI/CGS
Magnetic field intensity	H	Oe (Oersted)	Am^{-1}	79.58 $[10^3/(4\pi)]$	12.57×10^{-3} $(4\pi/10^3)$
Magnetic moment	m	–	Am^2	10^{-3}	10^3
Magnetic flux	Φ	Mx (Maxwell)	Weber	10^{-8}	10^8
Magnetic induction	B	Gauss	T (tesla)	10^{-4}	10^4
Magnetic susceptibility	ϰ	–	–	12.57 (4π)	79.58×10^{-3} $[1/(4\pi)]$
Magnetization	J	–	Am^{-1}	10^3	10^{-3}

magnetic flux-Weber/m^2 (instead of A/m^2), although the CGS system uses units of the same size and dimension. The Weber/m^2 unit, called a tesla, is the unit of magnetic polarization P. Thus, there are two distinct values used in the SI for magnetic intensity-magnetization J and magnetic polarization P. The units of magnetic polarization

IT$=0.7958 \times 10^3$ units in CGS or
I nT$=0.7958 \times 10^{-6}$ units CGS.

This magnetic polarization, expressed in T or nT, and not the magnetization is used in some European countries to describe the rock magnetic state (see, for instance, Bucha 1983). This conflicts with the accepted practice in magnetic survey and magnetic mapping, where this value, related to magnetic induction, is always determined as magnetization J. On the other hand, the transition into SI intensity units in magnetic mapping is practically impossible, because the reduction factor contains the coefficient 4π. Thus, in magnetic cartography (and magnetic prospecting) the geomagnetic field is characterized by its induction, the units of which are related by the coefficient that is a multiple of 10 in the CGS and SI systems. A similar difficulty in the case of magnetization arises by the use of SI units for the magnetic susceptibility ϰ.

In this book we follow the first variant of determination of the magnetic moment of the current loop, corresponding to the units of the SI magnetic values. The Earth's magnetic field is characterized by units of magnetic induction (T) and laboratory fields by their intensity (A/m). This decision appears unavoidable and emphasizes, the fact, that the best way out of the situation lies in a return to the CGS system.

1.2 Principal Properties of the Geomagnetic Field

1.2.1 Elements of the Earth's Magnetism

The Earth's magnetic field is characterized by a geomagnetic field vector (or induction B_T) and its components which are called the geomagnetic elements

(Yanovsky 1978). The angle between the vector H_T and a horizontal plane is called the magnetic inclination and is designated I. The angle between the horizontal projection of vector H_T and the direction to the geographic north is called the magnetic declination D. The relationships between the geomagnetic elements are shown in Fig. 1.1.

$$\begin{aligned} &H = H_T \cos I;\ Z = H_T \sin I;\ I = \operatorname{arctg}(Z/H);\\ &X = H \cos D;\ Y = H \sin D;\ D = \operatorname{arctg}(Y/X);\\ &H_T^2 = H^2 + Z^2 = X^2 + Y^2 + Z^2, \end{aligned} \tag{1.1}$$

where H, X, Y and Z are the horizontal, northern, eastern and vertical components of the vector H_T, respectively.

Changes of the magnetic field at the Earth's surface are easily followed in isomagnetic maps, which can be consulted in the works of Rikitake (1966) and Yanovsky (1978). The maps of isogonic lines (lines of the same magnetic declinations) and of isoclinic lines (lines of the same magnetic inclinations) are in common use. The line along which the inclination equals zero is called the magnetic equator; the point where $I = +90°$ is called the northern magnetic pole, and the point where $I = -90°$ is called the southern magnetic pole.

The magnetic field of the Earth (the magnetic induction) is usually expressed in gauss units of the CGS system which are related to the T units of the SI system in the following way:

$$1\ G = 10^{-4} T.$$

If the values of the magnetic field are small, the old unit, gamma, is still used:

$$1\ \text{gamma} = 10^{-5}\ G = 10^{-9}\ T = 1\ nT.$$

The maximal value of the geomagnetic field on the Earth's surface is about 0.7×10^{-4}T in the area of the southern magnetic pole.

In 1839 Gauss made the first mathematic analysis of the geomagnetic field. Its best, simplest approximation appears to be the field of the geocentric dipole with an inclination to the Earth's rotational axis of 11.5°. The axis of this dipole, if

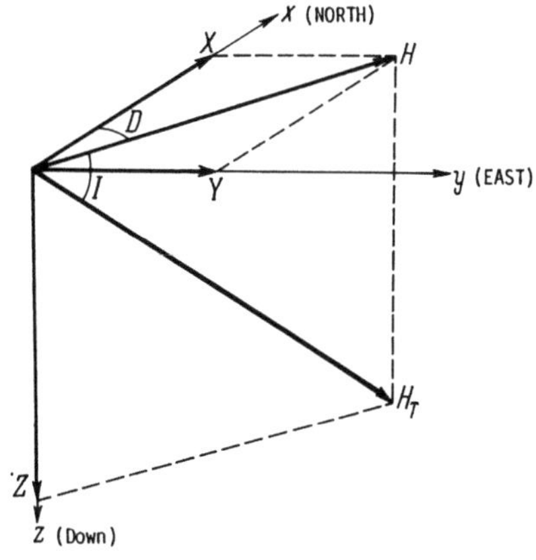

Fig. 1.1. The main geomagnetic elements (Yanovsky 1978). *D* declination; *I* inclination; *H* and *Z* horizontal and vertical components; *X* and *Y* northwards and eastwards components

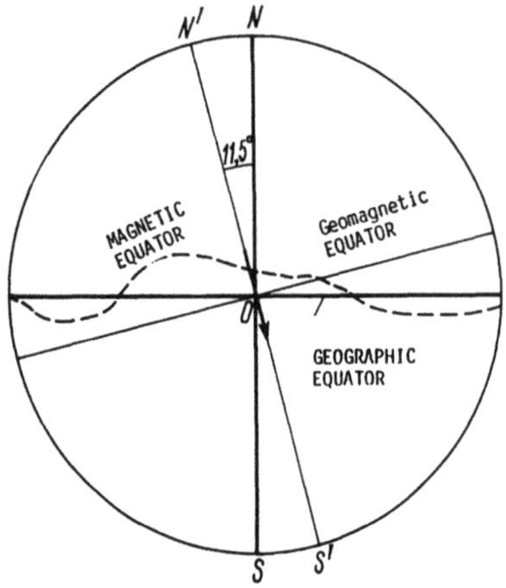

Fig. 1.2. The relationship between geographic, geomagnetic and magnetic coordinates. (After McElhinny 1973). Geographic poles: *N* northern, *S* southern; geomagnetic poles: *N′* northern, *S′* southern

continued up to the Earth's surface, cuts it at two points: 78.5°N, 290°E and 78.5°S, 110°E (Fig. 1.2). These points are called the northern and southern geomagnetic poles respectively. The locations of the geomagnetic and magnetic poles are different.

In 1634 Hellibrand noticed that the magnetic inclination changes at a given point with time. In the interval from 1580 to 1634 the inclination in London changed by 7.2°. This change exceeded the experimental errors. This phenomenon was called a secular variation of the geomagnetic field. Later, similar variations were recognized for other elements of the Earth's magnetism. Distribution of the secular variations on the Earth's surface is usually represented in the maps of isopors, i.e. the lines that connect the points with the same annual change of a given geomagnetic element (Yanovsky 1978). In these isoporic maps the closeness of lines mark areas around foci, where the secular variations occur at a maximum rate. These foci are not stable, but drift westward at a rate of about 0.2° of longitude per year.

1.2.2 Origin of the Geomagnetic Field

The magnetic field observed over the Earth's surface is composed of three components with respect to the effects they produce: the "internal" field, the source of which is within the Earth; the "external" field with its source outside the Earth; and the "vorticity" field due to electrical currents on the Earth's surface. Assuming that the "vorticity" field is negligibly small, then the magnetic field at the Earth's surface can be determined as a gradient of a certain scalar function U satisfying the Laplace equation:

$$\Delta U = 0.$$

The function U in this case is expressed as the spherical harmonic progression (Rikitake 1966; Yanovsky 1978):

$$U = R \sum_{h=1}^{\infty} \sum_{m=o}^{n} P_n^m(\cos\theta) \left\{ \left[(g\hat{e})_n^m \left(\frac{r}{R}\right)^n + (g\hat{i}) \left(\frac{R}{r}\right)^{n+1} \right] \cos(m\lambda) + \left[(h\hat{e})_n^m \left(\frac{r}{R}\right)^n + (h\hat{i})_n^m \left(\frac{R}{r}\right)^{n+1} \right] \sin(m\lambda) \right\}, \tag{1.2}$$

where R is the Earth's radius; θ, λ are the magnetic colatitude and longitude; r is a distance from the Earth's centre: $P_n^m(\cos\theta)$ is the associated Legendre polynomial of the n degree and m order; g and h are the Gauss coefficients, with the indexes e and i that designate the external and internal sources respectively.

The distribution over the Earth's surface of the Z and X (or Y) components must be known in order to calculate the Gauss coefficients. The estimates show that the coefficients calculated form the X distribution are similar to the coefficients calculated from the Y distribution. Thus, the validity of the initial supposition that the "vorticity" field is negligibly small is substantiated. Schmidt (1899) showed that the "external" field is also small compared to the internal one. Thus, the source of the major portion of the geomagnetic field is within the Earth, a finding that is vitally important for paleomagnetology.

There have been many theories to explain the sources of the magnetic field of the Earth. The reasons for their untenability are beyond the scope of this book, but the interested reader can consult, for instance, Rikitake (1986). At present, magnetohydrodynamic theories are regarded as satisfactory. However, mention will first be made of the theory related to the gyromagnetic effect. This effect concerns the observation that a rotating body magnetizes along the direction of the rotational axis. It has been shown, however, that only 10^{-10} of the Earth's magnetic field might be caused by such a gyromagnetic effect. However, dynamo theories require the presence of a weak initial poloidal (see later) magnetic field, and the gyromagnetic effect might provide this "initiating" field.

The physical basis of the dynamo theories is the presence of a conducting fluid core in the Earth and the Earth's rotation. The action of the magnetic hydrodynamo is roughly similar to the operation of the mechanism shown in Fig. 1.3. A conducting disk rotates in a weak field B_0 directed along the rotational axis. According to Faraday's law of induction an electromagnetic field appears in a disk with the direction away from its centre toward the periphery. Thus, a potential difference arises between the brush and the rotational axis (point 0′) and a current flows along the coil KO′. The magnetic field B_0, created by the coil, is parallel to the initial field B_0 and amplifies it.

The processes operating in the Earth are certainly much more complicated. The Earth's fluid core represents an almost homogeneous mass, in which it is difficult to recognize the different current loops. Cowling (1934) developed a proof of the theorem concerning the impossibility of generating a magnetic hydrodynamo when the fluid motion has an axis of symmetry. Thus, the fact that the equations describing the magnetic hydrodynamo must be three-dimensional also implies a particular complexity in any theory of the generation of the Earth's magnetic field.

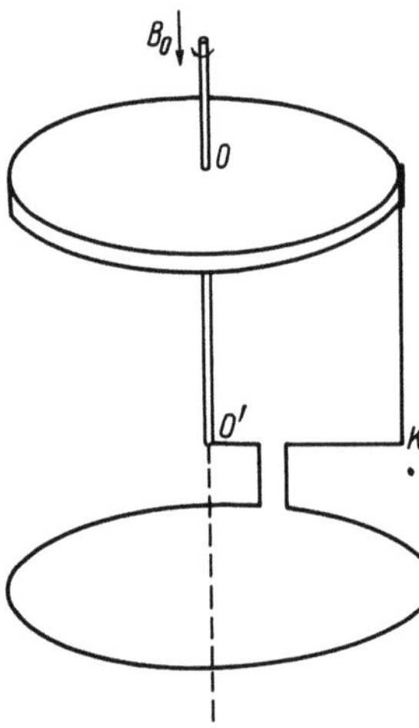

Fig. 1.3. Simple disk dynamo. (After Yanovsky 1978)

For a magnetic field of poloidal type (i.e. like the field of a dipole bar magnet) to be generated in the Earth's fluid core, at least two different types of inner motion are required. The first type of motion results in a conversion of a weak initial poloidal field into a strong (up to 0.5 T) toroidal field. The second type of motion converts the toroidal field into a poloidal field that amplifies the initial field. At present, we have many so-called kinematic models of the hydrodynamo in which the velocity field is defined, but not its origin.

First, these theories explain the origin of the toroidal field B_T. It appears to be due to the "freezing-in" of the magnetic lines of force into a highly conducting material and it is easily qualified: during the relative movement of a magnetic field and material with low electric resistance, the strong currents should be generate in this material by induction; however, this does not occur because the magnetic lines of force are carried along with the material. Bullard (1949) suggested that rotation in the fluid core is heterogeneous: the internal and external parts rotating relative to another. In this model, the toroidal field arose due to freezing-in of the magnetic lines of force (Fig. 1.4). Other explanations have been suggested but these are in poor agreement with the experimental data.

The conversion of the toroidal field into a poloidal one is more difficult to explain. It is clear that this conversion cannot take place during axially symmetric motion (as shown by Cowling). Thus, the theories explaining this phase in the

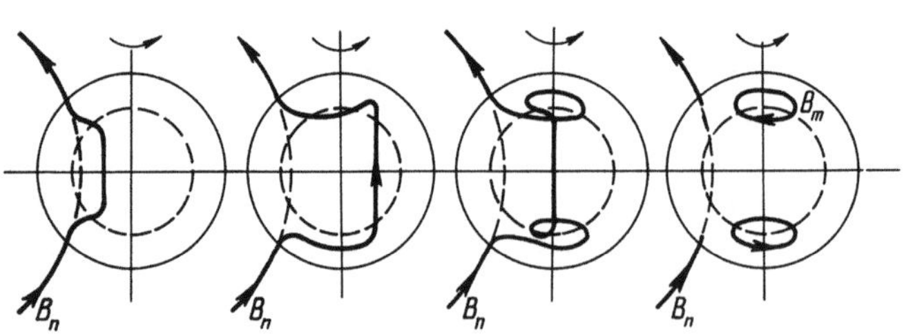

Fig. 1.4. The origin of the toroidal field. (After Yanovsky 1978)

dynamo operation generally verify the presence of convective motion in the core. The convective cells deformed by the Coriolis force, thus form toroidal field lines in such a way that a large-scale toroidal current is generated, and finally the poloidal magnetic field is formed. Two groups of theories can be recognized. The first group includes mechanisms operating under the influence of large-scale, laminar slow motions (in comparison with the rate of differential rotation in the core) existing within the core. The second group comprises theories of small-scale turbulence. More details of these theories can be found in Jacobs (1975).

As to the causes of the motions within the core resulting in dynamo action, Malkus (1963) suggested inertial effects, which appear from the precession of the Earth's axis. A model of the precession dynamo was created in terms of inertia, although its details have not yet been elaborated. Dolginov (1977) demonstrated that only this model easily explains the relationship between the parameters of the magnetic fields in the planets of the solar system. Besides, an assumption of the same order of the inertia force, the electromagnetic force and the Coriolis force results in fairly satisfactory estimation of the values of the magnetic field and of the westward drift

It must be noted, however, that Rochester et al. (1975) examined the mathematical and physical inaccuracy in the conclusions of Malkus and showed that the precession energy is at least two orders less than that required for the dynamo to operate. Thus, the probability of dynamo generation under the effect of precession is extremely low (Jacobs 1975).

Heat convection within the Earth's core might also generate the required pattern of velocities. Braginsky (1964), however, showed that in this case the heat flow from the core is incredibly great, which is related to the fact that the efficiency of the heat "engine" in the core is very low, and its power must be correspondingly greater. Braginsky maintained that convection in the core is due to the emersion of a light component (silicon) that is being permanently continually crystallized in the inner core. From the existence of convection in the core, according to the law of conservation of the angular momentum, the presence of heterogeneous rotation within it directly follows, which is necessary for the appearance of a strong toroidal field.

The model of Braginsky (1978) is one of the most elaborate models for the generation of the geomagnetic field. So-called MAC waves, in which the magnetic Archimedean and Coriolis forces are placed in equilibrium, play an important role in this model. An advantage of this model is that it not only explains the generation of the main field, but also explains the secular variations with a periodicity of about 10^3 years, as this is the same as for MAC waves. It also provides an explanation of why the geomagnetic dipole axis has a small angle relative to the Earth's rotational axis, although this angle and its direction is random. Hence, it is natural to assume that the direction of the geomagnetic axis, time averaged for $t \gg 10^3$ years, coincides with the direction of the Earth's rotational axis. This supposition agrees with both the theoretical and observed paleomagnetic data; it is known as the principle of the central axial dipole (see Sect. 1.2.3).

Notably nothing has yet been said about the polarity of the Earth's magnetic field, which might be normal or reversed in any of these dynamo theories. Furthermore, Rikitake (1966) constructed the so-called double disk model, which can generate the magnetic field either parallel or antiparallel to the axis of disk rotation. Thus, it has been proved both theoretically and experimentally that the Earth's magnetic field in past epochs could have had normal (similar to the actual polarity) as well as reversed polarity. Thus, reversals of the geomagnetic field are feasible.

1.2.3 Principle of the Central Axial Dipole and the Paleomagnetic Field

The poloidal magnetic field exhibits a rotation symmetry relative to the dipole axis; the magnetic meridians and isoclines of the central dipole field on the Earth's surface are similar to the meridians and parallels of the geographic grid. Thus, the values of declination, D, and inclination, I, of the central dipole field uniquely determine the situation of the geomagnetic pole from any single point of the Earth's surface.

The geographic coordinates, latitude Φ and longitude Λ, of the geomagnetic pole can be calculated from the following formulae, if the values of declination D and inclination I are known at a point with geographic coordinates, latitude φ and longitude λ:

$$\begin{aligned} &\Phi = \arcsin(\sin\varphi \sin\varphi_m + \cos\varphi \cos\varphi_m \cos D);\\ &\Lambda = \psi + \lambda, \text{ when } \sin\varphi_m \geqslant \sin\varphi \cdot \sin\Phi\\ &\Lambda = 180 - (\psi - \lambda), \text{ when } \sin\varphi_m < \sin\varphi \cdot \sin\Phi, \end{aligned} \tag{1.3}$$

where

$$\psi = \arcsin(\cos\varphi_m \cdot \sin D / \cos\Phi)$$

and

$$\varphi_m = \text{arctg}(\tfrac{1}{2}\,\text{tg} I), \text{ the geomagnetic latitude at a site.}$$

The recalculation of the D and I values at a site with the coordinates φ and λ with the known coordinates of the pole Φ and Λ can be made from formulae:

$$\begin{aligned} &I = \text{arctg}(2\,\text{tg}\,\varphi_m)\\ &D = D', \text{ when } \sin\Phi \geqslant \sin\varphi \cdot \sin\varphi_m\\ &D = 180° - D', \text{ when } \sin\Phi < \sin\varphi \cdot \sin\varphi_m, \end{aligned} \tag{1.4}$$

where

$$\begin{aligned} &\varphi_m = \arcsin[\sin\varphi \cdot \sin\Phi + \cos\varphi \cdot \cos\Phi \cdot \cos(\Lambda - \lambda)]\\ &D' = \arcsin[\sin(\Lambda - \lambda)\cos\Phi / \cos\varphi_m)]. \end{aligned}$$

The graphic relations between the angles and the arcs, determining the situations of the geomagnetic pole P, are shown in Fig. 1.5.

The presence of a nondipole field results from the fact that the geomagnetic pole coordinates, calculated from the magnetic inclination and declination in different points of the Earth's surface by formulae (1.3), vary and so the "geomagnetic poles" scatter around the true geomagnetic pole. These are the virtual geomagnetic poles, the scatter of which can be considered as a measure of deviation of the observed geomagnetic pole from the that of the central dipole. This pattern of distribution of the geomagnetic field represents an instant image of the permanent process of its changes (secular variations) and should be regarded as one of its possible various states.

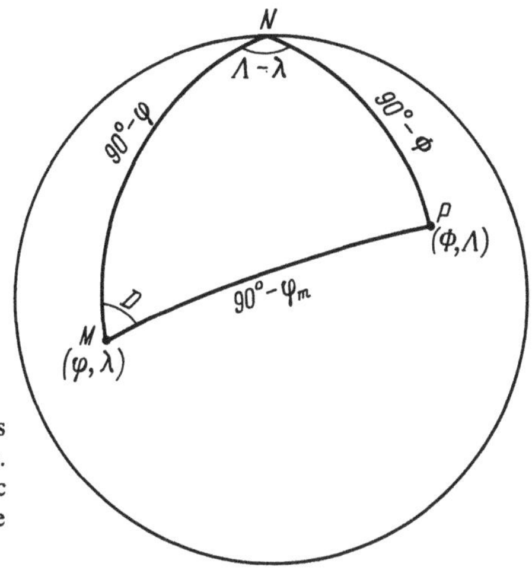

Fig. 1.5. The relationships between angles and arcs that determine the position of the geomagnetic pole. (After Khramov and Sholpo 1967). φ, λ geographic coordinates; φ_m geomagnetic latitude; Φ, Λ pole coordinates

Long series of observations of the Earth's magnetic elements (up to 400 years) at geomagnetic observatories and archeomagnetic studies[1] have revealed that the secular geomagnetic variations have a cyclical nature and can be mainly represented by the precession of the geomagnetic axis (i.e. by the rotation of the equatorial component of the dipole relative to the axis) and by the drift of the nondipole field. Furthermore, not only the equatorial dipole field, but also the nondipole field, are to a great extent excluded in time-averaging (Burlatskaya 1983). This time-averaged geomagnetic field is called the paleomagnetic field at a given site, with corresponding poles, which to the paleomagnetic poles can be calculated from formulae (1.3).

1.2.4 Models of the Geomagnetic Field

The Earth's magnetism can also be studied by other means, e.g. by a mathematical modelling of the geomagnetic field. In this method the origin of the geomagnetic field is left open and a simplified model is selected which might best explain the available experimental data.

One of these models was proposed by Cox (1962), (Fig. 1.6). It assumes the presence of a central dipole and several radial dipoles situated at the core-mantle boundary. The magnetic moment, M, of the central dipole changes sinusoidally, while the radial dipoles have constant magnetic moments, m, of different signs. Hurwitz (see Alldredge and Hurwitz 1964) showed that the dipole moment obtained from a spherical harmonic analysis of the magnetic field produced by several dipoles within a spherical shell is the vector sum of these dipoles. Cox proposed that the field polarity is subjected to reversal when $M'_a/M_a < -1$ (where M_a and M'_a are the axial components of the central dipole and the vector sum of the

[1]Archeomagnetic studies, unlike paleomagnetic studies, deal with the examination of the geomagnetic field history and not the natural objects, i.e., rocks, but with the examination of the remains of the materials of ancient cultures, mainly baked bricks, pottery, baked heaths, ancient fire places, etc. (Burlatskaya 1965; Creer et al. 1983).

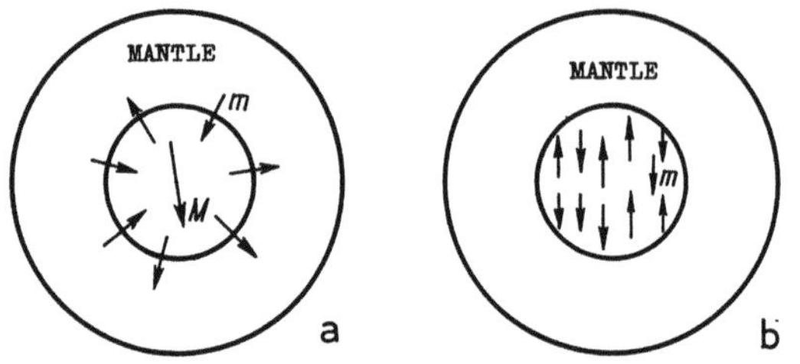

Fig. 1.6a,b. The mathematical models of the geomagnetic field. **a** Model of Cox. (After McElhinny 1973); **b** Model of Kono. *M* central dipole; *m* second-order dipoles

radial dipoles respectively) or when a complete axial component, $M = M_a + M'_a$, changes its sign. Alldredge and Hurwitz (1964) proved that the observed field is in good agreement with this model if eight radial dipoles are present.

Kono (1972) found that the Cox model agrees poorly with the results of direct determinations of the magnetic field intensity in past epochs or with paleointensity determinations. He suggested an alternative model of the geomagnetic field which assumes the presence of a great number of similar dipoles in the core, which are parallel or antiparallel to the Earth's rotational axis. From time to time these dipoles undergo spontaneous reversal. Hence, the polarity of the geomagnetic field is determined by the algebraic sum of their scalar values. The number of dipoles in the model varies widely (up to 80). Kono's model easily explains data on paleointensity, but is inconsistent with the existing distribution of the nondipole field and of its secular variations.

1.3 General Data on Rock Magnetism

1.3.1 Principal Magnetic Properties of Materials

All materials can be divided into three groups according to their magnetic properties: (1) diamagnetic; (2) paramagnetic; (3) ferromagnetic.

A peculiar aspect of diamagnetic materials is their repulsion by a magnetic field, which can be explained as follows. Atoms of diamagnetic materials do not have constant magnetic moments, i.e. all of the magnetic moments inside a single atom are equilibrated in such a way as to produce a resultant magnetic moment, in the atom, equal to zero. When such an atom is placed in a magnetic field, weak currents generated in the atom, due to induction, are directed according to Lenz's law so as to prevent an increase in the magnetic field and, thus, a repulsion effect is observed.

Atoms of paramagnetic materials possess individual magnetic moments. When an applied magnetic field is absent, the resultant magnetic moment of such a material equals zero because the moments of separate atoms are randomly distributed. When an applied magnetic field is present, the atom moments are largely oriented along this field and the material obtains a magnetic moment which is parallel to the applied field. A diamagnetic effect in such materials also exists, but its detection is difficult against the background of a stronger paramagnetic effect.

Ferromagnetics, representing the third group, are the materials with atomic magnetic moments that are not random, but are ordered. This ordering is of a quantum mechanical nature and the energy that maintains this ordering is called the exchange energy. This phenomenon has been examined in many works on ferromagnetism [see, for example, Vonsovsky, (1971)]. Ferromagnetics can be divided into four classes according to the type of ordering.

The first class includes the materials with magnetic moments of atoms oriented parallel to another (Fig. 1.7a). This ordering is observed only below a definite temperature for a given material, called the Curie point. These materials are called ferromagnetics (s.str.). In the second class of materials (Fig. 1.7b) the antiparallel orientation of atomic magnetic moments appears to be more efficient. These materials do not possess a macroscopic magnetic moment. The ordering is disturbed above a certain temperature, called the Néel point. These materials are antiferromagnetic and generally represent a subclass of the third class of ferrimagnetics (Fig. 1.7c). Different values of magnetic moments are characteristic of different types of atoms in ferrimagnetic materials and thus, in the presence of an antiferromagnetic ordering, these materials possess a resultant moment. Finally, the fourth class includes materials with almost antiparallel atomic moments, i.e. the angle between them is close to 180° (Fig. 1.7d). Therefore, they have a small resultant moment which is in perpendicular direction to the antiferromagnetic ordering. These materials are called weak ferromagnetics. The materials of the two last classes, ferrimagnetic and weak ferromagnetics, are the principal carriers of magnetism in rocks.

From the description given above it follows that ferromagnetic material is spontaneously magnetized: if it is heated above its Curie temperature and is then cooled in the absence of a magnetic field, it should have a macroscopic magnetic moment. Actually, this does not occur due to the existence of the so-called domain structure. How is this structure formed? Let us examine a ferromagnetic body (Fig. 1.8a). It is assumed that all atomic magnetic moments of this body are parallel and the resultant magnetic moment has the direction shown in this figure. However, this is inefficient from the viewpoint of energy, since the magnetostatic energy is high:

$$U_{ms} = N(J_s^2/2), \tag{1.5}$$

where J_s is magnetization, N is the demagnetizing factor.

This energy equals zero when $N = 0$, i.e. when the magnetic flux is closed inside the body. If the body is divided into two parts with opposite directions of magnetization (Fig. 1.8b), the magnetostatic energy substantially decreases and a boundary appears between these two areas or domains. This boundary possesses an energy, called the domain wall energy. Subsequent division of the body can be done in two ways. One way is shown in Fig. 1.8c; in this case an optimal quantity of domains is due to the balance between the magnetostatic energy and the energy

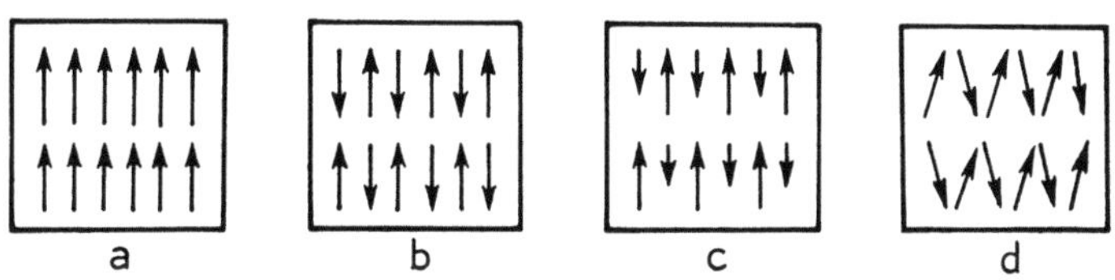

Fig. 1.7a-d. Classification of ferromagnetics. (After Kittel 1956). **a** ferromagnetic; **b** antiferromagnetic; **c** ferrimagnetic; **d** weak ferromagnetic

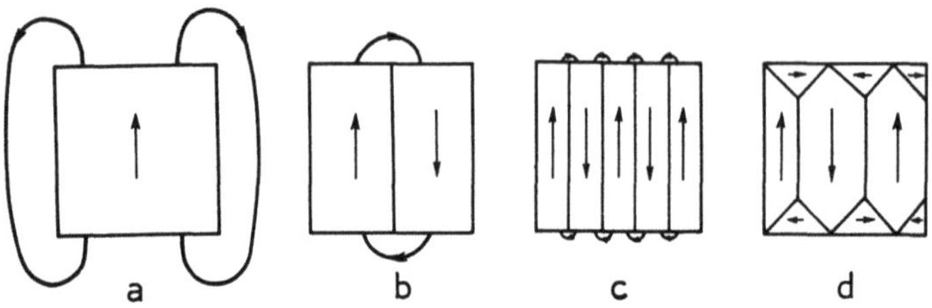

Fig. 1.8a-d. Domain structure formation in a ferromagnetic. (After Yanovsky 1978) Structure: **a** single-domain; **b** two-domain; **c** multidomain; **d** multidomain with "blocking" domains

of the domain walls. The second method is associated with the formation of "blocking" domains (Fig. 1.8d) leading to a quicker decrease of magnetostatic energy, but favouring an increase in the magnetic anisotropic energy.

Thus, in addition to an isotropic exchange interaction between the magnetic moments of atoms there is a weaker magnetic dipole-dipole interaction that depends on the orientation of moments relative to the straight line connecting them. In a crystal the situation of these straight lines is fixed. Thus, depending on the orientation of the magnetic moment of a ferromagnetic, this energy will change. In addition, the energy of spin-orbital interaction within the atom also depends on the orientation of a magnetic moment. Both interactions thus determine the energy of the magnetic crystallographic anisotropy. The magnetization direction of a sample, to which this minimal energy corresponds, is called the axis of easy magnetization of a sample. Certain ferromagnetic substances possess one axis of this type, others have two or three axes. On inspection of Fig. 1.8 it is noteworthy that the configuration and size of the domains in any given sample are determined by the minimum of the sum of all types of the energy enumerated. Domain structure is also characteristic of ferrimagnetics, weak ferromagnetics and antiferromagnetics, although the explanation of their domain structure is somewhat different.

Rocks are a special object of study and so it is necessary to examine the magnetism of macroscopically small particles. It is possible that a domain wall is not formed in such a particle, for it is not efficient from the viewpoint of energy. The thickness of the domain wall becomes commensurable with the particle size. Every magnetic material has a critical diameter d_0, thus if a particle has a diameter $d<d_0$, then it is in a single-domain state. This critical diameter is particularly larger in the weak ferromagnetics because of their low magnetization. Thus, d_0 can be up to ≈ 1.5 mm in hematite, which plays an important role in rock magnetism and paleomagnetism. Virtually, it means that all hematite occurring in rocks has a single-domain structure. At the same time, a ferrimagnetic material, magnetite, an important mineral for paleomagnetism, has $d_0 = 0.03 \div 3$ μm, i.e. it can be in a multidomain state.

Let us examine a sample of ferromagnetic material in a demagnetized state (domain magnetizations are mutually compensated). The process of magnetization of the sample is shown in Fig. 1.9. In weak fields a movement of the domain walls occurs; the domains with their magnetic direction close to that of the external field, grow at the expense of other domains. In stronger fields, when a sample acquires a single-domain state, a rotation of magnetization vectors starts relative to the direction of the applied field. Then, the saturation magnetization appears: under continuous increase of the applied field a slow linear growth of magnetization, a "paraprocess", is registered, which is due to the fact that the atomic magnetic moments, disoriented under the action of thermal agitation, become more and more efficiently aligned by the applied field.

Now let us decrease the external field, the demagnetization curve would then lie above the magnetization curve (Fig. 1.10) and at $H=0$ the magnetization $J=J_r \neq 0$. J_r is called the isothermal remanent magnetization. For the magnetization to equal zero, it is necessary to apply a field of an opposite direction H_c; this field is called the coercive force. In this case we continue to increase the applied field in this direction, again reaching saturation. Then a similar process occurs in the opposite direction. A complete cycle, i.e. magnetic hysteresis, is shown in Fig. 1.10.

1.3.2 Magnetic Properties of Rocks

Rocks are specialised objects of study. As a rule they contain some grains of ferro- and ferrimagnetic minerals dispersed in a para- or diamagnetic matrix. Because of the low concentration of these grains, the magnetization of rocks is very weak, as a rule, and amounts usually to ($10^{-3}-1$) A/m. Since these grains are very small they possess a single-domain or pseudosingle-domain (with few domains) structure. Only these grains yield the principal paleomagnetic information due to their high magnetic coercitivity which allows the preservation of the remanent magnetization despite environmental effects. Thus, the simplest magnetic model of a rock is an assemblage of non-interacting single-domain particles.

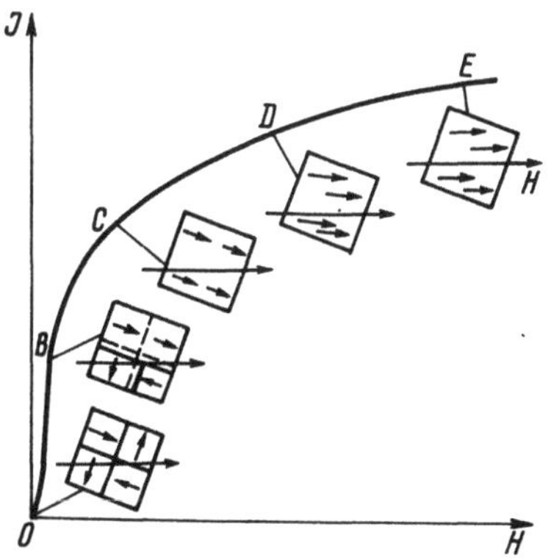

Fig. 1.9

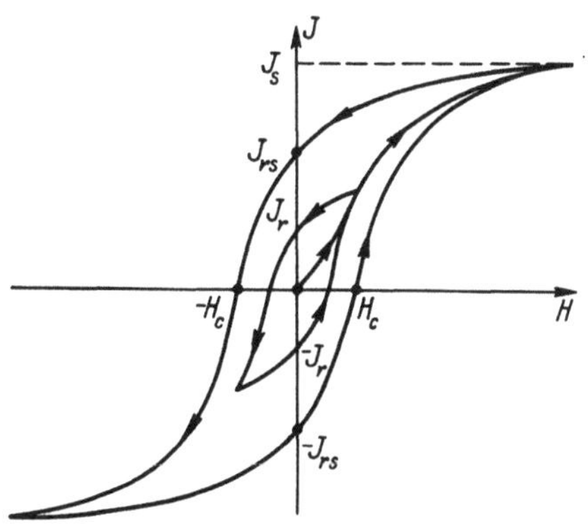

Fig. 1.10

Fig. 1.9. Magnetisation process according to domain theory. (After Irving 1964). *OB* Rayleigh's area; *OC* displacement of domain walls; *CE* rotation of spontaneous magnetization

Fig. 1.10. Magnetic hysteresis. (After Nagata 1961). J_s saturation magnetization; J_r isothermal remanent magnetization (IRM); J_{rs} saturation IRM; H_c coercivity

1.3.3 Assemblage of Single-Domain, Noninteracting Particles

The theory of magnetization of such an assemblage was proposed by Néel (1955). In spite of simplification, the conclusions drawn from this theory are in good agreement with the experimental data (Nagata 1961; McElhinny 1973).

Let us examine an assemblage of single-domain, uniaxial ferromagnetic grains. Their axes are randomly oriented and the whole assemblage thus possesses zero magnetization. With an external field applied in any direction the assemblage would acquire an induced magnetization associated with those grains that are axis-oriented in this direction. Thus, the assemblage as a whole is magnetically isotropic. The grain behaviour in the external field depends on its orientation relative to this field. With the field parallel to the grain axis, the magnetic behaviour is characterized by a rectangular hysteretic loop (Fig. 1.11a). The other extreme position is the external field perpendicular to the grain axis (Fig. 1.11b). With $H>H_c$ and $H<-H_c$, the magnetization equals J_s and $-J_s$, respectively. With $|H|<H_c$ the magnetization shows a linear dependence on the field intensity. Behaviour of the magnetization of the whole assemblage corresponds to an intermediate case (Fig. 1.11c).

The magnetization energy of a single-domain grain is

$$E = K_a V \sin^2 \theta, \tag{1.6}$$

where V is the grain volume; θ is the angle between the magnetic moment and the axis of a weak magnetization; K_a is a constant of anisotropy which is due to three factors: magnetocrystalline anisotropy, shape anisotropy and anisotropy rising from mechanical stress.

A simple relation between the coercive force and the saturation magnetization is expressed as

$$H_c = 2K_a / J_s. \tag{1.7}$$

For magnetocrystalline anisotropy:

$$K_a = K_{a1}; \qquad H_c = 2K_{a1} / J_s; \tag{1.8}$$

for shape anisotropy:

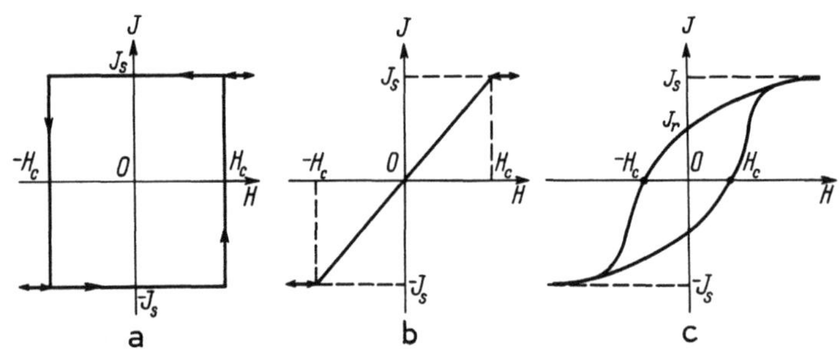

Fig. 1.11a-c. Hysteresis loops for single-domain particles. (After Nagata 1961). Directions of external magnetic field: **a** parallel to the grain axis; **b** perpendicular to the grain axis; **c** intermediate direction. J_s saturation magnetization; J_r IRM; H_c coercivity

$$K_a = (½)(N_b - N_a)J_s^2; \qquad H_c = (N_b - N_a)J_s, \tag{1.9}$$

where N_a and N_b are the demagnetization factors along the short and long axes of an ellipsoid;

for the anisotropy related with stress:

$$K_a = (^3/_2)\lambda\sigma; \qquad H_c = 3\lambda\sigma/J_s, \tag{1.10}$$

where λ is a mean coefficient of magnetostriction; σ is the amplitude of external stress.

At present, we have no satisfactory theory for an assemblage of single-domain, interacting particles, though certain studies in this field have been conducted (Dunlop 1969; Shcherbakov, and Shcherbakova 1977).

1.3.4 Magnetic Viscosity

As seen from Eq. (1.6) two orientations can exist in the magnetic moment of a single-domain grain in the absence of the applied field, thus corresponding to the energy minimum: $\theta = 0°$ and $\theta = 180°$. These positions are separated by a potential barrier. Maximum of energy corresponds to the position $\theta = \pm 90°$ when the potential barrier is

$$E_r = VK_a. \tag{1.11}$$

This barrier can be overcome as a result of thermal fluctuations. The energy of such fluctuations is known to be

$$E_t = kT, \tag{1.12}$$

where k is Boltzmann's constant; T is the thermodynamic temperature (Kelvin).

Consequently, the magnetic moment can spontaneously change its orientation from one direction to the other when $E_t > E_r$.

An assemblage of single-domain grains can now be examined (Yanovsky 1978; McElhinny 1973). It is assumed that it has a general magnetic moment M_o, i.e. the particle moments are generally oriented along the vector M_o. At a given temperature T a certain quantity of particles of the volumes which meet the condition $E_t > E_r$ will always be observed. With a dominant orientation present the majority of these grains exhibit magnetic moments directed at acute angles to M_o. Thus, a spontaneous change in the orientation of this magnetic moment, resulting from thermal fluctuations, will produce a decrease of the resultant magnetic moment of the assemblage. With the passage of time the magnetic moment of the assemblage will obey the following formula:

$$M_t = M_o \exp(-t/\tau), \tag{1.13}$$

where τ is the relaxation time, i.e. the time for the magnetic moment of the assemblage to decrease by a factor e.

The relaxation time can be calculated from the relationship:

$$\tau = (1/C)\exp(E_r/E_t) = (1/C)\exp\,[VK_a/(kT)]. \tag{1.14}$$

Substituting the expression of anisotropy for a constant K_a we obtain:

$$\tau=(1/C)\exp[VH_cJ_s/(2kT)], \quad (1.15)$$

where C is the coefficient with a dimension of frequency equal to 10^{10} Hz.

If the relaxation time is short (for instance, less than 10^3 s), the assemblage loses its magnetization as rapidly as it is acquired. The grains making up this assemblage are said to be superparamagnetic as their behaviour in the magnetic field closely resembles that of paramagnetic particles.

A change in magnetization with time is called the magnetic viscosity. The magnetic viscosity also exists in multidomain particles, when it is related to the movement of domain walls. The relaxation time can be constrained not only by thermal fluctuations, but also by the diffusion of point defects into the boundary layer that produces stabilization of the domain walls. As single-domain particles are known to be the principle carriers of paleomagnetic information, the diffusion viscosity is not discussed here.

The relaxation time, as seen in Eq. (1.15), decreases, the higher the temperature and the lower the particle volume. For every grain of a volume V there is a temperature at which the relaxation time becomes short (less than 10^3 s), called the blocking temperature T_b. Similarly, there is a blocking particle, diameter d_b, for a given temperature.

It is postulated that the anisotropic constant changes only slightly with temperature change. So

$$T_1\ln(C\tau_1)=T_2\ln(C\tau_2). \quad (1.16)$$

Consequently, maintaining of a sample at a temperature T_1 for a time τ_1 equals the effect of a higher temperature T_2 for a shorter time τ_2. If $C=10^{10}$ Hz is substituted into Eq. (1.16), then the heating of a sample to 150°C during 1 m.y. equals its heating to 450°C during 1000s. This fact is fundamental in temperature cleaning.

1.4 Types of Remanent Magnetization of Rocks

1.4.1 Remanent, Induced and Viscous Magnetizations

Remanent magnetization is of exceptional importance in paleomagnetology. In nature the total remanent magnetization of a rock is called its natural remanent magnetization (NRM) and is designated as J_n. In addition to remanent magnetization a rock also possesses an induced magnetization

$$J_i=\varkappa H_T. \quad (1.17)$$

The relationship J_n/J_i, which is called the Koenigsberger ratio, or factor Q, plays an important role in rock magnetism

$$Q=J_n/(\varkappa H_T). \quad (1.18)$$

The magnetization of a rock, as for any ferromagnetic, depends not only on its properties and the value of any applied steady field, but also on such factors as: time, temperature, mechanical stress, alternating magnetic field, and chemical changes. These factors alone, in the absence of a steady magnetic field, can only destroy the remanent magnetization, but cannot create it. In the presence of a steady field any of these factors might favour the generation of an additional magnetization superimposed on the isothermal magnetization J_r (or IRM) which was discussed in Sect. 1.3.1. In weak fields, such as studied in paleomagnetology, this addition might be very great compared to J_r. Depending on which of the factors affect the magnetization, this process proceeds according to certain laws and bears a corresponding name (a special type of magnetization). Here, the major types of magnetization processes and remanent magnetizations are examined.

Among the factors enumerated above, time is a common factor for all rocks, thus determining the type of rock magnetization. Therefore, the process of so-called viscous magnetization, due to the phenomenon of magnetic viscosity, is known to be of universal occurrence (see Sect. 1.3.4). Remanent magnetization appearing in a rock after a long duration in a magnetic field is called the viscous remanent magnetization (VRM) and is designated as J_{rv}. The magnetic viscosity is mostly related with a weak magnetizing field. The relaxation time of viscous magnetization in rocks is very long compared to that of ferromagnetics used in industry: the processes of growth of viscous magnetization, and its decay at $H=0$, lasts for extremely long time periods. The viscous magnetization of rocks acquired in the geomagnetic field over several months might exceed hundred fold the isothermal remanent magnetization in the same field (Khramov and Sholpo 1967). The relationship of J_{rv} and time in the weak fields can be expressed by the following empirical formula

$$J^a_{rv2} - J^a_{rv1} = S(\lg t - \lg t_o) - S'\lg t', \tag{1.19}$$

where J_{rv1} is the remanent magnetization acquired by a sample until the time t_o; t is the duration of after the field is turned off; J_{rv2} is the remanent magnetization appearing when the process of magnetization is renewed until time $t>t_o$; S and S′ are the "coefficients of viscosity" (commonly $S' \approx S$) related to the Rayleigh constant b by $S = S_v(b/2)^a$; S_v is a characteristic value for a given ferromagnetic and a given temperature; a is a power within the limits of 0.5–1.

This expression is not valid at $t \to \infty$ and thus, it does not give the finite limit of J_{rv}. However, in practice it is useful not only for the laboratory time scale, but also for the geologic time scale (Khramov and Sholpo 1967).

1.4.2 Thermoremanent Magnetization

All ferromagnetic materials heated above the Curie temperature lose their ferromagnetic properties. When this material is cooled down to room temperature, it acquires a remanent magnetization, the direction of which coincides with that of the external field. The thermoremanent magnetization (TRM), denoted as J_{rt}, is typical of the magnetization associated with the cooling of volcanic rocks. It was noted that, in general, the remanent magnetization of these rocks exhibits certain peculiar properties. First, the direction of this remanent magnetization often differs from the direction of the recent geomagnetic field. And, generally, it remains rather stable under different laboratory tests. Second, factor Q in these rocks usually lies between 2 and 10, but often exceeds 100. However, this is not related to the fact that the ancient geomagnetic field was much higher than the actual one (Nagata 1961). It has been established that remanent magnetization appears in the process of rock cooling and that its direction agrees with that of the geomagnetic field during cooling. The TRM is an important process for carrying paleomagnetic information due to its high resistance to external effects.

Experimentally it was shown that thermoremanent magnetization appears not only upon cooling of a sample from the Curie point down to room temperature, but also over any other temperature interval below the Curie point (Nagata 1961; Day 1977). The remanent magnetization acquired upon cooling of a simple in a field H from a temperature T_1 down to T_2 is called the partial thermoremanent magnetization (PTRM) and is designated as $J_{T_1H}^{T_2}$. Thellier (1941) experimentally established three relationships which were called Thellier's three laws. The formula illustrating the first law is

$$\sum_{i=1} J_{T_iH}^{T_{i+1}} = J_{T_1H}^{T_n} \quad \text{where} \quad T_1 \geqslant T_2 \geqslant \ldots \geqslant T_n. \tag{1.20}$$

It follows from this law that a complete TRM is equal to the sum of partial magnetizations acquired over the total temperature interval from the Curie point to room temperature.

If a sample possessing a TRM is heated to T_1 and is then cooled to room temperature in the absence of any external magnetic field, then its magnetization decreases. If the same process is repeated for the temperature $T_2 > T_1$, $T_3 > T_2 \ldots$, $T_n > T_{n-1}$, then, according to Thellier's second law, the result will be similar to the one obtained upon heating the sample directly to T_n and cooling it down to room temperature when the magnetic field is absent. Thellier's third law follows from the first two: a magnetization $J_{T_1H}^{T_2}$ does not change upon heating to T and subsequent cooling of a sample in the absence of the external field if $T < T_2$; it vanishes completely if $T > T_1$; it partially vanishes if $T_2 < T < T_1$.

The dependence of partial magnetization upon the temperature range in which it was acquired (Fig. 1.12) indicates that the maximum change in total thermoremanent magnetization occurs at a temperature close to, but still with a certain interval from the Curie point T_C; this temperature is the blocking temperature T_b.

Some studies have been published in which it is stated that Thellier's laws are valid only for the single-domain grains (Bol'shakov and Shcherbakova 1979). Furthermore, the authors of these works argue that the mere fact of the invalid nature of Thelliers's laws indicates a multidomain nature of the carriers of magnetization.

A simplified, theoretical model of TRM is the Néel model for single-domain particles (Néel 1955). The coercive force of any of these particles is calculated from Eq. (1.7). Thus, in a field H, a magnetization is acquired only by those particles for which $2K_a/J_s < H$. Upon increased temperature the coercive force decreases.

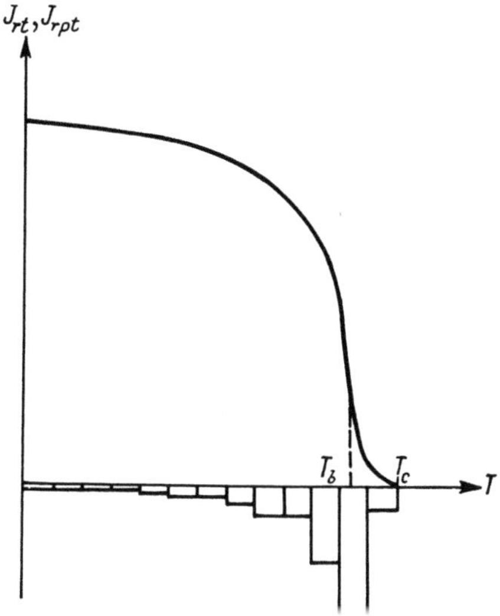

Fig. 1.12. Decrease of thermoremanence with heating and hystogram of partial thermoremanences. T_b blocking temperature; T_c Curie point

Near the Curie point it becomes so weak that all of the particles might acquire magnetization under the effect of an extremely weak field. Then, the acquired TRM equals J_s and is independent from the applied field. This does not occur, however, due to thermal fluctuations. The following expression is given for the TRM according to the Néel theory:

$$J_{rt} = J_s(T_o)\text{th}\,[VJ_s(T_b)H/(kT_b)], \quad (1.21)$$

where T_o is room temperature.

The Néel theory is rather idealized: all particles are of the same volume and of the same blocking temperature. Nonetheless, this theory is useful because it explains the existence of blocking temperature, from which Thellier's laws follow. The coercive force at room temperature is much higher than it is at the blocking temperature. Consequently, more particles are capable of being magnetized at the blocking temperature than at room temperature in the same field, thus explaining why the TRM greatly exceeds the isothermal magnetization acquired in the same field. It also explains the high stability of TRM under the effect of an alternating magnetic field. The resistance of TRM to heating is also due to a much higher blocking temperature in comparison with room temperature.

A rock contains ferromagnetic grains of different sizes, thus involving different blocking temperatures. If, during cooling, the temperature decreases below the blocking point for a given grain, its relaxation time increases sharply and the grain magnetization becomes "frozen-in". Thus, any changes occurring in the external magnetic field direction below T_b do not affect the direction of grain magnetization. It follows from Eq. (1.16) that if the relaxation time of a grain is $\tau = 10^3$ at a temperature $T = T_b = 530°C$, then at $T = 30°$ the relaxation time increases to 10^{17} years. Even the grains of relatively low blocking temperatures (about 300°C) yield relaxation times of the order of 10^9 years at room temperature. Therefore, TRM has a high stability throughout geological time.

The exceptional importance of single-domain particles in paleomagnetology is well known. However, multidomain and pseudosingle-domain particles in particular, which also appear to be carriers of paleomagnetic information (Day 1977), must also be considered. Multidomain particles also have blocking temperatures, but in this case the process of blocking is different; it is related to the displacement of the domain walls. Several theories have attempted to explain the origin of TRM in multidomain particles. Unfortunately, they cannot be considered satisfactory; a review of these theories can be found in the work of Day (1977). Thus, according to McElhinny (1973) for magnetite grains with a diameter d–20 μm in a pseudosingle-domain state

$$J_{rt} = \left(\frac{m}{V}\right)\left[\frac{\cos\alpha \ln \mathrm{ch}\,(2\,\beta\,m)}{2\,\beta\,m}\right] , \tag{1.22}$$

where

$\beta = [(H\cos\alpha)/(kT_b)]\,[J_s(T_b)/J_s(T_o)]$;
$\cos\alpha = 0.85$; $m = 4\times 10^{-4} d^2$ (d is in cm).

The expression in square brackets is practically constant for the grains in a pseudosingle-domain state and it follows that $J_{rt} \approx d^{-1}$. The experiment, however, gives $J_{rt} \approx d^{-s}$ where $S \approx 0.7$ (Day 1977). In the multidomain grains TRM does not depend on the grain size:

$$J_{rt} \approx 0.125H. \tag{1.23}$$

1.4.3 Detrital Remanent Magnetization

This type of magnetization is known to appear in sedimentary rocks during their formation. The external magnetic field produces an orientational effect on the magnetic particles during deposition. As a result, a statistical alignment of the magnetic moments of these particles is observed along the direction of the external field. This condition is called detrital remanent magnetization (DRM) and is commonly designated as J_{ro}.

Nagata (1961) proposed the following mechanism for the origin of DRM. An assemblage of N similar spherical particles with magnetic moments m is examined. At time $t_1 = 0$ their orientation is random. At $t_2 = t$, the particles reach the bottom of a sedimentary basin. The equation of particle movement is as follows (the inertia member of the second order is negligible):

$$\lambda \cdot (d\theta/dt) + mH\sin\theta = 0, \tag{1.24}$$

where θ is an angle between the magnetic moment m and the external field H; λ is a coefficient proportional to the dynamic viscosity of the medium.

The solution of this equation is

$$\mathrm{tg}(\theta/2) = \mathrm{tg}(\theta_o/2)\exp(-mHt/\lambda), \tag{1.25}$$

where

$\theta = \theta_o$ at $t = t_1 = 0$.

As the initial distribution is random, the number of particles with magnetic moments situated in the angle range from θ_o to $\theta_o + d\theta_o$ will be $dN = (N/2)\sin\theta_o d\theta_o$. As the projection of the magnetic moment of every particle in the direction of the applied field H equals $m \cdot \cos\theta$, the magnetic moment of a sediment is

$$M = \frac{Nm}{2} \int_{0}^{\pi} \cos\theta \sin\theta_o \, d\theta_o . \tag{1.26}$$

Substituting here θ from Eq. (1.25), and taking into consideration that $\mathrm{cth}\,\chi - \chi \cdot \mathrm{cosech}^2\,\chi \approx \mathrm{th}(2\chi/3)$ and solving the equation for magnetization, we obtain

$$J_{ro} = mn\,\mathrm{th}\,[2mHt/(3\,\lambda)], \tag{1.27}$$

where n is the number of particles per unit volume.

It thus follows that if the time of deposition is suffciently long ($t \to \infty$), then $J_{ro} \to mn$. In reality, this value is always less because of disorientation effects. Brownian motion represents such an effect for small particles. Khramov (1968) took this effect into consideration. An assemblage of weakly interacting particles behaves as a paramagnetic gas. Using a parameter, K, the precision parameter, to determine the grade of particle orientation, we obtain

$$K = mH/(kT), \tag{1.28}$$

where k is Boltzmann's constant, so

$$J_{ro} = \sigma J_{rm}(\mathrm{cth}K - 1/K), \tag{1.29}$$

where $\sigma = NV/V_s$ is the volume concentration of particles; V_s is the sediment volume; V is the volume of a magnetic particle; J_{rm} is the particle magnetization.

If $K > 3$, i.e. the magnetic energy dominates over the energy of the thermal fluctuations, then

$$J_{ro} = \sigma\,[J_{rm} - kT/(VH)].$$

Thus, in this case detrital magnetization is proportional to the concentration of a single particle with the correcting term $kT/(VH)$ included.

When $K < 1$, then $\mathrm{cth}K - 1/K \approx K/3$ and

$$J_{ro} = J_{rm}^2\,\sigma VH/(3kT). \tag{1.30}$$

Thus, when the magnetic energy of particles is less than the energy of thermal fluctuations, the magnetization J_{ro} is proportional to the intensity of the field H, particle concentration, particle volume and the square of its magnetization. Therefore, if a sediment contains different particles, the ones with a higher value of J_{rm}^2 V will yield the highest contribution to magnetization. These are mainly the hematite particles which might be sufficiently large thus keeping their single-domain nature, i.e. they are magnetized to saturation. And again we emphasize the significance of single-domain particles in paleomagnetology.

Stacey (1972) made an attempt to assess the contribution of particles with different magnetic moments to the detrital magnetization. With a uniform distribution of the particle magnetic moments, starting from 0 to the maximum value m_{max}, the detrital magnetization will be

$$J_{ro} = \sigma J_{rm}(1/x)\,\ln[(\mathrm{sh}x)/x], \tag{1.31}$$

where $x = m_{max}\,H(kT)^{-1}$.

With small x this expression gives $J_{ro} = (1/6)\,\sigma J_{rm} \cdot m_{max}\ H(kT)^{-1}$, and $m_{max} = 2J_{rm}V$ will again give Eq. (1.30). Thus, in the first approximation the distribution of particles according to their magnetic moments changes neither the form of the function $J_{ro} = f(J_{rm}, H, T, V)$, nor its value. The properties of this magnetization (the hysteretic nature, for instance), however, might differ from those of similar magnetizations created by other particles with equal magnetic moments.

This model of detrital magnetization also seems applicable to larger particles, when a disorientation of the magnetic moments is caused not by thermal fluctuations, but by motions in the aqueous medium.

If the kT value is substituted for the energy of motion of the aqueous medium, formulae (1.28)–(1.31) remain valid for this case. They are also applicable for the models which examine other types of randomization energy, for instance, gravitational energy (Burakov and Ryabushkin 1975).

Experimental studies (Khramov 1968; Stacey 1972) demonstrated the validity of the relationships $J_{ro} \sim H$, $J_{ro} \sim \sigma$ and $J_{ro} \sim 1/T$, thus proving the applicability of the model examined in the studies of the natural processes of detrital magnetization. At the same time, Nagata's model had limited application as theoretically and experimentally (Verosub 1977), the typical time of particle orientation in water did not exceed 1 s and the particle assemblage, while precipitating in water, acquired its stationary state given by formula (1.29) or (1.31).

Johnson et al. (1948) found that, during the redeposition of clays, a good consistency could be observed between the magnetic and external field declination, but the inclination of magnetization appeared to be 20° less than the field inclination. King (1955) examined this effect in detail and called it the "inclination error". King obtained the following empirical formula:

$$\mathrm{tg} I_0 = f\,\mathrm{tg} I_f, \tag{1.32}$$

where I_0 is the inclination of magnetization; I_f is the field inclination; f is a constant.

King determined $f \approx 0.4$ for the sediments studied. He explained this "inclination error" by the presence of spherical and disk-like particles in the magnetic fraction of a sediment, the magnetic moments of which are situated in the plane of the disks. The magnetic moments of the spherical particles thus acquire orientation along the geomagnetic field, but the disk-shaped particles lie horizontally and thereby make the inclination lower (Fig. 1.13). However, even when all of the sedimentary particles are spherical, an inclination error is still observed (King 1955). Such inclination errors have been suggested to occur when the spherical particles hitting the bottom and then roll into the nearest depression (Fig. 1.14). In this process particles 1 and 2 acquire inclination errors of different signs that are mutually compensated. Particles 3 and 4 have declination errors of different signs that are also mutually compensated, but the inclination errors are of the same sign, so that a resultant inclination errors occurs. Probably, in different types of sediments, such inclination errors might appear due to both shape anisotropy and as a result of rolling.

King also determined in laboratory studies that the orientation of sedimentary magnetization is affected by the tilt of the underlying surface. He called this effect a "deposition error", i.e. the difference between the inclination observed between deposition on a horizontal and a tilted surface. Hamilton and King (1964) expe-

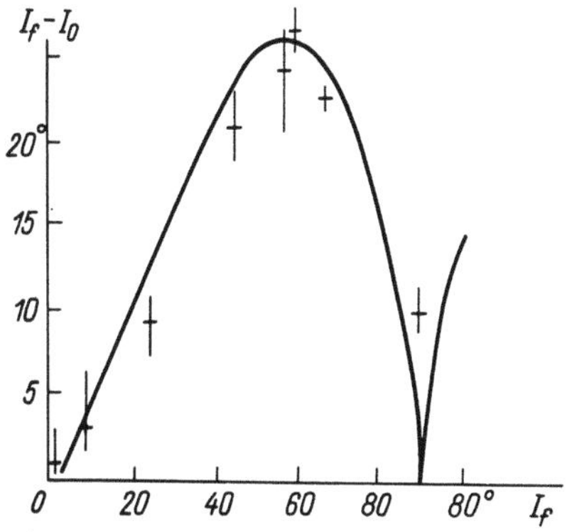

Fig. 1.13

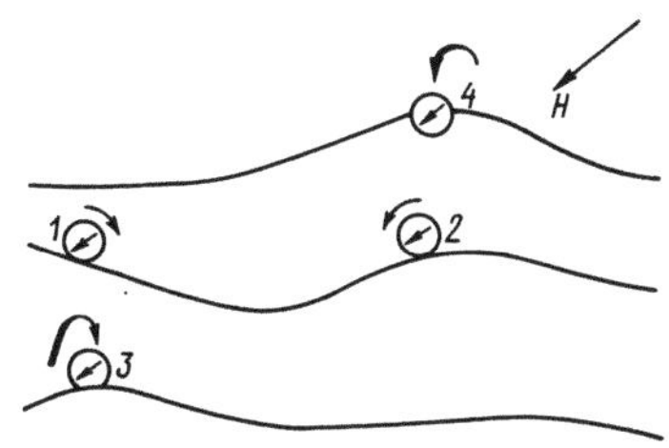

Fig. 1.14

Fig. 1.13. Variation of an inclination error with an inclination of external field. (After Nagata 1961). I_f inclination of external field; I_0 inclination of remanence

Fig. 1.14. Inclination error of a sequence of spherical particles. (After Verosub 1977)

rimentally confirmed that the deposition error is approximately equal to the tilt of the bottom. King proposed a model of magnetization of spherical particles depositing on such a tilted surface (King 1955). According to this model, the deposition error at small angles of the bottom tilt amounts approximately to 10°; it is thus consistent with experimental data. Furthermore, King found that the bottom currents also affect the direction of sediment magnetization.

The process of depositional magnetization combines the mechanism of detrital magnetization together with the effects of the three "errors". The magnetization acquired by the sediment by this process is called depositional or detrital magnetization (SDRM).

Graham (1949) and later Irving (1957-1967) discussed whether magnetization (or remagnetization) of a sediment occurs after its deposition, but prior to its lithification. Irving studied the Precambrian red sandstones and established the detrital nature of their magnetization and confirmed that the magnetization was acquired after the formation of slump folds. He suggested that after deposition the magnetic particles in certain sediments are allowed to rotate in the cavities filled with water. He called this process postdepositional magnetization (PDRM).

Irving and Major (1964) reported the occurrence of postdepositional magnetization when they produced a stable magnetization in wet synthetic sediments containing quartz and magnetite. However, the particle size and proportions of magnetite and quartz varied widely. Furthermore, the inclination of the external field was considerably changed (from 3° to 90°). Notable is that there was no significant difference registered between the magnetization direction and the field direction. Irving proposed a hypothesis that the water concentration characteristic for every type of sediment may be critical. When the water concentration in a sediment exceeds this limit, the magnetic particles can change their orientation. As soon as the water concentration is below this critical level, the magnetization becomes "locked-in".

Numerous magnetization studies of varved clays demonstrated the presence of inclination and deposition errors, and errors due to currents. It was found

(Verosub 1977) that this type of sediment is characterized by detrital magnetization. After studying the deep-sea sediments (Opdyke and Henry 1969), good agreement was found between the mean magnetization direction in the column and the field of a geocentric axial dipole. It was also determined that the inclination error is absent in this type of sediment. Thus, these factors, together with the ubiquitous presence of benthos, which loosens the upper beds of deep-sea sediments, led to the conclusion that their magnetization is of a post-depositional nature. The sediments of lakes and marginal seas might exhibit both types of detrital magnetization according to a review by Verosub (1977).

Experimental deposition studies of fine-grained silts confirmed the presence of both types of processes, which are characterized by the different relationships of magnetization J_{ro}, depending on the time of deposition and the magnetic field (Barton et al. 1980). During the time in which a dilute suspension precipitates, the magnetization J_{ro} reaches its maximum in 2 days. Its dependence on the magnetic field shows a divergence from linearity at H = 80 A/m and saturation appears in fields of the order of 10^3 A/m. Here, the dependence J_{ro} (H) shows good agreement with Eq. (1.31). During the time in which the concentrated suspension precipitates, the magnetization J_{ro} stabilizes in several minutes and the dependence J_{ro} (H) exhibits a linear character up to at least H = 800 A/m. Barton et al. (1980) suggested that the first process simulates the SDRM and the second one simulates the magnetization appearing as a result of submarine slumping and benthos activity. It is of interest that in both cases the inclination error appears only in the fields which are weaker than the terrestrial one and that the postdepositional DRM formed 24 h after the beginning of precipitation produced only 10–30% of the depositional magnetization.

Thus, even if the detrital nature of remanent magnetization is known, it is not sufficient to obtain valid information on the direction of the ancient geomagnetic field at the time of sediment formation. If this magnetization is of a depositional nature, one must consider the inclination and deposition errors and also the error related to currents. Postdepositional magnetization does not seem to invoke these errors, but the time of its formation has not been precisely determined. This fact is most important when studying paleosecular variations. In addition to remanent magnetization, other carriers of "memory" of the ancient geomagnetic field exist in sediments. For instance, the actual orientation of particle axes and the degree of their grouping yield information on the direction and intensity of this field and this information is the basis of the method of "elongated particles" (Pechersky et al. 1975).

1.4.4 Chemical Remanent Magnetization

When a rock is being formed, magnetic minerals are being formed and transformed at the same time and also later. If this occurs at high temperature (above the Curie point of a mineral), then upon cooling a rock acquires a thermoremanent magnetization. If this process occurs below the Curie point of a mineral, then the cooling only partially affects the remanent magnetization. Thus, three situations might occur: (1) a magnetic mineral is transformed into a non-magnetic one; (2) a magnetic mineral generates from a solution or from a non-magnetic mineral;

(3) transformation of one magnetic mineral into another. Similar processes can produce a remanent magnetization at room temperature without any cooling.

A reaction of the first type would result only in the disappearance of the total remanent magnetization, or a part of it. Under reactions of the two last types a new kind of remanent magnetization is acquired, which is called chemical remanent magnetization (CRM) and is designated as J_{rc}. Haigh (1958) proposed a theoretical model according to which a chemical reaction is regarded as a process of nucleation of a new magnetic material. This model applies the Néel theory of thermal fluctuations of single-domain, noninteracting particles. With this model, it is possible to determine the occurrence of CRM as a result of the second type of reaction and partially of the third type.

Haigh's mechanism resembles that of TRM acquisition, differing only in the constant temperature and the variance in grain size. Proceeding from Eq. (1.14) with the relaxation time assumed to be $\tau = 10^3$ s, i.e. $\ln(C\tau) \approx 30$, we obtain the blocking volume V_b when the particle passes from a superparamagnetic into a single-domain state

$$V_b = 30\, kT/K_a. \tag{1.33}$$

The chemical magnetization acquired by red sediments during hematite formation is of special importance in paleomagnetology. For hematite $K_a = 125$ j/m^3, thus the blocking diameter is $d_b \approx 1 \mu m$ (Haigh 1958), which agrees with experimental data.

Stacey (1963) obtained an expression for the CRM of single-domain grains forming at a temperature T:

$$J_{rc} = J_s \mathrm{th}\, [V_b J_s H/(kT)]. \tag{1.34}$$

According to its magnetic characteristics, CRM is analogous with TRM, which is feasible, because magnetizations of these types appear in a similar manner.

It must be noted that Haigh's model does not seem to be useful in all reactions of the third type. Thus, the demagnetizing fields of the original grains and the exchange interaction need to be taken into consideration. According to the type of exchange interaction, the relation of its energy with the energy of magnetostatic interaction and the external field, a new direction of magnetization can occur. It might be parallel or nonparallel to the previous direction or it might coincide with the direction of a new geomagnetic field. At present, we have no rigorous theory of CRM origin either for this type of reaction or for other types. A very important fact has been established upon observations and experiments: single-phase oxidation of ferromagnetic materials in oceanic basalts does not change the direction of their magnetization (Pechersky et al. 1975).

1.4.5 Pressure Effects Upon Remanent Magnetization

During their evolution all rocks are recurrently subjected to hydrostatic or uniaxial pressure. This is likely to affect their remanent magnetizations.

Kume (1962) demonstrated that magnetization decreases by 50–90% in magnetite, maghemite and hematite samples isothermally magnetized in the field of 0.5T and then subjected to a hydrostatic compression of 10×10^8 Pa (this pressure is typical of a depth of 30 km). Girdler (1963) obtained similar results on samples with TRM. The magnetization directions did not appear to change at all.

Stott and Stacey (1960) heated samples of igneous rocks above their Curie point and cooled them in a field of 6 mT under an uniaxial pressure of 0.5×10^8 and 1×10^8 Pa. It appeared that the direction of magnetizations acquired is consistent with the direction of the external field and with the magnetization direction of duplicate specimens cooled under the same conditions, but in the absence of pressure. However, in thermomagnetization of anisotropic rocks, the pressure effect results in a significant discrepancy between the directions of the acquired TRM and of the external field. Besides, it has been established that initially isotropic magnetic rocks might transform to anisotropic rocks under the effect of uniaxial pressure.

Vlasov et al. (1961) recognized that pressure also affects DRM. It was established that the magnetization inclination decreased 6° under the effect of 0.4×10^8 Pa pressure in a sediment with magnetic particles of 1 μm size; and it decreased 10° after applying a pressure of 3.3×10^8 Pa. In a sediment with smaller particles ($\sim$0.02 μm), however, pressure practically does not affect the direction of remanent magnetization.

Blow and Hamilton (1978), while studying finely dispersed sediments, found that a decrease in the magnetization inclination might also occur upon compaction.

Domen (1962) established that ferromagnetics acquire a remanent magnetization under static pressure, which was called the piezoremanent magnetization. Another type of remanent magnetization, dynamic magnetization, appears under repeated loading and impacts. The piezoremanent and dynamic magnetizations appear as a result of a decrease in the level of energy barriers under external stresses. Both types of magnetization might be registered in rocks, but their role is not yet clear. It is noteworthy, however, that their resistivity to external effects seems to be quite insignificant (Yanovsky 1978).

1.4.6 Isothermal and Anhysteretic Remanent Magnetizations

It was stated above that a rock sample placed in a steady magnetic field appears to be magnetized when this field is removed. This remanent magnetization is called isothermal remanent magnetization (IRM) and is designated J_r. In weak magnetic fields, such as the magnetic field of the Earth, the IRM value is proportional to the square of the field intensity:

$$J_r = bH^2/2. \tag{1.35}$$

This relationship is called the Rayleigh law and b is the Rayleigh constant. The IRM acquired in the Earth's magnetic field is much weaker than any other type of magnetization, thus, IRM is of minor importance in paleomagnetology. A lightning strike is the only natural phenomenon that creates a strong isothermal remanent magnetization. Only in laboratory tests does IRM play an important role.

A rock sample in an alternating magnetic field with a slowly decreasing amplitude down to zero and in the absence of a steady magnetic field becomes demagnetized. This state is called the zero state (there is also an absolute zero state that results from the heating of a sample above the Curie point and its subsequent cooling in the absence of a magnetic field). If this process occurs in the presence of a steady magnetic field, the sample acquires a remanent magnetization that greatly exceeds the isothermal one appearing in the same steady field. This magnetization is called an anhysteretic remanent magnetization (ARM) and is designated J_{ri}. Anhysteretic magnetization might also appear in the case of a lightning strike. In paleomagnetic studies natural ARM is merely a disturbance; in laboratory tests on the magnetic properties of rocks, however, it might be of great importance.

As a graphic example of the magnetization processes, the Preisach-Néel diagram is commonly used (Néel 1955; Sholpo 1977). In multidomain grains displacement of the domain wall does not occur constantly, but proceeds in irreversible jumps related to the overcoming of energy barriers. Every jump is equivalent to remagnetization of an imaginary particle characterized by a rectangular hysteretic cycle (Fig. 1.15). A rock sample might be regarded as a set of such particles. Suppose the coordinate plane ab is divided into two half-planes by a line a = b(0'0") (Fig. 1.16). The particles which have $b<a$ belong to the upper left half-plane and consequently, these are characterized by a reversible magnetization process. These particles, after removal of the field, become demagnetized and eventually have no magnetic memory. The remaining particles from either the first or third quadrants have magnetic memory, because, irrespective of the amplitude and direction of the external field, they acquire a similar state after the field is removed. Thus, only the particles of the second quadrant carry information on the external magnetic field (Fig. 1.17a).

If a sample is placed in an alternating magnetic field with a large initial amplitude and this field is gradually diminished to zero, then the particles with $/a/>/b/$ would acquire a "negative" magnetization, and the particles with $/a/</b/$ would acquire a "positive" one (the direction is conventional), i.e. the sample would be in a demagnetized or zero state (Fig. 1.17b). If this sample is placed in a steady magnetic field H, then the particles with $a<H$ would become remagnetized. Magnetization of an assemblage is proportional to the square of a triangle OHK (Fig. 1.17c) and depends on the particle density in the diagram, i.e. on the particle distribution in the critical fields a and b. In the area of weak fields (of the order of the Earth's field), this density is considered to be constant and thus, the remanent magnetization is proportional to H^2, in accordance with Rayleigh's law.

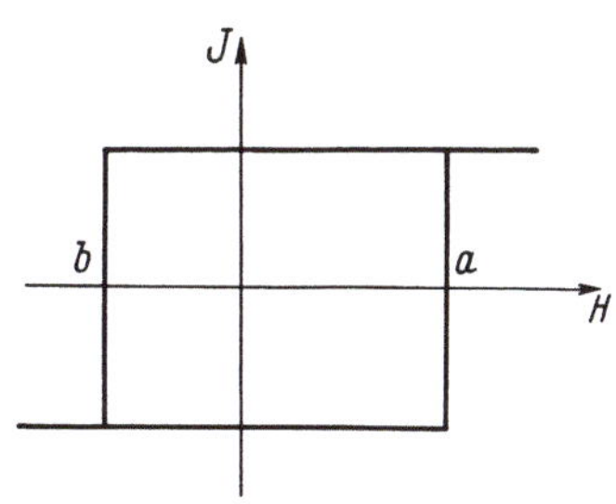

Fig. 1.15. Elementary hysteresis cycle

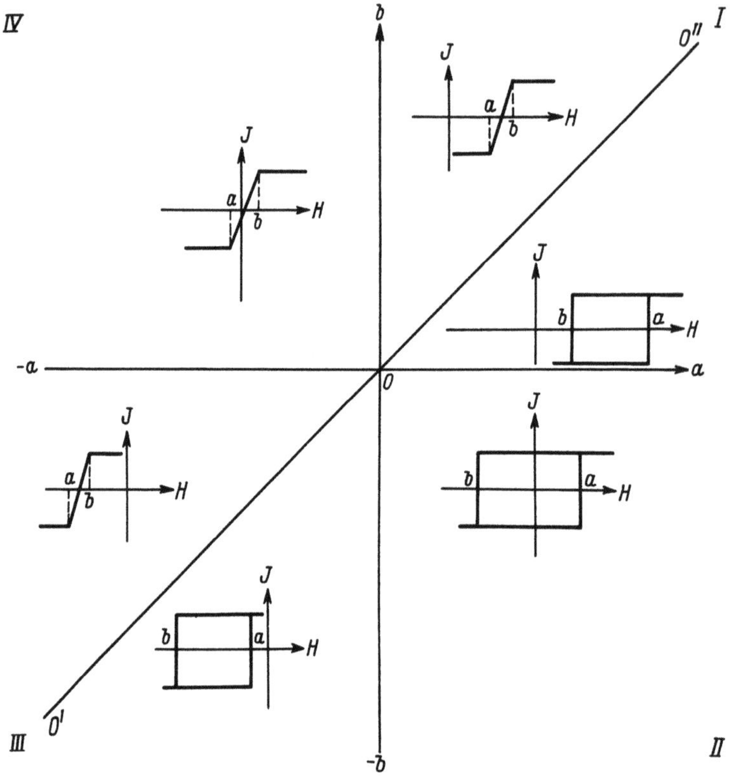

Fig. 1.16. Distribution of elementary hysteresis cycles on ab plane

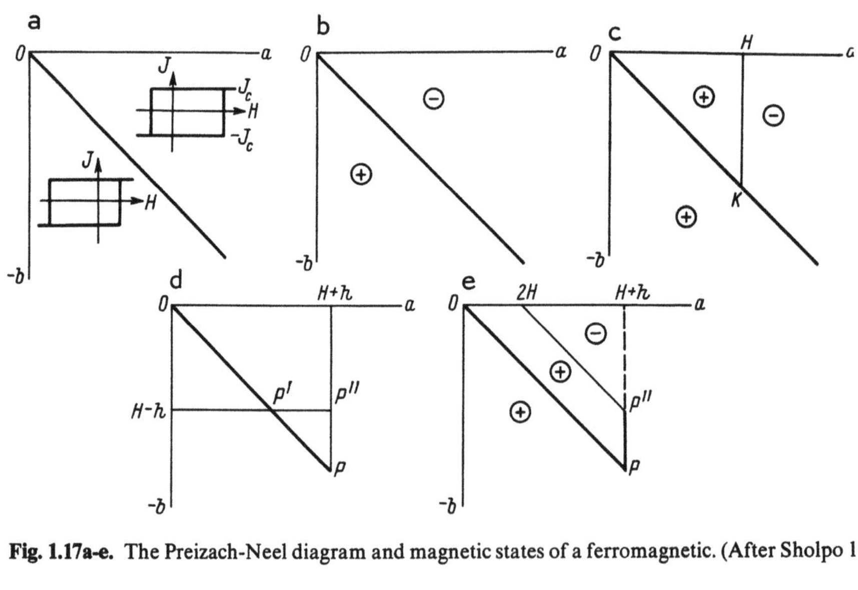

Fig. 1.17a-e. The Preizach-Neel diagram and magnetic states of a ferromagnetic. (After Sholpo 1977)

If a sample is placed simultaneously in both a steady and alternating field and the amplitude of the latter is gradually diminished to zero, then during the positive half of the period, the particles of the triangle 0(H + h)P and during the negative half of the period those of the triangle 0(H–h)P' (Fig. 1.17d) become remagnetized. Thus, during one period the triangle PP'P" becomes magnetized in a conventionally positive direction. When the amplitude of the alternating field decreases, the triangle would be transformed into a band (Fig. 1.17e). The anhysteretic magnetization is then proportional to the area of the band 0(2H) P"P. Consequently, simultaneous action of a steady field H with an alternating one is equivalent to a movement to the right of the diagonal in the Preisach-Néel diagram by a distance of a = H. Similar reactions can occur when the action of a steady field is accompanied by some other process that results in magnetization of a sample (cooling, orientation of sedimentary particles, chemical alterations, pressure).

The establishment of particle distributions in the critical fields facilitates the diagnosis of the structure of ferromagnetic materials present in a rock; single-domain, noninteracting particles exhibit symmetrical hysteretic cycles, i.e. they group in the area of the diagonal in the Preisach-Néel diagram.

1.4.7 Stability of Remanent Magnetization

The stability, or resistivity, of any remanent magnetization denotes the ability to preserve its value and direction (Khramov and Sholpo 1967). Remanent magnetization might change under the effect of various factors: steady and alternating magnetic fields, temperature variations, chemical and mineralogical transformations, mechanical stresses and relaxation phenomena. Thus, changes may be due to any agents that provide additional magnetization of ferromagnetic materials, thus resulting in the occurrence of any type of remanent magnetization. If a steady field is absent, all of these effects cause a decrease in the remanent magnetization and if they are very intensive, the remanent magnetization disappears completely (Khramov and Sholpo 1967).

Differences in the types of magnetization are due to the fact that the energy barriers in a ferromagnetic sample are distinguished not only by their size, but also by their peculiar physical properties. Thus, the same barrier may be more or less sensitive to one or another additional effect (alternating field, temperature, time, etc). It means that there are distinct populations of barriers more easily overcoming by the action of certain additional effect. This selectivity and different values of J_{rt}, J_{ri}, J_{rc}, J_{ro}. etc., can be obtained for the same field H. It also explains the varying resistance of these magnetizations relative to the same additional effects in the absence of an external field. The total thermoremanent magnetization J_{rt} increases most quickly with an increase in field, that is, after the action of a given steady field H, it has the highest value. Anhysteretic magnetization J_{ri} increases at a slightly lower rate and the increase of isothermal J_r is the lowest. The remaining types of magnetization can all be represented by curves situated between J_r and J_{ri} and approaching the first one with a gradual decrease of any incremental effects (Yanovsky 1978).

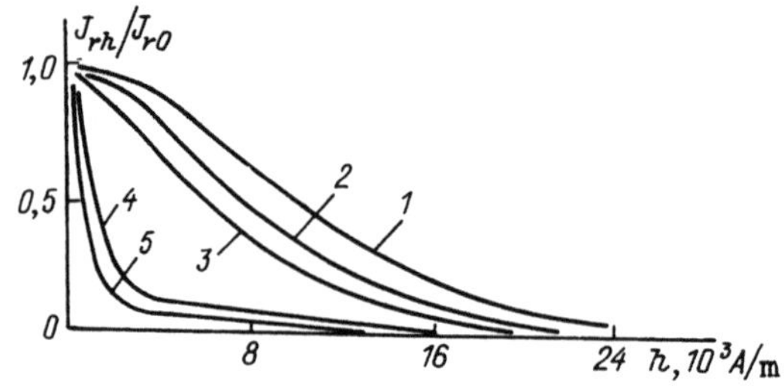

Fig. 1.18. The alternating field demagnetization curves for different types of remanence of equal initial values. (Khramov and Sholpo 1967). *1* Thermoremanent magnetization; *2* anhysteretic remanent magnetization; *3* viscous magnetization; *4* dynamic and *5* isothermal remanent magnetization

Remanent magnetizations of different types, equal in value and thus obtained at different amplitudes of a steady field H, are situated in the same order according to their resistance to the action of an alternating magnetic field h (Fig. 1.18). The total thermoremanent magnetization J_{rt} appears to be the most resistant. The stability of the chemical magnetization J_{rc} is not definitely known, but it is reported to vary over a rather wide range and in the case of single-domain grains it approaches the stability of J_{rt}. The anhysteretic magnetization J_{ri} is less resistant. The stability of the viscous magnetization J_{rv} steadily increases from the time of its formation. Since the upper limit of magnetization J_{rv} cannot be obtained experimentally, the behaviour of its corresponding curve cannot be determined. The values of J_{rv} obtainable in laboratory tests generally form curves of demagnetization by an alternating field that pass below the curves of J_{ri}. The curves of decay of the isothermal magnetization J_r are situated still lower and disappear completely in the alternating field with an initial amplitude equal to the field in which this magnetization was formed. For any other types of magnetization, the demagnetizing, alternating field greatly exceeds this value (Khramov and Sholpo 1967).

Resistance of remanent magnetization to heating is shown by the curves of temperature demagnetization. The thermoremanent and chemical magnetizations exhibit the highest stability, the isothermal and viscous ones the lowest.

The stability of remanent magnetization through time has not been thoroughly investigated, as the time intervals are necessarily limited to laboratory conditions and the spontaneous changes of magnetization thus obtained are quite insignificant. In accordance with the general statements on ferromagnetism, it can be concluded that the highest stability is also attributable to thermoremanent and chemical magnetizations (Khramov and Sholpo 1967; Yanovsky 1978).

Any sample of ferromagnetic material might possess several types of remanent magnetization at the same time, although generally with different directions. As the appearance of each magnetization is caused by a definite process, the resultant magnetization of a sample represents the sum of several magnetizations of various types. With the sample placed in an alternating field h and with an increasing amplitude, the less resistant components of magnetization, for instance, J_r or J_{rpt}, would be first destroyed, then J_{rv}, J_{ri} and finally J_{rc} and J_{rt}. The same might be stated about other demagnetizing effects. Figure 1.19 depicts the curve of demagnetization by an alternating field as an example. The resultant remanent magnetization equals zero when the thermoremanent and isothermal magnetizations, of the same value and of opposite directions, are added together.

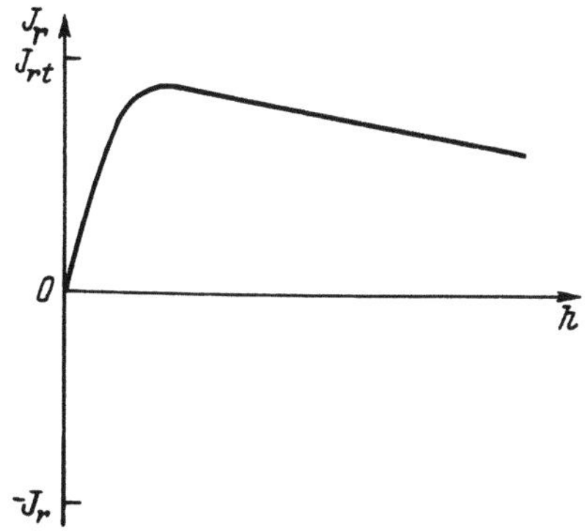

Fig. 1.19. Alternating field demagnetization curve of a specimen which carry thermoremanent and isothermal remanent magnetizations of equal values and opposite directions. (Khramov and Sholpo 1967)

1.4.8 Spectra of Demagnetization and the Principles of Magnetic Cleaning

The difference in stability of any magnetization to demagnetizing factors forms the basis for isolating components of NRM by magnetic cleaning. The principle of magnetic cleaning can be demonstrated in terms of the demagnetizing spectra dJ_{rx}/dy, where x is the type of magnetization; y is the demagnetizing factor. These spectra are related to the coercive spectra of the blocking temperature and relaxation time, i.e. with distributions of effective particles (single-domain grains or barriers in multidomain assemblages) according to the coercive forces, blocking temperatures and relaxation times in a given sample of a ferromagnetic material.

Suppose two different magnetizations J_{ra} and J_{rb} are subsequently created and destroyed in a sample by an alternating magnetic field and two curves of demagnetization $J_{ra}(h)$ and $J_{rb}(h)$ are plotted. According to these data their coercive spectra are drawn, i.e. the curves dJ_{ra}/dh and dJ_{rb}/dh. Now, if both magnetizations J_{ra} and J_{rb} are created simultaneously in a sample, then, when demagnetizing this sample, due to the additivity of different magnetizations, one obtains the coercive spectrum $dJ_r/dh = dJ_{ra}/dh + dJ_{rb}/dh$. Thus, the following cases are shown in Fig. 1.20.

1. The coercivity spectra J_{ra} and J_{rb} do not overlap; with the amplitude of an alternating field increasing to a certain value h_1, i.e. between the boundaries of these spectra, the magnetization J_{ra} is absolutely destroyed and J_{rb} is completely separated (Fig. 1.20a).
2. The coercivity spectra partially overlap; at $h \geqslant h_1$ the component J_{ra} is absolutely destroyed and J_{rb} is partially destroyed, i.e. it is possible to reconstruct the direction of magnetization J_{rb}, but not its magnitude (Fig. 1.20b).
3. The coercivity spectra overlap in such a way that at any value of h, only a partial isolation of components J_{ra} and J_{rb} is possible (Fig. 1.20c).
4. The coercivity spectra fully overlap; even a partial isolation of components J_{ra} and J_{rb} is impossible (Fig. 1.20d).

These can be observed during thermo- and chemical demagnetization, and during the maintenance of samples with two- or more component magnetizations.

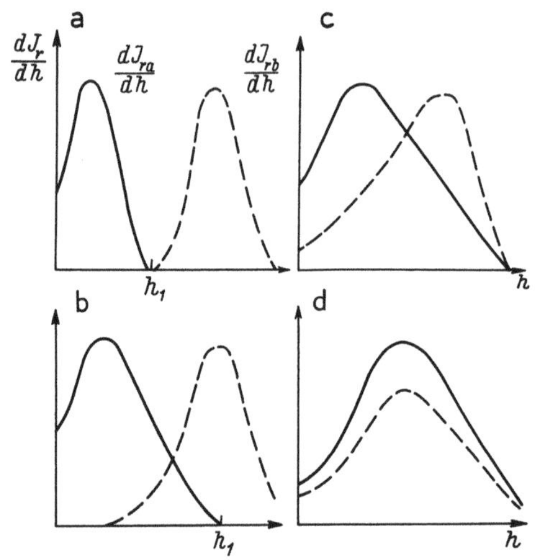

Fig. 1.20a-d. Demagnetization spectra for rocks with two components of magnetization. Four probable relations between two spectra: **a** non-overlapping spectra; **b** partially overlapping spectra with two "windows" ($h<h_0$ and $h>h_1$); **c** partially overlapping spectra without windows; **d** fully overlapping spectra

In complex magnetics (i.e. all rocks) the components of J_r with different spectra of demagnetization might appear not only when the rock has acquired different magnetizations, but also when two or more carriers of magnetization are present: the grains of different minerals (for instance, of magnetite and hematite) or different generations of a single mineral.

Partial demagnetization of a rock with the aim of isolating components of different stability in relation to this demagnetizing effect is called magnetic cleaning. From the above discussion it is evident that magnetic cleaning may be carried out in different ways and, thus, several types of cleaning are distinguished: alternating field, thermal, chemical or temporal cleaning.

Application of these different types of magnetic cleaning depends on whether the preferred cleaning appears useful and on the type of magnetization to be eliminated. In some cases it becomes necessary to use complex cleaning, i.e. to apply two or three types of cleaning to the same samples.

1.5 Ferromagnetic Properties of Rocks

1.5.1 Mineral-Carriers of the Ferromagnetic Properties of Rocks

Magnetic characteristics of igneous rocks and also of sedimentary and metamorphic rocks are mainly due to the accessory minerals that possess ferromagnetic properties. Magnetite and its varieties (titanomagnetite, manganomagnetite, etc.), maghemite, hematite and hemoilmenite belong to these minerals. Hydroxides of iron (goethite, hydrogoethite, hydrohematite, lepidocrocite) in sedimentary rocks and pyrrhotite in igneous and metamorphic rocks can also be quite important.

The iron oxide minerals in rocks that are most widely distributed belong to the three main series of solid solutions of the ternary system FeO-Fe_2O_3-TiO_2: titanomagnetite series Fe_3O_4-Fe_2TiO_4 (an inverse spinel structure of crystalline lattice), ilmenite-hematite Fe_2O_3-$FeTiO_3$ (a rhombohedral structure) and pseudobrookite Fe_2TiO_5-$FeTi_2O_5$ (an orthorhombic structure). Members of the first two series represent ferrimagnetic and weak ferromagnetic materials at room temperature (Nagata 1961).

In magmatic and commonly in metamorphic rocks the minerals of the titanomagnetite series are the main carriers of ferromagnetism. The crystalline lattice of all these minerals represents a cubic structure of an inverse spinel. Magnetite, Fe_3O_4, is an extreme member of the series with a typical saturation magnetization $J_s = 480 \times 10^3$ A/m, and a Curie point $T_C \approx 580°C$. Titanomagnetites are the intermediate members of the series forming solid solutions of magnetite and ulvöspinel Fe_2TiO_4. With increasing concentration of ulvöspinel, which is paramagnetic at room temperature ($T_C = -153°C$), the values of J_s and T_C decrease. However, the natural titanomagnetites rarely have J_s lower than 80×10^3 A/M and T_C lower than 100°C, because titanomagnetites of intermediate composition are metastable at room temperature and exsolve into two phases, one of which (with a lower concentration of ulvöspinel) maintains a higher magnetization J_s (Sholpo 1977).

Natural titanomagnetites are commonly of a composition departing from the line Fe_3O_4-Fe_2TiO_4 towards a higher concentration of oxygen and thus forming cation-deficient varieties (Verhoogen 1962). If this deviation is great, then these minerals are called titanomaghemites. Titanomaghemites, approaching the ilmenite-hematite series in their composition and representing the products of low-temperature oxidation of minerals of the titanomagnetite series, preserve an inverse spinel structure. Maghemite γ-Fe_2O_3 is an extreme member of the titanomaghemite series and also has an inverse spinel structure; $J_s = 415 \times 10^3$ A/m, $T_C = 740°C$. At high temperature maghemite is transformed into hematite (Pechersky et al. 1975; Sholpo 1977).

The ilmenite-hematite solid solution series is represented by minerals with much lower values of J_s compared to those of the minerals of the titanomagnetite series. Hematite, α-Fe_2O_3, is an extreme member of the series and a weak ferromagnetic with $J_s = 2.6 \times 10^3$ A/m and $T_C = 678°C$. The other extreme member of the series, ilmenite $FeTiO_3$, is a paramagnetic at room temperature. The intermediate members of the series, hemoilmenites (the solid solutions of hematite and ilmenite) with a concentration of ilmenite from 0 to 45 mol%, are also weak ferromagnetics, like hematite, and from 45 to 80 mol% are ferrimagnetics (at room temperature) with the maximal value of $J_s = 140 \times 10^3$ A/m at $T_C = 250°C$ (Pechersky et al. 1975).

In nature the following hydroxides of iron commonly occur: goethite, hydrogoethite, lepidocrocite, hydrolepidocrocite and hydrohematite. Goethite α-FeOOH and hydrogoethite are antiferromagnetics, but they often possess a remanent magnetization due to internal defects; for hydrogoethite $J_s \approx 0.5 \times 10^3$ A/m, for goethite $J_s \approx 4.8 \times 10^3$ A/m. Hydrogoethite is not stable and upon heating to 200°C it transforms into goethite and with further heating, into hematite. Lepidocrocite, γ-FeOOH, is antiferromagnetic and shows extremely weak ferromagnetic properties; at 150°–275°C it transforms into maghemite and then into hematite. Hydrohematite α-$Fe_2O_3 \cdot nH_2O$ is also antiferromagnetic with an extremely weak ferromagnetism, $J_s = 2$–700 A/m; at 250°C it transforms into hematite (Pechersky et al. 1975).

Ferromagnetics, in addition to a saturation magnetization J_s and a Curie point T_C, are characterized by hysteretic properties: an initial magnetic susceptibility $\varkappa_0$, a saturation field H_s, a saturation remanent magnetization J_{rs}, a coercive force H_c and a residual coercive force of maximal remanence H'_{cs}. These values, however, depend strongly on the grain size and shape, and on defects, inclusions, stresses, etc. present in the grains. Thus, the parameters, which are called structural-susceptible (enumerated above), should be used to characterize the rocks containing these minerals and not the minerals proper.

Mineral-carriers of the ferromagnetic properties in rocks are present in different forms (Khramov and Sholpo 1967). In magmatic and metamorphic rocks two forms of minerals are recognized: actual crystals, commonly with mineral intergrowths (for instance, magnetite and ilmenite) and as "ore-dust", a constituent that is unsuitable for optical diagnosis and which represents fine crystalline precipitations along the crystal margins and inside crystals of rock-forming minerals, and also in the noncrystallized portion (glass) of a rock. The division of grains of ferromagnetic minerals into three groups is possible according to the petrographic features of magmatic rocks, and this division is very important for paleomagnetism. The three groups are as follows: (1) idiomorphic grains of primary-magmatic minerals; (2) grains of secondary minerals formed at the stage of rock cooling (for instance, the products of high-temperature oxidation of titanomagnetite); (3) grains of secondary minerals formed during the processes that post-dated the formation of an initial rock over a long time interval.

Three forms are characteristic for the ferromagnetic carriers in sedimentary rocks. The first form encompasses clastogenic ore grains of magnetite and its varieties, martite and hematite (specularite): these grains belong to the sand-silt fraction (from 0.01 to 2 mm); grains of quartz and biotite flakes with magnetite inclusions also belong to this fraction. The second form includes clastogenic and authogenic hematite, maghemite (locally magnetite) dispersed in a rock; the particle size varies from micrometer fractions to several tens of micrometers, all belonging to the clay fraction. The third form includes various, late diagenetic and epigenetic compounds formed of iron hydroxides, products of oxidation and substitution by siderite, pyrite and magnetite.

1.5.2 Certain Magnetic Characteristics of Rocks

Magnetic properties of rocks, in comparison to those of continuous polycrystalline ferro- and ferrimagnetic materials, appear to be specific. This is due, first of all, to the fact that the accessory minerals are the carriers of ferromagnetism and their grains are dispersed in a "nonmagnetic" matrix formed of para- and diamagnetic minerals. The magnetic grains are arranged differently in size and shape, and their concentrations and relationships also differ. Thus, the magnetic parameters of rocks are diverse. For certain classes of rocks and magnetic parameters, it is only

Table 1.2. Relationships between ferrimagnetic constituent characteristics of rocks and their magnetic properties

Characteristic of ferrimagnetic	Magnetic properties[a]								
component of rocks	T_c	T_b	H_s	H_c	J_s	J_{rs}	λ	$\varkappa_0$	J_n
Composition	++	++	++	++	++	++	+	++	++
Percentage	–	–	+	++	++	++	+	++	++
Grain size	–	++	++	–	–	++	+	+	++
Grain shape	–	+	+	–	–	+	++	+	+
Distribution of grains in rocks	–	+	+	–	–	+	++	+	+

[a] ++ Denotes strong correlation; + weak correlation; – no correlation.

possible to determine the most typical and probable values, as for others, only the limits of parameter change can be indicated. Table 1.2 presents the relationships of grain characteristics and the magnetic properties of rocks.

Magnetic Susceptibility. Generally in magmatic, metamorphic and in relatively strongly magnetic sedimentary rocks, an initial magnetic susceptibility $\varkappa_0$ is mainly due to the concentration of minerals of the titanomagnetite series in which the value $\varkappa_0$ reaches 20 SI. The percentage of these minerals in magmatic rocks varies over a wide range, i.e. from traces (less than 0.01%) to 10%. It is related to the chemical composition of rock-forming magma and the thermodynamic conditions of crystallization and also to the intensity and nature of secondary rock alterations (Pechersky et al. 1975; Sholpo 1977).

Magnetic susceptibility of sedimentary and weakly magnetic crystalline rocks is created by a whole set of accessory minerals possessing ferromagnetic properties, whereby magnetite is again the principal mineral. The most widely distributed rock-forming minerals of sedimentary rocks are as follows: quartz, calcite, feldspars, gypsum, anhydrite and halite. These are diamagnetics with $\varkappa = -(0.2\text{–}0.8) \times 10^{-5}$ SI and contribute little to the magnetic susceptibility of a rock. The most important paramagnetic minerals are siderite, chlorite, pyrite, ilmenite, biotite and sometimes clay minerals. However, their magnetic characteristics are mainly due to admixtures, relics and newly formed compounds of iron oxide minerals possessing ferromagnetic properties. Thus, in natural siderite the admixtures of maghemite and hydroxides of iron are common; quartz grains, biotite and chlorite flakes contain small inclusions of magnetite; aggregates of clay minerals are usually impregnated with iron hydroxides. These inclusions and admixtures are responsible for the higher values of magnetic susceptibility and ferromagnetic properties of the enumerated minerals, their pure varieties belong to paramagnetics $\varkappa = 10^{-5}\text{–}10^{-3}$ SI.

This strong dependence of the magnetic susceptibility of rocks on the concentration and composition of accessory minerals results in a wide range of $\varkappa$ change. The histograms of $\varkappa$ for every lithologic variety are characterized by the principal maximum, which is shifted towards the lower values and has a long "tail" towards the higher values (Khramov 1958; Dortman 1976). Distributions of $\varkappa$, thus, differ greatly from the normal ones and they are closer to the lognormal type for sedimentary (Fig. 1.21) and magmatic rocks (Belokon' et al. 1973).

A direct functional dependence of the percentage of iron oxide minerals and of the magnetic susceptibility of rocks has not been registered, but a statistical dependence of this type exists. The statistical character of this dependence is explained by the fact that the magnetic susceptibility of a rock is affected not only by the ferromagnetic particle concentration, but also by the composition, size, shape and preservation of these particles.

Natural Remanent Magnetization. The principal law in the distribution of magnetization J_n is lognormal in magmatic rocks as it is for the magnetic susceptibility, which might be explained by the similar distribution of concentrations of the titanomagnetite minerals present in rocks as an "admixture" (Belokon' et al. 1973; Sholpo 1977). Sedimentary rocks are characterized by the same lognormal law of distribution of J_n for the same reason. These distributions are recorded not only in a narrow range of rocks, for instance, in the Upper Carboniferous red argillites of the Donbas (Fig. 1.22a), but also in a wider range of complexes, for instance, in the Upper Permian marls, clays and siltstones of the northern and eastern parts of the Russian Platform (Fig. 1.22b). The distribution of J_n in sedimentary rocks of the USSR area is also in accord with the lognormal law (Fig. 1.22c).

A direct relationship between the lithologies and magnetization J_n does not exist. The value of J_n depends greatly on the rock origin, and the percentage and form of iron oxides. Thus, the most probable value of J_n in red beds is one order higher than in other sedimentary rocks (for example, in the USSR formations 13×10^{-3} and 1.6×10^{-3} A/m, respectively). A relationship between magnetization J_n and rock colour (its red component intensity) is recorded within any variegated sequence. This relationship is not functional, but of a statistical nature; the correlation coefficient for variegated rocks of the western Turkmenistan, for instance, amounts to 0.57 (Khramov 1958). A decrease of J_n by the gradual removal of an iron oxide component has shown that the contribution of red pigment and ore grains in rock magnetization might vary considerably (Roy and Park 1974).

The particle type, to which the magnetization J_n of red beds is related, is exceptionally important in paleomagnetism, for it is closely related to the time of appearance of the initial magnetization of these rocks. The red beds are one of the most important objects in paleomagnetic studies due to their higher magnetization J_n and their resistivity to demagnetization effects.

In magmatic as well as in sedimentary rocks, heterogeneity of magnetization J_n is often observed even if the rocks are fine-grained and homogeneous in colour and composition. Great variations of J_n within the same flow, sheet or sample might be explained by the rather small number of grains that create the magnetization J_n of a rock or by their concentration as beds, aggregates, etc.

Magnetic Anisotropy (Khramov and Sholpo 1967; Sholpo 1977). Different rock structures are responsible for an anisotropy λ of their physical properties, including the anisotropy of magnetic susceptibility $\lambda\varkappa$. Thus, the magnetic susceptibility is not a scalar, but a tensor. There are two types of magnetic anisotropy: crystallographic and shape anisotropy. Crystallographic, magnetic anisotropy is caused by

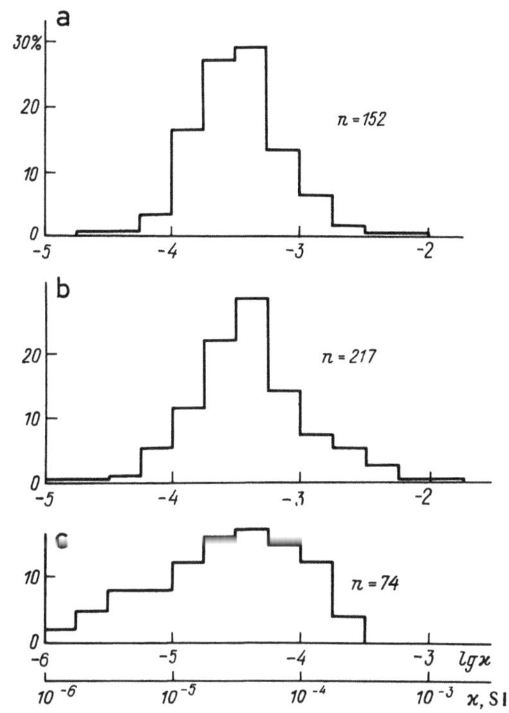

Fig. 1.21a-c. Log-normal distribution of magnetic susceptibility for sedimentary rocks. **a** The Upper Carboniferous red argillites of the Donets Basin; **b** the Upper Permian marls, clays and siltstones of the northern and eastern parts of the Russian platform; **c** mean values for 74 sedimentary formations in USSR territory. (Distribution of mean values, after Ozerskaya, pers. commun.)

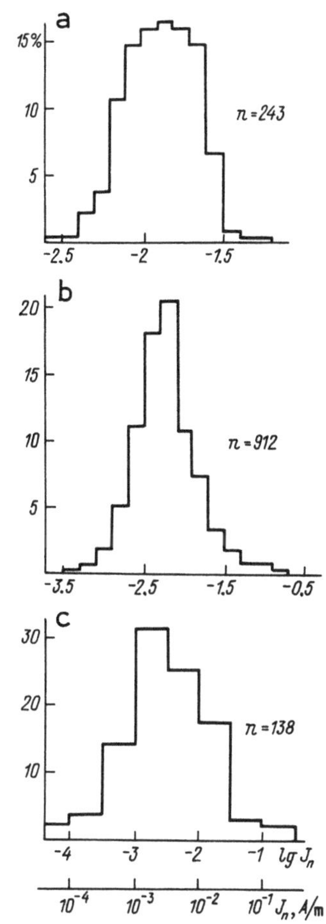

Fig. 1.22a-c. Log-normal distribution of NRM for sedimentary rocks. **a,b** The same groups of rock as **a** and **b** in Fig. 1.21; **c** mean values for 138 formations from USSR sedimentary basins

the presence in a ferromagnetic crystal of the so-called axes of "easy" and "hard" magnetization that represent specific crystallographic axes differing for crystals of different lattice. Upon magnetization of a monocrystal, the magnetization vector departs from the magnetizing field direction towards the nearest axis of an easy magnetization. Thus, if the axes of ferromagnetic crystals of a rock display a preferred orientation, the remanent magnetizations associated with them would not coincide with the direction of the ancient geomagnetic field from the very beginning.

Shape anisotropy is of great importance in rocks, since it is related to the fact that magnetization of a ferromagnetic body occurs not only in the applied field, but also in the field created by the body itself. If the body shape is nonisometric, the magnetization vector would deviate from the direction of the field H towards the long axis of the body. This deflection might be very great if magnetization values are high and the body is extremely elongated. Anisotropy of magnetic properties due to shape effects can be of two modes: macro- and microscopic. Macroscopic anisotropy is related to the shape of a geologic body proper; microscopic anisotropy is due to oriented, elongated or flattened ferromagnetic grains enclosed within rocks. Thus, the bedded structure of strongly magnetized rocks might explain the consistency of the remanent magnetization in the direction, not of an external magnetic field, but of the bed strike as in the KMA area (Krutikhovskaya et al. 1964). In relatively strongly, magnetized igneous rocks, the marginal parts of flows and intrusions might show magnetization with abruptly distorted directions. This effect is enhanced by the fact that the blocks could be mobile in the marginal parts at relatively low temperature when a major portion of the thermoremanent magnetization had already been acquired.

When a sedimentary rock is being formed, the ferromagnetic, nonisometric particles could become components. The long axes of these particles, due to shape anisotropy, would be the principal directions for their remanent magnetization. These particles tend to be situated with their long axes along the current direction. In stagnant water they are mostly situated in a horizontal plane, and while depositing they show a tendency to turn with their long axes in the direction nearest to the horizontal. Subsequent rock compaction continues to affect the magnetization direction, resulting in a some inconsistency between the direction of magnetization J_n in a rock and the direction of the Earth's field at the time of rock formation.

Anisotropy of magnetic susceptibility is usually estimated by the value

$$\lambda_\varkappa = \frac{\varkappa_{max} - \varkappa_{min}}{2\bar{\varkappa}} = \frac{\varkappa_{max} - \varkappa_{min}}{\varkappa_{max} + \varkappa_{min}} \qquad (1.36)$$

or

$$A_\varkappa = \varkappa_{min}/\varkappa_{max}.$$

The relationships of these values are given in the formulae

$$\lambda_\varkappa = \frac{1 - A_\varkappa}{1 + A_\varkappa}\,; \qquad A_\varkappa = \frac{1 - \lambda_\varkappa}{1 + \lambda_\varkappa}\,;\ \text{where} \qquad (1.37)$$

the values $\lambda_\varkappa = 0$ and $A_\varkappa = 1$ correspond to a magnetically isotropic rock and $\lambda_\varkappa = 1$ and $A_\varkappa = 0$, to a maximally anisotropic rock.

Anisotropy of magnetic susceptibility is high most often in metamorphic rocks and in high-grade sedimentary rocks, where it reaches $\lambda_\varkappa = 0.75$ (Sholpo 1977). An unaltered, sedimentary rock exhibits much lower anisotropy, and depends mainly on the tectonic situation of the area. In sedimentary rocks from fold zones, the anisotropy of magnetic susceptibility is appreciably higher than in similar rocks from platform areas. Thus, for the Permian and Triassic rocks of the Verkhoyansk fold area $\lambda_\varkappa = 0.05$–0.15 (Dortman 1976) and for the sedimentary cover of the Siberian platform $\lambda_\varkappa \leq 0.05$.

In addition to the initial magnetic susceptibility $\varkappa_0$ (at $H \to 0$), magnetic anisotropy is commonly characterized by another limiting value, i.e. the saturation remanent magnetization J_{rs} (at $H \to H_s$). Sholpo (1977) reported that the estimation of anisotropy of magnetization J_r in weak fields (in the Rayleigh region) has certain advantages. The main advantages are: the high value of λ_r in comparison to $\lambda_\varkappa$ in the same sample and the absence of effects of paramagnetic minerals that appreciably contribute to the magnetic susceptibility in weakly magnetized rocks and, thus, diminish the anisotropy of the ferromagnetic component. The relationship $A_r \leqslant A_\varkappa^3$, given in the work of Sholpo (1977), provides an advantage in the study of magnetic anisotropy using the J_r value, which is generally obtained in the fields $H \approx 1 \times 10^3$ A/m. This relationship and consequently the reliability of magnetic anisotropic characteristics of a rock of only one of the values $A_\varkappa$ or A_r is correct, only when the contributions of different grains into parameters $\varkappa$ and J_r are equal. Strictly speaking, it is correct only for a particular case. Furthermore, $A_\varkappa$ (at $H = 40$ A/m) and A_r (at $H \approx 1 \times 10^3$ A/m) characterize the magnetic anisotropy of the magnetically soft component of a rock. Therefore, the anisotropy of the saturation remanent magnetization J_{rs} must be estimated in order to determine the magnetic anisotropy due to magnetically hard, small grains.

Hysteretic Properties. In sedimentary rocks these properties account mainly for the size and relative concentration of grains of magnetite, hematite (specularite) and pigment particles of hematite and hydroxides present in the clay fraction of rock. Red beds exhibit the most typical hysteretic properties. The varities that do not contain magnetite have an almost rectangular hysteresis loop and their saturation remanent magnetization is nearly equal to that of the saturation magnetization ($J_{rs} \approx J_s$). Generally, $J_{rs} = (3\text{–}60)$ A/m; $H'_{cs} = (0.1\text{–}0.6) \times 10^6$ A/m; $J_n : J_s = 10^{-4}\text{–}10^{-3}$ (Nagata 1961; Khramov et al. 1974). The lower limit of H_{cs} refers to the rock rich in specularite, the upper limit is typical of red beds, the magnetic properties of which are entirely responsible for the presence of iron pigment. In general, saturation in the fields of the order of 10^6 A/m is not achieved (Fig. 1.23). A destructive field of natural remanent magnetization (coercive force of NRM) of red beds is represented by $H'_c = (2\text{–}16) \times 10^3$ A/m, factor $Q = 1\text{–}50$. Parameters of magnetic hardness of other sedimentary rocks usually reach the lower limits of the values indicated. Nearly all groups of sediments show a direct correlative relationship between the two pairs of values: Q and H'_c, and J_{nv} and $\varkappa$ (Sholpo 1977).

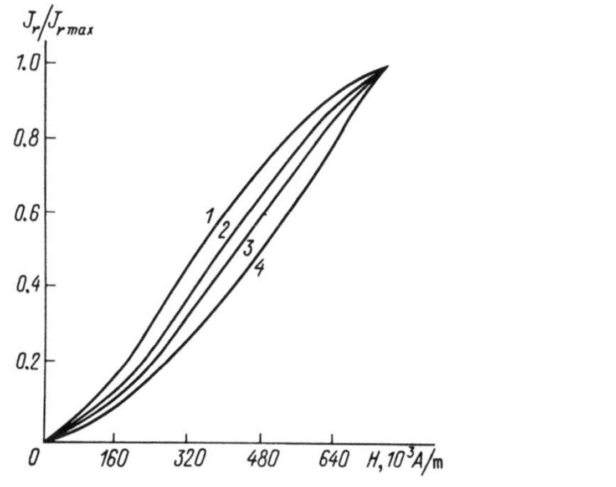

Fig. 1.23. Isotermal magnetization of red-coloured clays and claystones. *1,2* The Donets Basin (*1* Upper Carboniferous; *2* Lower Permian); *3,4* southern Siberian plate (*3* Ordovician; *4* Upper Cambrian)

Thermomagnetic Characteristics. The determination of the mineral-carriers of ferromagnetic properties, includes the examination of Curie points, the temperature relationship of saturation magnetization $J_s(T)$ and $J_{rs}(T)$, and also the changes of these values, and the saturation destructive field H'_{cs} of the isothermal magnetization of rock samples after thermal maintenance. The methodology in the determination of minerals, according to their magnetic transformations, have been reported in detail (Pechersky et al. 1975; Burov and Yasonov 1979).

When studying a magmatic rock, it is most important to determine the members of the titanomagnetite series and their alteration products. The composition of these products depends on the natural processes responsible for their formation. Single-phase oxidation, commonly occurring under hypergenesis, and multiphase oxidation, typical of the autometamorphism at high temperatures, have been distinguished (Pechersky et al. 1975; Gapeyev and Pechersky 1983). A cation-deficient titanomagnetite (titanomaghemite) is formed during single-phase oxidation. Its presence in a rock is revealed by an increase of J_s after the first heating of a sample to the Curie temperature and further elevation of this temperature to $T_C>500°C$ during the second heating. This heterophase oxidation results in the formation of hemoilmenite lamella that divide grains into cells filled with titanomagnetite, impoverished in titanium, or magnetite. Since the heterophase system formed prevents further magnetic transformations, its magnetic characteristics slowly change during repeated heating; the Curie points and the saturation magnetization diminish slightly (Gapeyev and Pechersky 1983).

Special aspects of sedimentary rocks as carriers of ferromagnetic properties are mainly related to rather small particles of weakly magnetized, but highly coercive iron oxides and hydroxides, and ferruginous minerals unresistant to heating. Therefore, the thermomagnetic characteristics of sedimentary rocks are quite complicated. A differential method (differential thermomagnetic analysis: DTMA) has now been elaborated and widely used, and thus, the relations dJ_s/dT and dJ_{rs}/dT can be obtained and compared. The distinction of magnetic transformations in sedimentary rocks is easier, when compared to the examination of integral curves (Burov and Yasonov 1979).

1.5.3 Natural Remanent Magnetization J_n of Rocks

Generally, the natural remanent magnetization of a rock is a complex phenomenon. This complexity accounts for a variety of physicochemical factors that affect a rock during its formation and subsequent evolution. Thus, different types of magnetizations might simultaneously characterize one and the same rock in different proportions. Various types of magnetization might appear in a rock at different times and, thus, could yield various directions in accordance with the changes of the Earth's magnetic field through time. In general, a rock contains several kinds of ferromagnetic grain and each type of magnetization appears to be related to corresponding grains. Thus, while examining J_n we face the simultaneous presence of vectors of the remanent magnetization of different origin and stability that exist irrespective of one another and change according to their own laws (Khramov and Sholpo 1967; Yanovsky 1978).

All rocks, substantially heated during formation or at a later time, had to acquire thermomagnetization while cooling in the Earth's magnetic field. Total thermomagnetization is shown by a rock, when its ferromagnetic minerals has been heated above the Curie point. All effusive and intrusive rocks are just this type of rock. The total thermoremanent magnetization of rocks might reach 100 $\varkappa_0 H_T$. Thermomagnetization is known to be the most stable type of magnetization. The relaxation time for magnetization J_{rt} in most cases theoretically exceeds 10^9 and even 10^{17} years (see Sect. 1.4.2). Thus, magnetization J_{rt} would not be noticeably affected by time alone or by the geomagnetic field changes after it is formed. After cooling, such a rock preserves its magnetization for a long period and this magnetization yields the direction of the field in which it was cooled.

The natural remanent magnetization of clastic sediments arises as a detrital magnetization of small ferromagnetic particles (see Sect. 1.4.3). The smallest of these, which are single-domain size, possess a spontaneous magnetization of this ferromagnetic material. Larger particles, consisting of several domains, preserve the magnetization they acquired when present in the parent rock, i.e. the thermoremanent or chemical magnetization. Many rocks could acquire an initial remanent magnetization as a result of chemical magnetization. In certain cases the moment of appearance of magnetization J_{rc} might be precisely fixed and then these rocks can be used for paleomagnetic examinations. These are chemogenic carbonate rocks, syngenetic concretions, old soils-laterite (bauxite) and other formations. Certain iron ores of hydrothermal and sedimentary origin also fall into this group. The relaxation time of these types of magnetization is likely to be as long as that of thermoremanent magnetization.

The unavoidable superposition of viscous magnetization needs to be taken into account as its formation in a rock is especially favourable. Magnetic viscosity is relatively intense in such weak magnetic fields as the Earth's field; the time in which the rocks are in this field amounts to hundred thousands and millions of years. Therefore, the natural remanent magnetization J_n always has a viscous magnetization component J_{rv} approximately directed along the present Earth's field. The value of J_{rv} might be attained in certain rocks $(0.6–0.8)\varkappa_0 H_T$, but sometimes it amounts only to a small portion of the magnetization J_n (Khramov and Sholpo 1967). Among the secondary components of a different type of magnetization, the most common is chemical magnetization, because certain physicochemical processes occur within a rock, resulting in alternations of the primary composition and structure. These processes include the formation of new minerals and the destruction of primary minerals during oxidation and reduction, recrystallization, desintegration of solid solutions, etc.

If a rock is subjected to a second heating below the Curie point during its evolution, then a partial thermomagnetization J_{rpt} is observed. During tectogenesis, rocks might suffer from mechanical stress of varying intensity, which induces the appearance of dynamic magnetization. And finally, the remanent magnetization of some samples might include another component which is created by the magnetic field accompanying the electric discharge produced by lightning strikes. This is isothermal magnetization, but the magnetizing field exceeds the Earth's field more than hundred fold.

During alteration, metamorphic rocks acquire a remanent magnetization, the nature of which depends on the type of metamorphism. This magnetization might be chemical, thermoremanent, partial or dynamic. It either strongly distorts the premetamorphic magnetization or replaces it completely. Metamorphic rocks are useful in paleomagnetic studies when their magnetization is stable and provides information about the magnetic field at the time of metamorphism.

Thus, the natural remanent magnetization of a rock generally includes several magnetizations that appeared at various periods in its geologic history and will exhibit different directions and values.

Furthermore, it should be recalled that anisotropy is another cause of deviation between the direction of magnetization J_n and the direction of the geomagnetic field at the time of its origin (see Sect. 1.5.2). Magnetization of a rock occurs strictly along the direction of an applied field only when it has an isotropic structure. In reality, the anisotropy of magnetic properties is commonly observed, particularly in rocks with a clearly expressed structure and also when a rock has been subjected to unilateral stress. Effects of anisotropy are evident in dynamometamorphism and secondary magnetization of altered rocks, but also in initial detrital magnetization when they are associated with the occurrence of an inclination error (see Sect. 1.4.3).

1.5.4 Physical and Chronologic Components of the NRM; Paleomagnetic Stability of Rocks

The acquisition of remanent magnetization in rocks is discussed here in broad terms. At the first stage of rock formation (deposition of sediments, lava cooling, precipitation from solutions, etc.), a thermoremanent, detrital or chemical magnetization is acquired, which might be called the initial magnetization J_n^s. This magnetization inevitably decreases with time because of magnetic and structural ageing of a rock and because of the gradual decomposition of the primary minerals. The part of the initial magnetization preserved until the present is called the primary magnetization J_n^o.

For solving certain tasks while studying paleosecular variations in particular, two components in the primary magnetization need to be distinguished: the "rapid" and "slow" components (Baag and Helsley 1974a); the former is called synchronous magnetization. This magnetization is a component of magnetization J_n^o, the age of which fully agrees with that of the rock. It might be, for instance, a thermoremanent magnetization J_{rt}, acquired during the first cooling of a lava, or a depositional detrital magnetization of a sediment. The slow component of the primary magnetization is part of magnetization J_n^o, the age of which is practically the same as the age of the rock. However, this component was formed at a relatively slow rate, lagging behind the processes of lava outflow and sedimentation. The slow component might be, for instance, postdepositional detrital and chemical magnetizations that are formed under gradual compaction and dehydration of a sediment, or the low-temperature components of magnetization J_{rt} of slowly cooled intrusions.

All the components of magnetization J_n acquired after the formation of the initial magnetization and preserved up to the present are called the secondary magnetization J_n^h. A component that is recent in geologic terms, with a direction practically coincident with that of a local geomagnetic field, might be separated from J_n^h. This component is the recent magnetization J_n^p, partially resulting from the process of present weathering, and of a chemical nature, and partially of viscous origin. The remaining, older part of secondary magnetization is called the metachronous or ancient secondary magnetization J_n^m. The sum $J_n^o + J_n^m$ is called ancient magnetization J_n^a. All four components of magnetization J_n, i.e. two components of J_n^o and two of J_n^h, are the chronologic components of the natural remanent magnetization. The prefolding and postfolding magnetizations isolated by geometrical and statistical analysis of the vector distributions of J_n can also be attributed to different chronologic components (see Sect. 2.3). All of these components differ in the time of their appearance, but not in their mode of formation, i.e. not in the type of magnetization.

Components of J_n differing in the mode of formation (in the type of magnetization) and in the association of mineral-carriers of these magnetizations are called the physical components. The physical components, generally, are not consistent with the chronologic ones and the separation of the latter is the main task in any paleomagnetic study. Thus, a chemical magnetization of small hematite grains in a sediment might arise at any stage of rock evolution, from early diagenesis to recent epigenesis. Similarly a thermoremanent magnetization of igneous rock might be the result of secondary heating, etc. Relationships between the chronologic and physical components are illustrated in Fig. 1.24, which also shows the possible types of magnetization of grains that later became part of the composition of metamorphic and sedimentary rocks and that produced another chronologic component, the relict magnetization J_n^r.

Stable and unstable components isolated during magnetic cleaning (see Sect. 2.4) make up the third system of components of magnetization J_n. Earlier it was demonstrated (see Sect. 1.4.8) that these components are not generally comparable with the physical components of vector J_n. The component of magnetization J_n, isolated during magnetic cleaning and used in subsequent analysis, is commonly called the characteristic magnetization and, depending on the type of cleaning, is designated as J_{nt}, J_{nh}, J_{nT}, J_{nc}.

The impossibility of identification of components of magnetization J_n, isolated by magnetic cleaning, and the chronologic components of J_n (which also refers to the physical components in most cases) by any direct and suitable means necessitated the elaboration of a technique of stepwise treatment in order to determine the direction of the ancient magnetization J_n^a. This is part of the direct task of paleomagnetism, i.e. finding the intensity and direction of the geomagnetic field at a certain moment in the geologic past. As the moment of the initial stage of rock formation can be determined most precisely, the direct task of paleomagnetism mainly involves the examination of primary magnetization J_n^o. The following evaluations and determinations must be made: (1) to evaluate a proportion of primary magnetization J_n^o in magnetization J_n of a rock; (2) to find the intensity and direction of magnetization J_n^o; (3) to determine the nature of magnetization J_n^o and

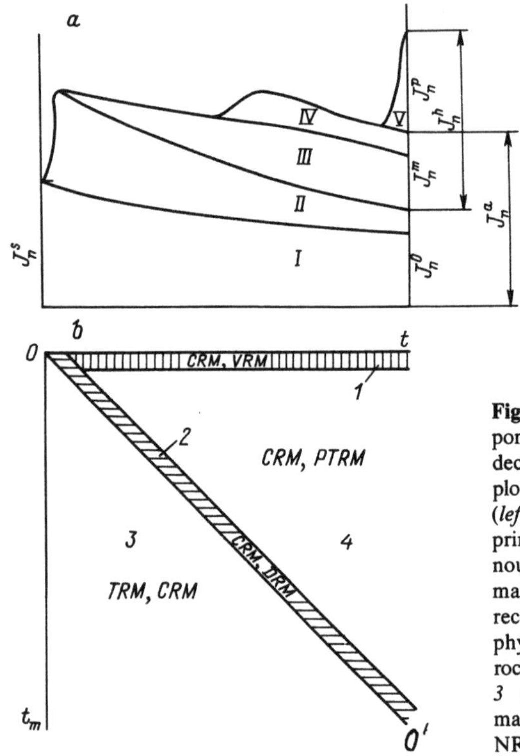

Fig. 1.24a,b. Physical and chronologic components of NRM. **a** Scheme of creation and decay of NRM components. Geological time is plotted on the abscissa from the origin of rock (*left*) to the present (*right*). Components of primary magnetization J_n^o: *I* rapid (synchronous); *II* slow. Components of secondary magnetization J_n^h: *III, IV* metachronous J_n^m; *V* recent J_n^p. **b** Relation between chronologic and physical components of NRM of sedimentary rocks. Areas of remanence: *1* recent; *2* primary; *3* relict magnetizations; 00′ synchronous magnetization line; t age of a rock, t_m age of NRM components

the way this magnetization is related to the ancient geomagnetic field H_a; (4) to determine magnetization J_n^o and the type of relationship with the field H_a, whereby the intensity and direction of the field H_a can also be determined.

The first step is called the determination of the paleomagnetic stability of a rock; the third one is the determination of the significance of natural remanent magnetization. The paleostability is the "purity" of the primary magnetization which depends on the resistance of the primary magnetization to the processes of destruction and of the secondary one, and on the intensity of these processes in any particular rock. Paleomagnetic stability is a basic physical factor that is responsible for the effective preservation of the initial direction of natural remanent magnetization of a rock. The significance of natural remanent magnetization means a correspondence between the direction of primary magnetization and the direction of the ancient geomagnetic field at the moment of rock formation in situ and the possibility of determination of the intensity of this field.

A measure of paleomagnetic stability is given in the relationship of a vector length of primary magnetization J_n^o to the sum of lengths of primary and secondary magnetizations:

$$S = J_n^o/(J_n^o + J_n^h). \tag{1.38}$$

This stability measure might also be estimated through the angles between the vectors J_n, J_n^o, J_n^h:

$$S = \sin\Delta/(\sin\Delta + \sin\delta), \quad (1.39)$$

where Δ is an angle between J_n and J_n^h; δ is an angle between J_n and J_n^o. The expressions (1.38) and (1.39) are evidently identical.

The measure of paleomagnetic stability is a nondimensional positive value revealing the proportion of the primary magnetization in the total remanent magnetization of a rock. At $S=0$ a rock is absolutely unstable and at $S=1$ it is absolutely stable.

A measure of preservation of initial magnetization is given in the relationship of the primary magnetization J_n^o to the initial magnetization J_n^s: $Q_{ns} = J_n^o/J_n^s$.

The value S is commonly called "stability" and Q_{ns} is the "ageing factor".

1.5.5 Problem of Acquisition of Reversed Magnetization in a Rock

Magnetic examinations have shown that many rocks have a natural remanent magnetization, the direction of which is almost opposite to the Earth's present magnetic field of site locality. The reversed magnetization of rocks is a common phenomenon: approximately half of the paleomagnetically stable rocks exhibit reversed magnetization (Khramov and Sholpo 1967; McElhinny 1971). The first data obtained on rock and ore reversed magnetization posed the question of its origin: whether it was acquired in a geomagnetic field of reversed polarity (Mercanton 1926; 1932) or whether it can be explained by peculiarities in composition, genesis and occurrence of geologic bodies.

At the beginning of the 1950's two hypotheses on the acquisition of reversed magnetization in rocks were developed (Néel 1951; Hospers 1953). According to the first hypothesis of self-reversals, reversed magnetization can be explained by the physicochemical processes causing magnetization of a rock to be acquired in a direction opposite to that of the geomagnetic field. The second hypothesis of reversals does not reject the possibility of self-reversals, but it argues in favour of the existence of a geomagnetic field during rock formation with a direction opposite to that of the present field and considers this as the principal cause of reversed magnetization. This hypothesis suggests recurrent reversals of the geomagnetic field.

The problem of the occurrence of geomagnetic field reversals is of exceptional importance in the theory of origin of geomagnetism. This is also important for stratigraphy and geochronology. If the boundaries of normally and reversely magnetized strata actually correspond to the moment of polarity change in the geomagnetic field, these boundaries are then synchronous at all sites on the Earth's surface. In this case it is an effective method in geology for synchronizing different events in geologic history in regions separated by many thousands of kilometers.

Laboratory experiments have confirmed the possibility of self-reversal of magnetization in rocks. A reversed, thermal magnetization occurred in these tests within definite temperature ranges of cooling, while in other ranges the acquired magnetization was parallel to the applied field. The proportions of these partial magnetizations affected the total thermoremanent magnetization in such a way that it appeared to be reversed (complete self-reversal) or normal, though decreases (partial self-reversal). Néel (1951, 1955) was the first to consider theoretically mechanisms of self-reversal in rocks. These mechanisms are divided into two groups: (1) those acting in single-phase systems; (2) those typical of two- and or more phase-systems.

Group 1. The spontaneous magnetization of ferrimagnetic materials is the difference in magnetization caused by two sublattices A and B, that is $J_s = J_{sA} - J_{sB}$ (Néel 1951, 1955). If the temperature changes of magnetizations J_{sA} and J_{sB} are different, then the magnetization during a temperature change can be positive as well as negative. If at high temperatures $J_{sA} - J_{sB} > 0$ and at low temperature $J_{sA} - J_{sB} < 0$, then upon thermomagnetization of this ferrimagnetic material, a magnetization opposite to the field applied, might appear, i.e. self-reversal might occur.

A difference in the temperature relationship of magnetizations J_{sA} and J_{sB} might be produced by a change in cation distribution in sublattices A and B during and after cooling of the ferrimagnetic material (Néel 1955). This can also occur in substituted magnetites when nonmagnetic ions (Mg^{2+}, Al^{3+}) diffuse from one sublattice into the other (Verhoogen 1956). This mechanism seems to have operated in a magnesian titanomagnetite from the Maymecha-Kotuy region (Gusev 1962). After thermal treatment at 800°C in a neutral medium, the resultant magnetization J_{rt} was parallel to the applied field during cooling. A self-reversal of J_{rt} was registered 10–40 days after thermomagnetization. The possibility of a similar mechanism was also shown (theoretically and experimentally) for a single-phase oxidation of homogenic titanomagnetite. In this particular case the vacancies that appear in the octahedral sublattice B play the role of nonmagnetic ions, appearing as a result of oxidation of Fe^{2+} to Fe^{3+}, which leads to the over-balance of the magnetic moment of the sublattice A. Partial self-reversal of the same type was later observed in the Kola Peninsula ores (Bol'shakov et al. 1973) and the Yakutian traps (Trukhin et al. 1984b).

Group 2. Mechanisms of self-reversals in two- or more phase systems can be divided into two types with respect to the nature of the interaction between the phases across their boundary.

The magnetostatic mechanism of self-reversal might operate if the B phase, in a two-phase system, magnetizes in a demagnetizing field of grains of A phase. Calculations reveal that for this mechanism to operate, the A and B grains must be quite close and must have a distinct boundary (nonmagnetic lamella) between them and the grains must also be very small, i.e. of the order of single-domain grain size (Néel 1955; Nagata 1961). Furthermore, an asynchronous magnetization of grains of different phases in a rock is also essential. Since the magnetic field of every particle averaged for the total volume equals zero (due to the absence of magnetic charges), the magnetostatic interaction under isotropic distribution of grains of both phases would not accomplish a self-reversal of magnetization.

An exchange interaction in a two-phase system might be realized only with a crystallographic continuity of phases and the presence of common oxygen layers. A self-reversal might occur if $J_{sB} > J_{sA}$ and $T_{CA} > T_{CB}$. The most effective self-reversal occurs if the A phase is a weak ferromagnetic and the B phase is a ferrimagnetic, then $J_{sB} \gg J_{sA}$ (Nagata 1961). Dacite from Mt. Haruna (Japan) can exemplify this situation: the A phase is represented by hemoilmenite and the B phase by titanomagnetite; the reversal of J_{rt} was first confirmed in laboratory tests on this rock (Nagata 1953). In natural titanomagnetites of the Siberian traps, a partial self-reversal was recorded as a result of an exchange interaction at the boundary of phases formed after a high temperature baking (Zhilyayeva et al. 1970; Agabekov and Metallova 1972). The same mechanism was assumed to operate in order to produce partial self-reversals of rock magnetization elsewhere (Zhilyayeva et al. 1970; Mel'nikov and Khisina 1976).

In many cases the question of which mechanism produces any specific self-reversal remains open. Thus, the magnetization self-reversal in Mt. Haruna dacites might be related to the super-exchange interaction between the magnetic moments of the hemoilmenite phases of close composition (Nagata 1961). A similar mechanism is assumed to operate in synthetic hemoilmenites and magnesian hemoilmenites with the composition (1–X) $(FeMg)TiO_3 \cdot Fe_2O_3$ (Zvegintsev et al. 1974; Bugayev et al. 1972) and in natural rocks and minerals, which have been separated from them, when these minerals are of the same composition (Agabekov and Metallova 1972; Trukhin et al. 1984a).

Many more examples of laboratory tests of self-reversals of the TRM can be cited. They make up, however, just a small proportion of the results of TRM tests of rocks and they are not comparable with the quantity of reversely magnetized rocks in nature, which amounts to 50%. On the other hand, self-reversal processes are not always reproducible in laboratory tests if they are related to the irreversible changes in rocks. Thus, studies of various mechanisms of self-reversals have not as yet unambiguously solved the question of in which processes a reversed magnetization is occurs in nature.

More basic evidence of the relationships of reversed magnetization and petrological changes can be obtained by the systematic correlation of magnetic properties and the composition of ferromagnetics in normally and reversely magnetized rocks (Sholpo 1977). In this respect a detailed analysis of the correlation of the reversed polarity of J_n and the high deuteric oxidation in the sequences of basaltic lavas of the same type in Columbia is imperative (Wilson and Watkins 1967). This correlation appears to be valid in some regions, but sometimes it is quite poor or even absent (Mikhailova et al. 1970; Khaburzania et al. 1979; Glevasskaya 1983).

The confirmation of an absolutely different origin of reversed magnetization in rocks, i.e. a magnetization acquired in a geomagnetic field of reversed polarity, will be given in Sect. 3.1.

1.6 Distribution of the Directions of Natural Remanent Magnetization in Rocks

1.6.1 Initial Paleomagnetic Data and the Drawing of Magnetic Directions

The principal data, upon which all paleomagnetic determinations are based, represent the combination of many vectors of natural (or characteristic) remanent magnetization measured in oriented samples of the rocks. These vectors never agree exactly with each other and show a more or less wide natural dispersion around a mean direction, even when the samples are taken from a single exposure or from a single bed of a sediment or lava flow. For a statistical analysis of the distribution of directions of J_n, each vector is assigned an equal weight, irrespective of values J_n and each direction of magnetization J_n is represented as a vector of a unit length j_n. Thus, a point on the sphere of a unit radius, described around a common origin of all unit vectors as the centre of this sphere, corresponds to each direction of vector J_n. Each point represents the projection of a vector on this sphere.

The analysis of the distribution patterns of the direction of J_n is possible when a unit sphere with all points on it is projected onto a plane (Fig. 1.25). Usually an equidistant projection is used (Figs 1.26 and 1.27), which affords the least distortion of the projected area. An initially horizontal rock surface (the ancient system of coordinates) or a recent plane of horizon (the recent system of coordinates) is employed as the projection plane. The directions of the vectors J_n are expressed in terms of the angles of declination D and inclination I, just as the direction of the geomagnetic field in a point is estimated by the angles of magnetic declination and magnetic inclination. In general, a polar projection is used (see Fig. 1.26) to depict the distribution of directions of vector J_n; the lines of equal declinations are projected as radii; and the lines of equal inclinations are projected as concentric circles.

Many causes might be responsible for the dispersion of the vectors J_n: different composition and alteration grade of the rocks and, as a consequence, different relationships between the components of J_n that appeared at different time intervals in geomagnetic fields of various directions; disturbances in rock occurrences, some other types of deformation, etc. Even if the rocks studied are not deformed and are of exactly the same age and a single-component, primary magnetization has been reliably established, a dispersion will occur because of heat fluctuations, hydrodynamic disturbances and structural inhomogeneities in rock during its formation. All these factors prevent the primary magnetization from orienting exactly along the geomagnetic field. The dispersion produced by random

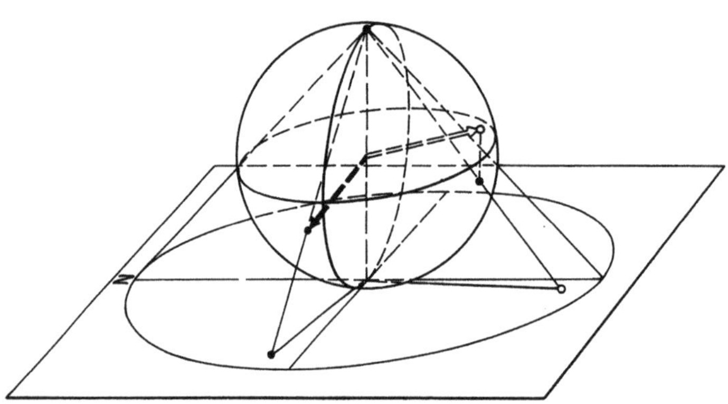

Fig. 1.25. Plotting of NRM directions on a plane. Equal angle stereographic projection. (Khramov and Sholpo 1967)

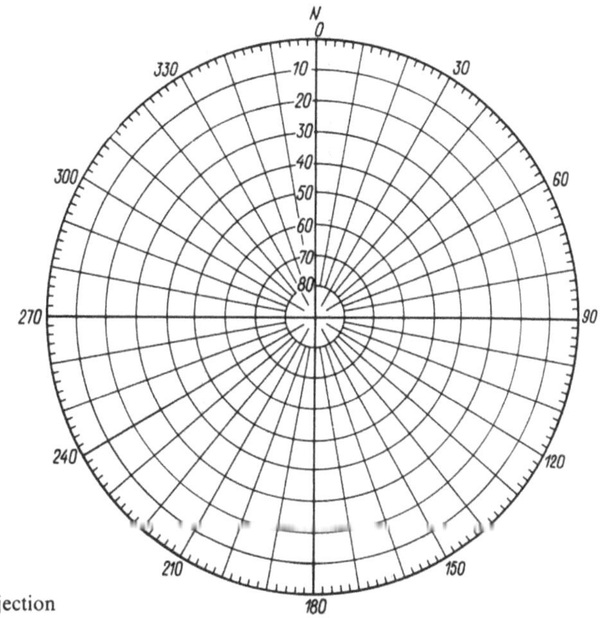

Fig. 1.26. Equidistant polar projection

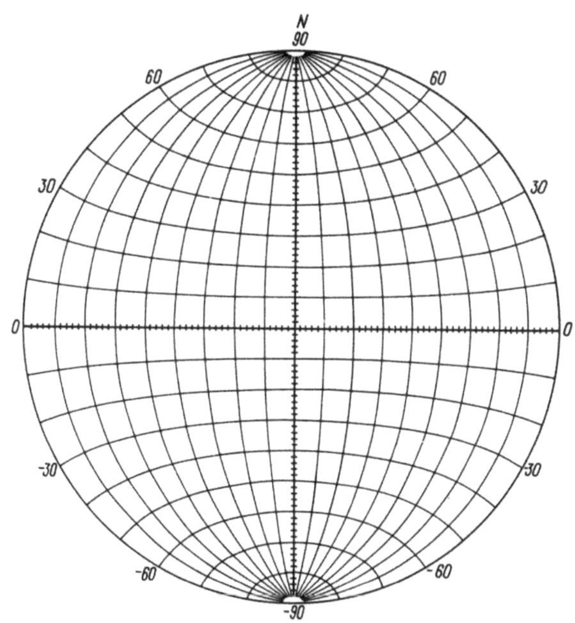

Fig. 1.27. Equidistant equatorial projection (the Kavraisky net)

errors in taking samples and in measuring their magnetization also exists. All of these variations in the samples and also in the dispersions of vectors J_n increase, if data are compared from different beds and flows. Furthermore, a dispersion may have been registered which is due to asynchroneity of the primary magnetization and to changes in the geomagnetic field throughout time.

1.6.2 Fisher's Distribution

It is easy to comprehend the general rule if the distribution of directions of vectors J_n under the effect of all of the factors enumerated above, if the random nature of almost all of these disturbances is taken into account. In particular, single-component magnetization should be considered. In this case the examined assemblage should fulfill the following requirements (Irving 1964; Khramov and Sholpo 1967): (1) the assemblage vectors are distributed with axial symmetry in the direction of the ancient geomagnetic field; (2) the vector density decreases with the increase of the angular distance ψ from the field direction; (3) the principle of this decrease is determined by the model of dipoles oriented in a magnetic field with random disturbances present, i.e. according to the model of detrital magnetization (see Sect. 1.4.3).

Suppose N vectors J_{ni} are given, their directions are expressed by the angles of declination D_i and inclination I_i ($i = 1,2,\ldots N$); unit vectors j_{ni} are calculated, thus their projections on the coordinate axes can be expressed as

$$x_i = \cos I_i \cos D_i;\ y_i = \cos I_i \sin D_i;\ z_i = \sin I_i. \qquad (1.40)$$

These N vectors, according to the model of detrital magnetization, might be regarded as a set of dipoles with equal magnetic moments m that did not reach complete orientation along the ancient geomagnetic field H due to disturbances of energy U_0. Then, the distribution of these dipoles, according to their magnetic energies $U = -mH\cos\psi$ (where ψ is the angle between m and H), should satisfy Maxwell-Boltzmann's law: the number of particles with energy within the limits from U to U + dU equals

$$dN = NC_0 \exp(U/U_0)\, dU.$$

Substituting $U = -mH\cos\psi$ and designating $C = mHC_0$ we obtain:

$$dN = NC\exp[(mH/U_0)\cos\psi]\sin\psi\, d\psi. \qquad (1.41)$$

The value $K = mH/U_0$ is called the precision parameter. This is a number indicating how many times the magnetic energy of vectors in the field H exceeds the energy of disturbances. The constant C is calculated from

$$\int_0^{\pi} C\exp(K\cos\psi)\sin\psi\, d\psi = 1,$$

where

$$C = K/(2\,\mathrm{sh}K).\ \text{Consequently,}$$
$$dN/N = [K/(2\,\mathrm{sh}K)]\exp(K\cos\psi)\sin\psi\, d\psi. \qquad (1.42)$$

The expression in the form of Eq. (1.42) is called Fisher's distribution (Fisher 1953). The value dN/N can be regarded as the probability $P_f(\psi)$ of the contact of vector j_{ni} into the ring (band) limited by the angles ψ and $\psi + d\psi$, counted away from the direction of the field H. In each interval the vectors should be uniformly distributed in all azimuthal angles φ, where $0 \leq \varphi < 2\pi$. Thus, the density of vectors j_{ni} in the unit of a solid angle, i.e. the part related to the ring area $dS = 2\pi \sin\psi\, d\psi$, is

$$p(\psi, \varphi) = p_f(\psi)\, p_a(\varphi) = [K/(4\pi\,\mathrm{sh}K)]\exp(K\cos\psi), \qquad (1.43)$$

where

$p_a(\varphi) = 1/(2\pi)$.

A projection of any of the dN unit vectors j_{ni} in the direction H equals $\cos\psi$, and the sum of these projections $dR_o = dN\cos\psi$. Thus taking into account Eq. (1.42) we obtain:

$$dR_o/N = [K/(2\,\mathrm{sh}K)] \exp(K\cos\psi) \sin\psi \cos\psi \, d\psi.$$

Integrating all angles ψ from 0 to π we obtain the sum R_o of the projections of all N vectors:

$$R_o = N\,(\mathrm{cth}K - 1/K), \qquad (1.44)$$

that is, a formula for detrital magnetization is obtained, differing from Eq. (1.29) only in the concentration of ferromagnetic particles, that here equals one.

At $K > 3$, the value $\mathrm{cth}K \approx 1$, where

$$K = N/(N - R_o). \qquad (1.45)$$

Thus, the distribution of directions of particle magnetic moments in a rock, as well as of vectors j_{ni} in the samples, (normalized vector sums of particle magnetic moments) obeys the same law (1.42), i.e. Fisher's distribution. By analogy it can be concluded that the vector sums Σj_{ni} ($i = 1,2,...N$) also satisfy Fisher's distribution. The precision parameter K of the vectors evidently increases at this summation.

The form of the function of distribution $p(\psi)$ depends on the precision parameter K (Fig. 1.28). If the value of K is large, the vectors j_{ni} are closely packed around the direction H and the function $p(\psi)$ exhibits a narrow peak; if $K = 0$, the vectors are distributed regularly in all directions and the plot $p(\psi)$ represents a horizontal straight line.

1.6.3 Observed Distributions of Magnetization J_n in Rocks

Observed distributions of the directions of magnetization J_n in rocks are, in general, consistent with Fisher's distribution, but they can also differ from it. In a rock body

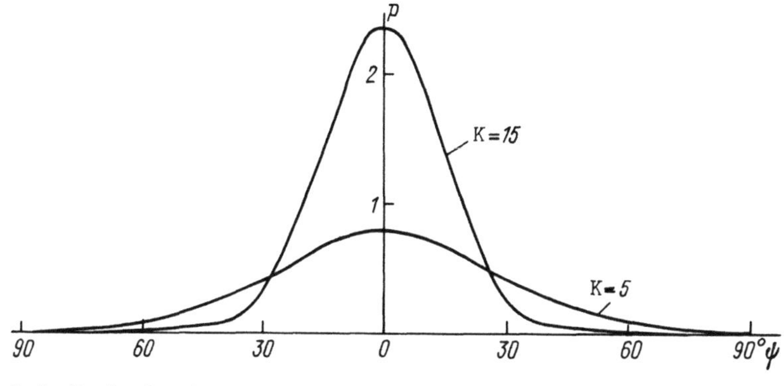

Fig. 1.28. Fisher's distribution function $p(\psi)$ for $K = 5$ and $K = 15$. (Khramov and Sholpo 1967)

or in a rock sequence, the directions of magnetization J_n satisfy Fisher's distribution, if the following requirements are fulfilled:

1. The rock contains only one component of magnetization J_n (a primary component or one of the components of secondary magnetization) or all of the components, except one, are randomly oriented.
2. During the acquisition of magnetization J_n of a given rock group, the geomagnetic field is kept steady (within the limits of determination precision) or its changes are of a random nature.
3. The anisotropy of the rocks does not greatly change the distribution of directions of vector J_n.

Naturally, not all of these conditions (or none of them) are always observed. Fisher's distribution is quite commonly violated in the presence of two components of magnetization J_n in a rock: the primary one is directed along the ancient geomagnetic field and the later, secondary component is directed along the present field.

Several examples of the observed distributions of magnetization J_n in rocks are illustrated in Fig. 1.29. Two of these agree with Fisher's distribution (Fig. 1.29a,c)

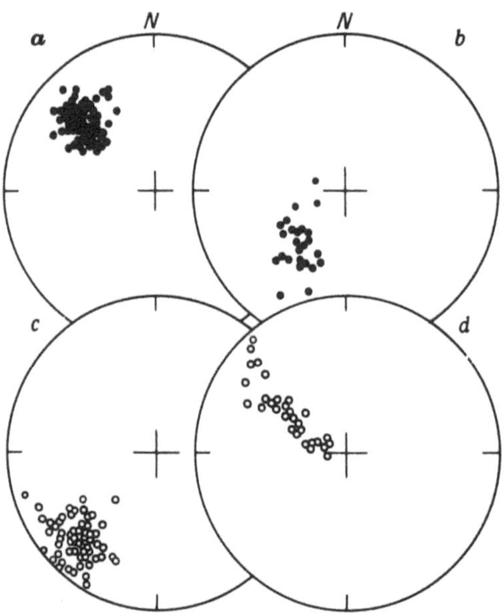

Fig. 1.29a-d. Examples of NRM direction distributions. (After Khramov and Sholpo 1967).**a** Site A_8, Torridonian sandstones, Scotland. (Irving 1957); **b** bed 6085, Middle Carboniferous Vereya clays, near Moscow; **c** Lower Permian clays, the Donbas; **d** Ordovician siltstones, Vikhoreva River (Angara basin)

and the other two do not (Fig. 1.29b,d). It is noteworthy that this inconsistency might appear only in an azimuthal set [variability of the function p_a in Eq. (1.43)] or only in a radial set [discrepancy between the function $p(\psi)$ and Fisher's distribution] or in both of these sets simultaneously.

CHAPTER 2

Methods and Techniques Used in Paleomagnetic Studies

2.1 Principles of the Investigation of Geologic Bodies; Collection of Oriented Samples

2.1.1 Objects of Paleomagnetic Studies

The paleomagnetic study of rocks of different composition and origin – sedimentary, effusive and intrusive complexes, effusive-sedimentary sequences and, in some cases, metamorphic complexes and ores – as the choice of the object to be studied is determined solely by the problem to be considered, namely whether it is a direct or indirect. Furthermore, there are general requirements, which narrows the group of bodies to be studied. These limits are mainly caused by: (1) weak rock magnetization, which cannot be accurately measured by means of modern magnetometers; (2) poorly preserved primary magnetization, which is difficult to isolate using the available methods; (3) incomplete paleomagnetic records for a given object.

The accuracy of any paleomagnetic determination depends not only on the establishment of ancient geomagnetic field components, but also on a fairly accurate estimation of the moment in time to which the data can be referred, i.e. the time at which the ancient magnetization was acquired. It is impossible to determine this time if the age of the rock is unknown, therefore, paleomagnetic objects should be reliably related to the geochronologic scale. If the study is aimed at the age determination of a rock, reliable standards for comparison are necessary, i.e. paleomagnetic data should be precisely located stratigraphically. Secondary components of rock magnetization can also be used in solving geologic problems. In particular, any stable, prefolding ancient magnetization can be used in the study of local and regional tectonic movements. In some cases the use of post-folding magnetization facilitates the determination of the upper age limit of the folding. However, in order to study rapid changes in the ancient geomagnetic field (secular variations and reversal morphology) the rapid component of the primary magnetization must be established. Moreover, the formation of a specific geologic object should provide a fairly comprehensive and continuous paleomagnetic record.

Thus, the suitability of the objects chosen for paleomagnetic study should be evaluated according to whether a direct or indirect problem in paleomagnetology or a specific geologic problem is to be solved.

2.1.2 Techniques Used in the Study of Geological Bodies

The sampling of geological bodies is primarily defined by the problem stated and objects to be used in paleomagnetic studies. There are different methods of sampling (Khramov and Sholpo 1967), namely:

1. *Uniform Sampling of a Formation from Two or Three Exposures.* One or two samples are taken at very site[2] and two or three specimens from each sample are measured magnetically. Such a procedure is used when the study is mainly aimed at a reliable determination of the mean direction of magnetization J_n in the body studied, whereby an exact direction of J_n at each site of the succession is not so important. This procedure is applied to homogeneous sediments, isolated beds, lava flows, intrusive sheets and dykes. The sampling interval is a function of the thickness of a geologic body and the minimum number of samples necessary for statistical analysis. Not less than 15–20 samples should be taken from every succession.

2. *Detailed Sampling of the Section.* This method is based on successive changes of the NRM vector (for example, to determine secular variations or geomagnetic reversals). The section should be sampled at many successive stratigraphic levels, and at short intervals between them, depending on the sedimentation rate. Not less than three samples should be taken at each precise stratigraphic level, and not less than eight oriented specimens prepared for each level.

3. *Selective Sampling of Certain Rock Types.* This technique is used when only a few rocks have preserved the primary magnetization. It does not differ from the above mentioned techniques, if adequate rocks are abundant. When only a few beds can be chosen, more samples should be used to obtain at least 15–20 specimens from each exposure.

4. *Sampling of a Body at Different Sites.* This technique is used in reconnaisance studies. It is recommended to prepare as many specimens as possible in the laboratory.

Each technique, except for series sampling, necessitates the use of special collections, mainly to determine the NRM components and the time of their formation. Samples are taken from folds (five to ten samples from each of two or three bands at different sites in a fold), pebbles from overlying deposits and baked rocks. Samples for petrographic, mineral and chemical analyses, as well as fauna and plant remains, are also usually required.

Depending on the problem, each sampling system has its own distinct characteristics. The first and second methods are applied in studies of the changes in time of the ancient geomagnetic field. The thickness of sections, sedimentation rate and the problem under consideration, determine the sample spacing. The second sampling system is used to study the changes in the fine structure in the ancient

[2]A sampling site of a section is the stratigraphic level within a bed, flow or succession, whereas within a nonstratified geologic body it is a segment whose dimensions along a sampling line are much less than the distance between two successive sampling sites (less than the sampling interval).

geomagnetic field. As a rule, the solution of this problem and the indirect problem of the detailed correlation of sections, necessitates continuous sampling (i.e. the distance between the centres of the samples equals the sample height, i.e. 2.5–5 cm, so that the entire section is sampled); the sampling interval increases in geosynclinical sections.

To solve problems of global tectonics, samples are taken from several sections (two or three) within an area of 100 × 100 km² within a single coherent block. The section must be uniformly sampled, as this is an obligatory requirement for determining the significance of the mean paleomagnetic direction. The choice of paleomagnetically defined stratigraphic intervals depends on the time for which reconstructions are to be made (period, epoch, age, etc.). The objects studied within a region should be chosen to minimize the effect of local tectonics, local magnetic anomalies, etc. It is very important in the case of local and regional tectonics to sample different slopes in a fold in order to determine the direction of the prefolding magnetization. The interval of sampling across the section in tectonic studies of any scale must allow the determination of the precision of the mean ancient magnetization vectors appropriate to the problem under consideration.

The study of geological bodies can also be related to their history and mode of occurrence.

Sampling of volcano-sedimentary strata is similar to that of sedimentary strata. Samples from lava flows are taken evenly along natural flow planes. As a rule, the primary magnetization of lava flows is better preserved in their marginal parts, however, these show a considerable scatter of NRM directions. Thus, approximately 15 samples need to be taken in order to obtain a reliable mean J_n^o direction for the flow. It is better to sample the intrusives along the profiles across the bodies, which enables the homogeneity of magnetization to be traced in the central and marginal parts. The number of samples from each intrusive body should not be less than 15–20. Uniform sampling of several profiles is preferable for metamorphic rocks in order to establish sites with preserved, relict (premetamorphic) magnetization of rocks. The number of samples should be adequate for the reliable determination of the direction of ancient magnetization. In solving stratigraphic problems, samples are taken uniformly downhole. The interval of sample spacing is the same as for exposed sedimentary strata.

2.1.3 Orientation of Samples and Specimen Preparation

Paleomagnetology deals with the vector values, so the samples for paleomagnetic study must be oriented in space. For this purpose one should choose a marker plane and calculate its azimuth and determine the tilt of the plane (Khramov and Sholpo 1967). Measurements are usually performed with a miner's magnetic compass. Geodetic (or theodolite) or astro-orientation (solar compass) are used only when the samples are of highly magnetized rocks ($J_n \geq 1$ A/m), or when the samples are from areas of extensive local magnetic anomalies, so that orientation by a miner's compass becomes impossible (Khramov and Sholpo; Collinson et al. 1967).

Oriented samples can be obtained in several ways, depending on their mode of occurrence, rock properties and the field and laboratory technique to be used. Any method should provide a restoration of the original horizontal plane position of the sample, i.e. when the marker plane was not originally horizontal (bedding plane in sedimentary rocks), the elements of the bedding (tectonic slope) should be known in order to recalculate the data in the coordinate system related to the primary horizontal plane. To study layered sedimentary rocks one should try to choose a bedding plane as a marker to avoid recalculation and the introduction of errors. Any joint surface can be used as a marker plane for non-layered hard rocks. A dip line is drawn on the marker plane, with an arrow pointing along the dip, and its azimuth and slope angle are measured and recorded (Fig. 2.1). The line of the magnetic meridian, with an arrow pointing to the north, is usually drawn on the bedding plane in the horizontal strata.

Different methods of preparing specimens for magnetic measurements are employed (Khramov and Sholpo 1967, Collinson et al. 1967):

1. A sample is chipped off the rock. The upper face, with an arrow on it, serves as a marker plane. The sample is shaped into an isometric form within a 2:1 ratio of the longest to shortest axis.
2. A sample is cut into similarly oriented cubes either manually (on the exposure for soft rocks) or using a diamond saw blade (for hard rocks). The orientation of cube and direction of the axis is illustrated in Fig. 2.2 for magnetic studies.
3. Cores are drilled out from samples perpendicular to the oriented plane. The cores are cut into cylinders of equal height using a diamond saw blade.
4. Similar cores are drilled at the exposure rather than taking samples. Punches, allowing the collection of cubes, are used for soft rocks at each site.

Preference is given to regularly shaped samples, as this improves the accuracy, when measured at the lower limit by a magnetometer, by an order of magnitude.

Fig. 2.1. Hand-sample orientation by means of magnetic compass and inclinometer. (After Khramov and Sholpo 1967)

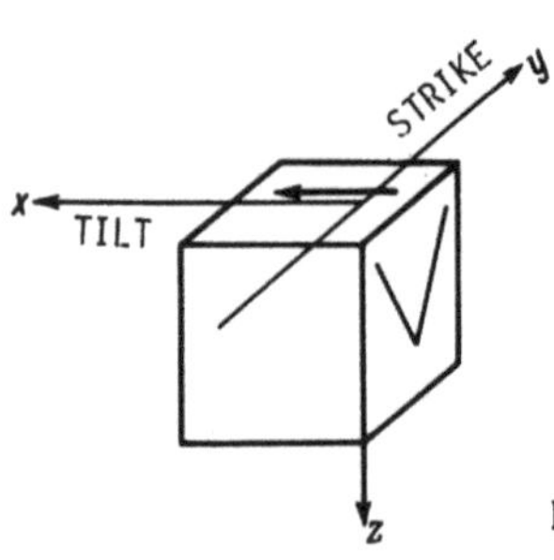

Fig. 2.2. Cubic specimen for magnetometric measurements

The dimension of the cubes depends on the magnetometer used. The cubes are usually 2 to 5 cm along the side. The cylinders are also of the same size, and the best accuracy is obtained with a height to diameter ratio of 7:8 (for example, 2.2 cm: 2.54 cm).

Cores from oriented boreholes can be used in paleomagnetic studies. There are magnetic methods for core orientation, based on a viscous or chemical component of J_n which is nearly always directed along the known present geomagnetic field direction at the site (Boronin 1979; Khramov 1984). However, other standard methods of orientation can also be used, which can be undertaken in the course of drilling. When cores are available, they should be marked "top-bottom". The oblique bedding of rock is a feature which helps to orient a core horizontally when the elements of its occurrence at the sampling site and the inclination of slant of the borehole are known. The top-bottom cores, kept at the sample store, are often disarranged. However, they can be reconstructed using the occurrence of fossils, the curve of the core end and other distinguishing features. The original horizontal is determined from the oblique bedding. In general, the determination of the core orientation depends upon the time, elapsed from the moment of drilling until the magnetic measurements. This interval should be minimized.

2.2 Magnetic Measurements

2.2.1 Equipment Used in Magnetic Measurements

Determinations of the natural remanent magnetization and magnetic susceptibility of samples, measured for purposes of magnetic exploration, have been carried out for several decades. Considerable documentation is available and the main principles of measurements and magnetometer design are well developed at present. Paleomagnetic studies impose heavy demands on the equipment due to the necessity of accurately measure low magnetic moments (well below these associated with magnetic exploration).

There is no universal instrument which can measure all the magnetic parameters over the entire range of their values.

To measure natural remanent magnetization and its components using magnetic cleaning, as well as artificial remanent magnetization, the following instruments are used: (1) ballistic magnetometer; (2) astatic magnetometer; (3) rock generator (spinner magnetometer); (4) magnetic modulation meter (flux gate magnetometer); (5) a combination of the third and fourth types; (6) cryogenic magnetometer, based on Josephson's effect. Instruments of type 1 and 2 can also be used to measure magnetic susceptibility $\varkappa$. The $\varkappa$-value and the total magnetization J and their dependencies J(H), J(T), $\varkappa$(H), $\varkappa$(T) are determined using magnetic balances of different design.

The description of instruments and their usage has been widely discussed in literature dealing with Earth and rock magnetism and in special paleomagnetic studies (Nagata 1961; Collinson et al. 1967; McElhinny 1973). Therefore, only the instruments used in the USSR for measurements of magnetization J_n and its components will be discussed here.

2.2.2 Astatic Magnetometers

All astatic magnetometers consist of an astatic magnetic system, suspended on a torsional fibre inside a casing. The fibre is linked to a head, thus providing a bias adjustment to the system. It also includes bars to locate a specimen in a known position relative to the magnet system and a counter to determine the torsional angle of resulting from the deviation from the steady position due to the magnetic field of a specimen. The simplest astatic system incorporates two magnets with equal and oppositely directed magnetic moments, rigidly fixed by a rod, a mirror and a support shackle. A small additional magnet, which one can rotate in two interperpendicular planes, is often used for final astatization. The main magnets are placed within the magnetic meridian plane.

An astatic magnetometer is based on the interaction between the specimen magnetic field and that of both or one of the magnets. The specimen is placed in such a way that only one component of its magnetic moment affects the system. As the specimen is turned through 180° around the axis, perpendicular to the component, the direction of the magnetic field of a specimen changes and the system turns through a certain angle which is measured by the counter. This angle is proportional to the magnetic moment component of the specimen. Figure 2.3 shows some possible positions for a specimen relative to the magnetic system and the components measured.

In position I the centre of the specimen is at the lower magnet level of the system and is situated east or west of the magnet centre. Hence, the system shows a bias from the original position affected by the specimen magnetic moment component, horizontal and perpendicular to the magnet axis, i.e. along the line, connecting the centres of the lower magnet and the specimen. Data averaging, measured east and west of the system, eliminates some effects of the vertical component of the magnetic moment and also those due to magnetic inhomogeneity. In position II the centre of a specimen lies along the continuation of the lower magnet axis. The horizontal component of the specimen magnetic moment, perpendicular to the line connecting the centres of a specimen and a magnet, makes the system deflect. In position III the centre of the specimen is under the lower magnet of the axis projection of the astatic system. The component measured, as in the first two cases, is horizontal and perpendicular to the magnetic meridian plane. Only the remanence (residual moment) of a specimen can be measured for these three cases, since the rotation of the system is produced by the horizontal component of the magnetic moment perpendicular to the magnetic meridian. In this direction, a specimen is not affected by induced magnetization.

In position IV, when the centre of a specimen is equidistant from both magnets and lies on the line perpendicular to their axis, the deviation depends on the vertical component of the magnetic moment

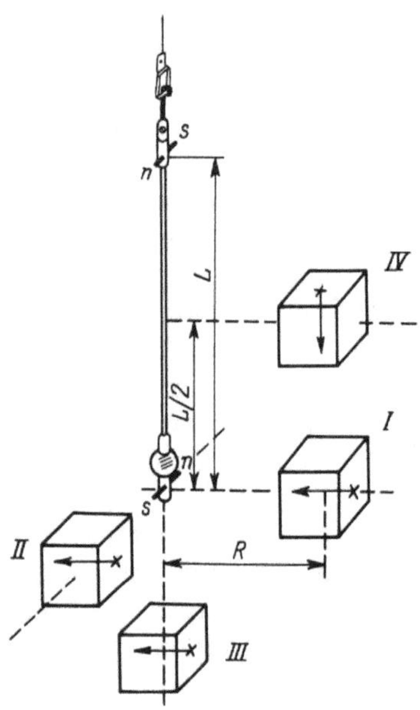

Fig. 2.3. Specimen positions relative to the astatic magnetometer. *Arrows* inside the cubes denote the measured component of magnetization. (Khramov and Sholpo 1967)

of a specimen. In this case the magnetic moment induced by the vertical constituent of the Earth's magnetic field is added to that of the remanent moment M_{rz}. When the specimen is rotated through 180° around the horizontal axis, its magnetic field resulting from the remanent moment, changes its direction and the magnetic field, due to the induced moment, retains its direction. Thus, at the beginning a reading n_1, proportional to $M_i + M_{rz}$, corresponds to the system deviation, and when the specimen turns, a new reading n_2 will be proportional to $M_i - M_{rz}$. Therefore, the effect of the magnetic moment M_i provides a distinct determination of these values. Position I is used in the astatic magnetometer designed by Yanovsky and Chernyshev, but it is possible to produce an induced moment using large Helmholtz coils, which form a uniform field (usually 0.05 mT) in a latitudinal direction.

The astatic system of the magnetometer which is set to unstable equilibrium, relative to the external field, provides the highest sensitivity. Such an equilibrium occurs when

$$m_1H_1 - m_2H_2 = 0$$

where m_1 and m_2 are the magnetic moments of the lower and upper magnets, respectively; H_1 and H_2 are the values of the horizontal component of the magnetic field in which the magnets are placed.

If $H_1 \neq H_2$, then even the addition of a small uniform field, i.e. noise (such as diurnal variation of the geomagnetic field), introduces changes into the equilibrium condition. Magnetic storms and industrial noise, producing a nonuniform field, greatly influence the equilibrium field. To decrease this effect a three-magnet system is used. It consists of an antiparallel middle magnet, the magnetic moment of which is twice that of the marginal magnets. This device is placed within large Helmholtz coils with self-compensation of the geomagnetic field.

Astatic magnetometers MA-21, MAL-036 (NPO "Rudgeofizika"), LAM-2, LAM-3 ("Geofizika" Enterprise, Brno, Czechoslovakia), VNIGRI and IZMIRAN systems are also used in paleomagnetic studies in the USSR and differ mainly in their mode of reading. The null method of automatic feedback is used in the MAL-036 (with a camera attachment), LAM-3 and IZMIRAN systems. A light beam deflection from the mirror on the astatic system (Fig. 2.3) is converted into an electrical signal by a differential photocell. The signal is amplified and fed to a feedback coil. The magnetic field of the coil compensates the primary magnetic field produced by a rock specimen. The electrical signal, which is proportional to the deflection angle, is then measured.

Magnetometers are generally calibrated by means of standard specimens with different magnetic moments or coils with known parameters.

2.2.3 Spinner Magnetometer

McNish was the first to use a rock generator, or spinner magnetometer, in the measurement of rock magnetization J_n (McNish and Johnson 1938). Since then, various types of rock generators have been described in the literature. There are several rock generators used in the USSR. Most of them, for instance the spinner magnetometer developed at the Institute of the Physics of the Earth of the USSR Academy of Sciences and the spinner magnetometer ION-1 of "Rudgeofizika" Scientific Enterprise, and based on the design developed by Dianov-Klokov (1960) and differ only in the mode of construction.

The operational principle of these devices is to measure the phase and amplitude of the alternating electromagnetic fied (emf) induced by a uniformly rotating specimen inside a pickup coil. The emf amplitude, resulting from the rotation of a specimen with a magnetic moment M, is proportional to the component M_i, which is the projection on a plane perpendicular to the axis of rotation. The emf phase, designated as ψ, forms an angle between component M_i and one of the axes. Thus, the magnitude of the vector M and angles a, j, characterizing its direction, can be determined when the specimen rotates successively around two interperpendicular axes. To estimate the error, a specimen is rotated around the x, y, and z axes.

To define the phase angle ψ a reference signal is required and, relative to this signal can then be measured by a phase detector. The reference signal is produced in different ways in rock generators of various types. In some cases the simultaneous rotation of a specimen with either a permanent magnet or with a coil occurs. Some rock generators employ a disk, half of which is painted black to deflect the light beam from the photocell during half the period. In the JR-3 and JR-4 rock generators ("Geofizika"

Enterprise, Brno) two reference signals, displaced in phase of 90°, are formed by two photocells. The "phase" and "quadrature" components of the signal are amplified separately by two amplifiers. Thus, components M_x and M_y in the horizontal plane, instead of M_i and phase ψ, can be easily measured and estimated.

2.2.4 Fluxgate Magnetometer

These magnetometers use two flux probes with oppositely directed axes. The system is balanced. The specimen to be studied is placed close to one of the flux probes. The magnetic field, induced by it then disturbs the balance. The effect of induced magnetization due to the geomagnetic field and that of irregular magnetization, are eliminated by multiple measurements of the sample in different positions. The measurements are similar to those made by the astatic magnetometer.

There are instruments which, in fact, are a combination of a rock generator and a fluxgate magnetometer (Foster 1966). They are more sensitive than magnetic modulation meters and allow unconsolidated rocks to be measured, which is not possible using a rock generator.

2.3 Field Methods Used in the Determination of Ancient Magnetization

2.3.1 Evaluation of the Paleomagnetic Stability of Rocks

Field or direct methods for the determination of the paleomagnetic stability of rocks were proposed by Brunhes and David (Brunhes 1906) and developed by Graham (Graham 1949). The principle of these methods is as follows: if blocks of rocks have changed their spatial orientation the time of the acquisition of the observed NRM, relative to this change, can be determined. The NRM components have been originated before block movements – paleomagnetically stable components – will have rotated together with the blocks, while unstable magnetizations will be parallel to each other despite the different rotation of the individual blocks.

Folding, and the redeposition of rocks as conglomerates, are commonly observed (Fig. 2.4). It is evident that the magnetization of layers III and IV preceded folding. Moreover, layer III was magnetized before the deposition of layer V. The magnetization of layer II postdated all of these events. When Graham tests, i.e. fold and conglomerate tests, are applied both to natural remanence and its components as revealed by magnetic cleaning, one can define the time of their

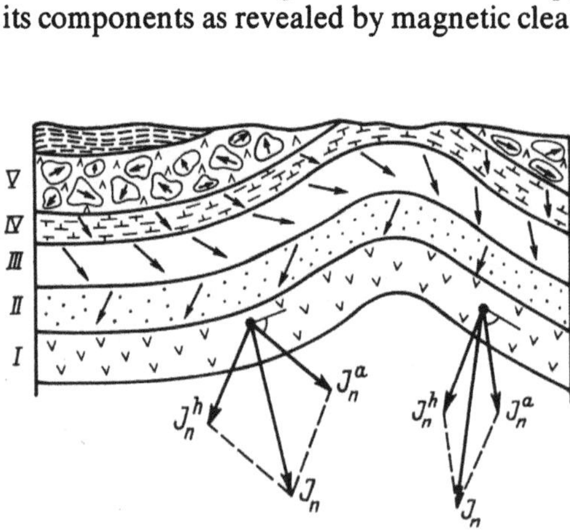

Fig. 2.4. NRM directions and relations between NRM components for rocks that have changed their initial position. (After Yanovsky 1978). The NRM of bed *III* and *VI* preceded the folding, moreover bed *III* was magnetized before deposition of the bed *V*. The magnetization of bed *II* is postfolding. The NRM of bed *I* has both prefolding J_n^a and postfolding J_n^h components

formation. If the natural remanence, or characteristic magnetization revealed by magnetic cleaning, consists of prefolding and postfolding components, then the directions at different points will be intermediate between those of layers II and III, i.e. similar to those of layer I. A method of component separation, based on geometry and statistical analysis, has been developed (Khramov and Sholpo 1967). All the methods of separation of J_n components and the assessment of their paleomagnetic stability are based on the use of two different coordinate systems. The axes of the first, present coordinate system are oriented as follows: x points along the present North, the true meridian of the site; y points East; and z is directed vertically downwards. Such a coordinate system is common for the present position of beds. Another system, i.e. the ancient coordinate system, is related to the beds and rotates with them during folding. Prior to folding, the axes of the ancient coordinate system had the same orientation as the present one. Since then, the system has rotated around the line of strike through an angle equal to the dip angle. The ancient coordinate system was originally the same in all horizontally bedded rocks. To calculate declination D_c and inclination I_c of a vector in the present coordinate system, and the values D_a and I_a in the ancient one, the following formulae are used (Zotkevitch 1966, with additions):

$$\begin{aligned} I_c &= \arcsin\,[\sin I_a \cos B + \cos I_a \sin B \cdot \cos(D_a - A)] \\ D_c &= A + \delta, \text{ if } \sin I_a < \sin I_c \cdot \cos B \\ D_c &= A + (180° - \delta), \text{ if } \sin I_a \geqslant \sin I_c \cos B, \end{aligned} \tag{2.1}$$

where A is the azimuth of the dip line of the bed, B is the angle of dip, angle δ is $\sin\delta = \sin(D_a - A)\cos I_a / \cos I_c$. The inverse recalculation is done by the same formulae, changing the "a" and "c" indices and the angle B by (−B).

The evaluation of paleomagnetic stability of rocks has the following basis. If samples of a similar age are taken from sections with different bedding, then the directions of the ancient (prefolding) component of magnetization J_n^a must be the same in the ancient coordinate system. Whereas the directions of postfolding magnetization J_n^h must be the same in the present coordinate system (Fig. 2.5). This suggests certain characteristic for the paleomagnetic stability of coeval rocks, i.e. the component composition of their NRM. These features can be summarized from stereoprojections of vectors J_n (Figs. 1.26 and 1.27), plotted in the ancient coordinate as follows:

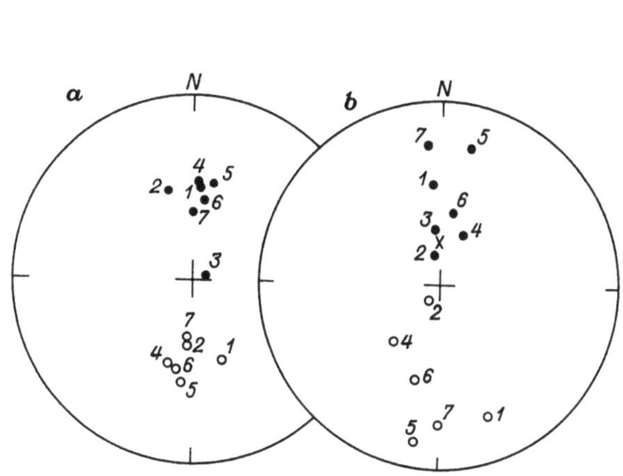

Fig. 2.5a,b. Mean NRM directions of the Pliocene sediments in different outcrops from western Turkmenia. (After Khramov and Sholpo 1967). **a** Ancient coordinate system; **b** Recent coordinate system: *1,2,4–7* rocks with prefolding magnetization

1. NRM projections form circular patterns, not including the projection of the total vector for the present geomagnetic field H_T, but including the mean direction of NRM from different sections, independent of the rock bedding (Fig. 2.6a,b). Such rocks are paleomagnetically stable and their magnetization is prefolding.
2. NRM projections form circular patterns, not including H_T, but the mean directions of J_n depend on the rock bedding. If in the present system these directions are similar to each other, then the magnetization is postfolding but ancient (for example, permo-Triassic as in Fig. 2.6c,d).
3. The area occupied by the NRM projections stretches towards the projection of H_T (Fig. 2.6e). Such rocks are paleomagnetically stable to a different degree. The relationship of ancient J_n^a and recent J_n^h magnetizations will be different in each sample.
4. The NRM projections form circular areas around the projection H_T (Fig. 2.6f). Here, we have complete instability of rocks and only recent magnetization.

Most of the observed distributions of magnetic directions for particular rock groups can be classified into one of the above distributions or their combinations (Fig. 2.6g,h). There is only one exception, i.e. when the NRM consists of three or more components that are comparable in value but considerably different in direction. The above features cannot be applied to the analysis of the paleomagnetic stability of horizontally lying young beds, where reversed magnetization is the only evidence for stability. If folds were formed by the rotation of beds, not only about a horizontal axis (for example, the tilting of the fold hinge of an anticline, or the superposition of a different folding or postfolding horizontal deformations), then the first feature, even in the case of complete paleomagnetic stability, will not be satisfied.

The contact method is very effective for the evaluation of paleomagnetic stability. The coincidence of NRM directions between the igneous rock proper and their contacts within the baked zone is the main criterion for paleomagnetic stability in igneous rock. This criterion is growing in importance, especially when such directions differ greatly from the NRM directions in the surrounding, unbaked rock and from that of the present geomagnetic field (Irving 1964).

2.3.2 Determination of the Degree of Paleomagnetic Stability

Quantitative assessment of paleomagnetic stability, i.e. computation of its degree, S, is performed by one of the following methods, namely the conglomerate and fold tests, or the plane of remagnetization method (Khramov and Sholpo 1967). The conglomerate and fold tests were proposed by Graham as far back as 1949. These tests explore the precision parameter of the NRM directions. Calculation of the precision parameter K for pebbles and fragments from pebblestone and conglomerates, resulting from the destruction of the rock studied, is performed using the conglomerate test[3]. If the magnetization J_n was acquired prior to conglomerate

[3] For $K \leq 3$ Eq. (1.44) should be used instead of Eq. (1.45) to compute K by the iteration method with initial value $K_0 = 10–20$ (Ryabushkin and Pevzner 1978).

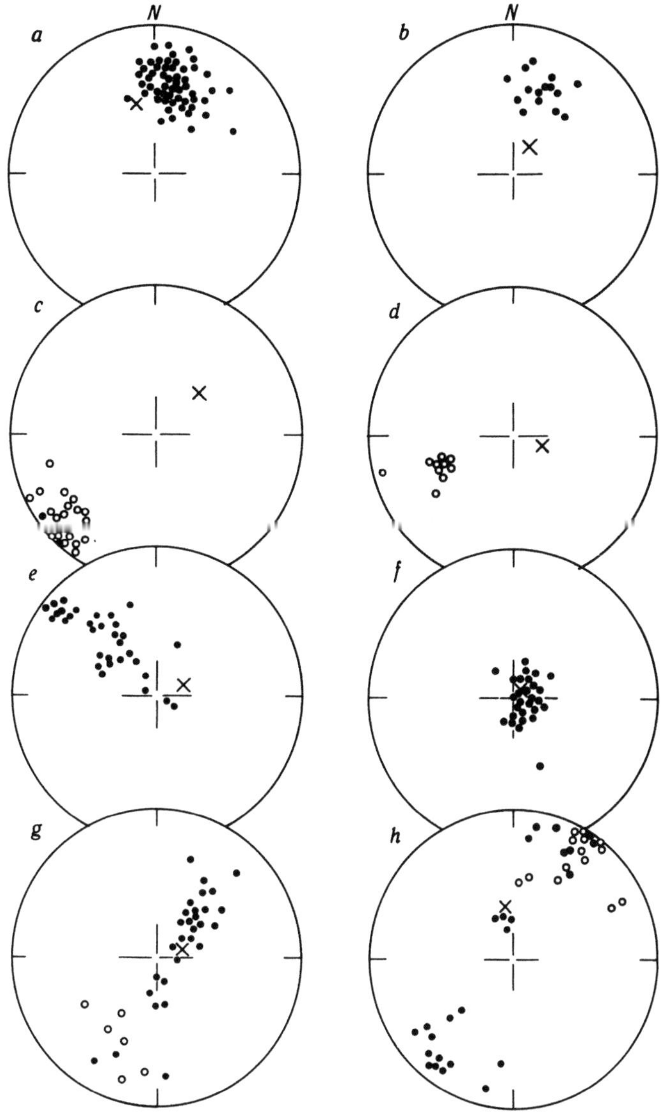

Fig. 2.6a-h. Distribution of the NRM directions of rocks of different paleomagnetic stability; ancient coordinate system. (After Khramov and Sholpo 1967). Clays from the upper part of the red bed sequence, Cheleken Peninsula, western Turkmenia: **a** northwest tilting strata, **b** southwest tilt; Ordovician red claystone from the southwestern part of GDR (Bucha et al. 1976): **c** northwestern tilting strata, **d** southeastern tilt; **e** Middle Ordovician red beds, Kulumbe River, Tunguska syneclise; **f** Middle Pliocene sandstones, Minor Balkhan, West Turkmenia; **g** Paleogene clays and marls, Minor Balkhan; **h** argillite and limestone, the lowermost Middle Carboniferous (d-suite) from the Donets Basin

formation, then the distribution of NRM vectors in pebbles is random since detrital component of magnetization of such "large" particles, i.e. pebbles, will be negligible (see Sect. 1.4.3). In this case the precision parameter K, calculated for the present coordinates, as well as for a pebblestone bed, will tend to zero. If the magnetization occurred after the deposition of pebble, the precision parameter will be high. The following value can serve as a measure of stability:

$$S = 1/(K^{1/2} + 1). \tag{2.2}$$

The fold test is based on the following principles. The directions of J_n are calculated in the ancient as well as in the present sample coordinates, taken from one and the same bed, but from different sites of the fold, or for a series of samples from similar and coeval rocks, having different bedding. If the precision parameter K_1, computed in the ancient coordinates, exceeds that of K_2 in the present coordinates, i.e. $K_1 > K_2$, it is evident that the magnetizatic component, formed prior to the folding and stable in time, is dominant (Fig. 2.5). Postfolding, secondary magnetization dominates when $K_1 < K_2$. A measure of stability S is given by approximate empirical formulae (Khramov and Sholpo 1967), which are useful only as rough estimations of S.

Unlike the test proposed by Graham, the remagnetization plane method necessitates a knowledge of the directions of both magnetizatic components. This method is, in fact, the plot of so-called planes of remagnetization and their traces on the surface of a unit sphere, i.e. circles of remagnetization. Considering J_n as the sum of vectors J_n^a and J_n^h, it becomes clear that these three vectors are always coplanar. Thus, the plane in which they lie is the plane of remagnetization. The degree of stability is calculated as a function of the angles between the directions of the Earth's recent field H, natural remanence J_n and the ancient magnetization J_n^a from formula (1.39). Hence, the directions of ancient magnetization are determined using the methods described in Section 2.33 and magnetic cleaning.

A series of samples, taken for the determination of the paleomagnetic stability using Graham's method, should give an approximation of the same paleomagnetic stability of the same rock (irrespective of its bedding). Therefore, the main justification of Graham's method is the assumption of similar paleomagnetic stability of folded or pebble-forming rocks, and those with gentle dips. This assumption well suits folded rocks, since it is quite obvious that conditions for the preservation of primary magnetization are not as favourable for pebbles as those in rocks in situ. One should take into account that the conglomerate test may give values too low for the measurement of rock stability.

The above methods can provide an assessment of the degree of the stability mean value only for a fairly large group of samples (not less than 40–50) to justify the use of statistical methods. However, in some cases (e.g., when the relationship between rock magnetic stability and lithology, the presence and the degree of secondary changes is studied), it is necessary to determine the degree of stability for every sample of the group. One can obtain this by constructing a circle of remagnetization. Here, apart from the NRM direction in a sample and that of remagnetization, the direction of ancient magnetization also needs to be known.

The mean direction of ancient magnetization for a given group of rocks can be taken as the ancient magnetization direction of a sample, hence, homogeneous magnetization is again assumed. A mean circle of remagnetization for the group of rocks discussed is a circle of remagnetization of the samples shown by plots on a stereographic projection. These will fall outside the circle due to the natural scatter in the NRM directions, but will cluster closely around it (Figs. 2.5 and 2.6). The determination of the degree of stability of a sample is then obtained from the projection of the NRM directions on the mean remagnetization plane (the plots are offset along the arcs of a great circle along the circle of remagnetization) and to the determination of angles Δ and δ. To calculate the degree of stability, Eq. (1.39) is employed.

2.3.3 Determination of the Direction of Ancient Magnetization

Estimation of rock paleomagnetic stability and the calculation of its measure S are not the same as the determination of the direction of ancient magnetization. To solve this problem, several methods have been developed, namely the methods of (i) intersection of remagnetization planes, (ii) displacement and (iii) reversal methods (Khramov and Sholpo 1967). The direction and intensity of ancient magnetization J_n^a and errors of their estimation can then be determined for the two-component case.

(i) The intersection of remagnetization planes method is based on the determination of the line of intersection between two (or more) planes, whose vectors J_n have been turned as a result of the postfolding, secondary magnetization in rocks of different bedding angles. Let us consider the two-component magnetization when the secondary magnetization is recent, i.e. it follows the direction of the present geomagnetic field H_T. Assume we have two series of NRM determinations for coeval rocks with different bedding and the mean directions of J_n and H_T are calculated in the ancient coordinate system for each series. Then, each pair of vectors J_n and H_T has its own plane of remagnetization in space within which the vector J_n^a (prefolding magnetization) should lie. Since the directions of J_n^a in the ancient coordinate system must be the same for all neighbouring coeval sites, the line of intersection of remagnetization planes, plotted according to this system, will represent the direction of J_n^a. (These are plotted diagrammatically on stereoprojection in Fig. 2.7).

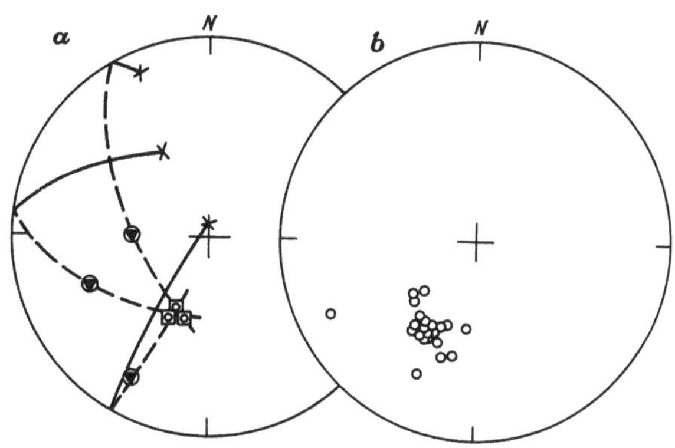

Fig. 2.7a,b. Determination of the prefolding magnetization by means of remagnetization planes intersection method. (After Khramov et al. 1974). **a** Mean NRM directions of the Tatarian stage red beds for three sites within the fold structure near the town of Bugulma and the intersection of the remagnetization circles; **b** ChRM directions for the specimens from the same sites, isolated by thermal cleaning

To obtain reliable results in the determination of the directions of J_n^{na} and to estimate the errors, we use the data from three or more sites of coeval rocks with different bedding. In the case of q number of sites, then the intersections of q number of planes of remagnetization give n = (½)q(q−1) directions. They can be considered as measured directions J_n^a, whose resultant vector R has a direction with angles D_R and I_R, a precision parameter K and a radius α_{95} of the confidence circle. When K≥10, the initial hypothesis will be correct, namely that vectors J_n^a and J_n^a are parallel in the ancient and present coordinate system, respectively, i.e. the secondary magnetization is recent. If the postfolding magnetization was formed in an ancient geomagnetic field, older than Recent but later than the folding, then its direction will differ greatly from that of the present field, and K>10. The direction of such metachronous magnetization J_n^m and, hence the direction of J_n^a can be determined using successive approximations. For this purpose, the direction of R and q number of J_n directions are converted into the present coordinate system and n number of J_n^m directions is determined by constructing remagnetization circles in order to calculate the resultant vector R_m. Then the vectors J_n and R_m are reconverted into the ancient coordinate system, and the second approximation for the directions of J_n^a is determined by constructing new circles of remagnetization. As a rule, it is sufficient to have only one or two steps.

(ii) The displacement method is based on the determination of an angle δ, through which the direction of NRM should be displaced along the remagnetization circle in order to obtain the J_n^a direction. The angle δ is determined by the measure of stability S, or by comparing NRM directions for normally and reversely magnetized rocks. When a succession comprises an alternation of normally and reversely magnetized rocks, similar in lithology and composition, then: (1) directions of primary magnetization of rocks in zones of different polarity differ by 180°, i.e. $J_{n+}^o = -J_{n-}^o$; (2) the mean measures of paleomagnetic stability normally and reversely magnetized rocks are equal, i.e. $S_+ = S_-$. In this case, when angles Δ_+ and Δ_- are known (the angles between the direction of the present field and vectors J_{n+} and J_{n-}, respectively), we can find the angle between J_{n+} and J_{n-}, i.e. δ_+, hence, we can determine the direction J_{n+}^o and J_{n-}^o, with a 180° difference (Fig. 2.8).

The above assumptions result in the equations:

$$\sin(\Delta_+ + \delta_+) = \sin(\Delta_- + \delta_-);$$
$$\sin\Delta_+ / (\sin\Delta_+ + \sin\delta_+) = \sin\Delta_- / (\sin\Delta_- + \sin\delta_-).$$

Solving the equation for δ_+ we obtain:

$$\mathrm{tg}\delta_+ = [\sin\Delta_+ \sin(\Delta_+ + \Delta_-)]/[\sin\Delta - \sin\Delta_+ \cos(\Delta_+ + \Delta_-)]. \quad (2.3)$$

Angle δ can be determined from Eq. (1.39), if the measure of stability S and, hence, the relationship between the ancient and recent magnetization, as well as the directions of H_T and J_n and the angle Δ between them, are known.

The displacement method is used to correct the mean NRM directions for both normally and reversely magnetized rocks or for the known mean value S of the set of samples studied. In this case, the angle δ is removed from the remagnetization circle, using a Kavraisky grid to determine the J_n^a direction.

(iii) Creer and Irving were the first to develop and use the reversal method (Creer et al. 1957). Thus, to calculate the mean direction of an axis, in the case of

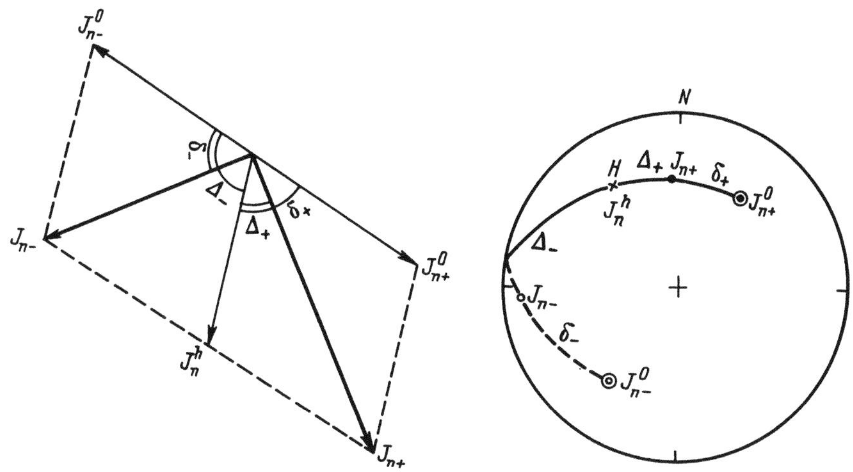

Fig. 2.8. Determination of the ancient magnetization direction by the displacement method. (Khramov and Sholpo 1967). J^o_{n+} and J^o_{n-} are normal and reversed primary magnetization of rocks, J^h_n is common secondary component, J_{n+} and J_{n-} are NRM directions of two rock groups. *Left* is vectorial diagram, *right* is stereoplot

normally and reversely magnetized rocks, along which the rocks were initially magnetized, the directions of reversely magnetized rocks are changed by 180°, i.e. angle $D' = D + 180°$ is taken instead of D and $I' = -I$. The unit vectors and their components are then derived from Eq. (1.40) and hence the direction of the resultant vector is defined, which is the direction of the ancient magnetization. It is clear that the influence of secondary magnetization components is eliminated, because its direction is, as a rule, the same in all rocks. This reversal method does not necessitate the use of any hypothesis about the direction of the secondary magnetization of rocks. For example, it was widely used in the studies of the southern margin of the Siberian platform (Khramov et al. 1974), where the direction of the secondary magnetization could not be definitely determined due to regional remagnetization of rocks by trap rocks of different age. The method is rather inaccurate, because one should add together the geometric sums of J_n vectors of normally and reversely magnetized rocks rather than the unit vectors (Fig. 2.9). However, some differences in lithology of normally and reversely magnetized rocks may also introduce errors, and this justifies the summation of the unit vectors.

2.3.4 Field Methods Used in the Estimation of the Synchroneity of Ancient Magnetization

Prefolding magnetization can be determined by the method of intersection of remagnetization planes, however, very often the sedimentation predates the folding by several million years. Thus, the metachronous magnetization could partly, if not completely, remain within the prefolding component. This can be demonstrated by pre-Paleozoic and Paleozoic rocks from the southern Urals and the northeastern Russian platform (Khramov et al. 1974).

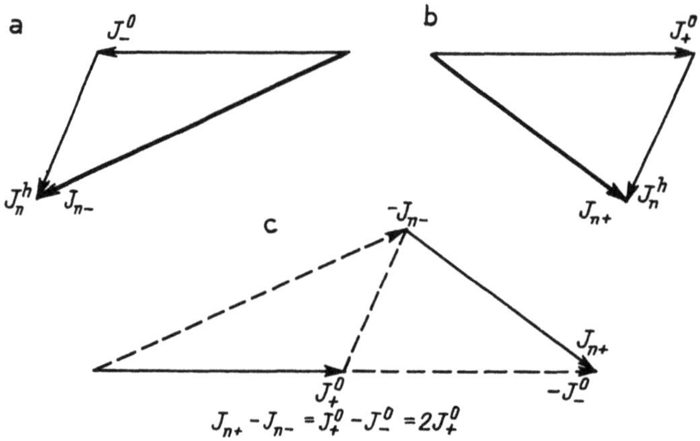

Fig. 2.9a-c. Determination of the ancient magnetization direction by the reversal method. (Khramov and Sholpo 1967). **a** NRM components for reversely magnetized rocks; **b** NRM components for normally magnetized rocks; **c** 180°-turning of reversed magnetization and obtaining the resultant vector. For definitions see Fig. 2.8

The reversal method, applied to N and R directions, resulting from the intersection of remagnetization circles, constructed for normally and reversely magnetized rocks, respectively, can be used to separate the prefolding magnetization components. If the ancient magnetization J_{n+}^a of some rocks include the primary normal J_{n+}^o and metachronous J_n^m magnetizations and the J_{n-}^o of other rocks have a primary reversed J_{n-}^o, equal and anti-parallel to J_{n+}^o and the same metachronous J_n^m magnetization, then it becomes evident that, like the reversals method, applied to the J_n directions, one may sum the vectors J_{n+}^a and J_{n-}^a to obtain the resultant vector, the direction of which is the direction of magnetization J_n^m and its length is $2J_n^m$. The vectorial addition of J_{n+}^a and the vector opposite to J_{n-}^a gives $2J_n^o$. In fact, here we also add together and subtract two normalized (unit) vectors, resulting from the geometrical addition of vectors, which corresponds to the intersection points of the N and R circles of remagnetization.

The conglomerate test can be applied easily in the determination of the direction and value of the ancient magnetization. Unlike the other methods, it considerably limits the possible time interval during which an isolated J_n component was formed. In the case of intraformational conglomerates, it is, in fact, a direct method of determining the synchronous magnetization J_n^c. The J_n components formed prior to the formation of conglomerates, due to random orientation, will have random directions in the coordinate system of the conglomerate bed (i.e. in the ancient system). Their geometrical sum, when there is a sufficient number of vectors, equals zero. Those components which were created after the formation of the conglomerate bed, are parallel to each other. Thus, the geometrical sum of the magnetization J_n of pebbles, divided by their number N, is equal to the mean asynchronous (secondary) magnetization and follows its direction. Gusev developed this method for an estimation of the viscous magnetization in conglomerates of the Quaternary terraces, in order to determine their age (Gusev 1970).

Since the intensity and direction of asynchronous magnetization, with respect to the underlying beds, is known, one can subtract them geometrically from the total magnetization of each sample. In some cases, the concentration of ferromagnetics can be accounted for by the introduction of a scaling factor of asynchronous magnetization with the values $\varkappa$, J_{ri} or J_{rs}.

The relationship between synchronous and asynchronous magnetization in pebbles can also be determined as follows: if we have N number of J_n^c vectors and they are random in the conglomerate, then their mean projections on any plane, normal to the vector of asynchronous magnetization, including J_n^m, will be equal to (½)J_n^c. Hence, designating the mean angle between the declination of vector J_n for each pebble and the J_n^m direction as angle θ, we may write

$$J_n^c/(2J_n^m) = \mathrm{tg}\theta$$

Taking into account that $\cos\theta = R/N$, where R is the geometric sum of unit vectors j_n, we obtain

$$J_n^c/J_n^m = 2(N^2/R^2 - 1)^{1/2}. \quad (2.4)$$

Using this relationship and the direction and value of J_n^m being known, we can determine the direction and value of synchronous magnetization J_n^c.

2.4 Magnetic Stability Tests and Magnetic Cleaning

2.4.1 General

All paleomagnetic laboratory studies are aimed at the determination of the primary magnetization J_n^o, representing the geomagnetic field pattern in rocks at the moment their formation. When the magnetization J_n^o is more stable to certain laboratory effects (time, temperature, variable alternating magnetic fields, chemical reagents), all secondary magnetizations are destroyed (or compensated) during the course of demagnetization. Thus, primary magnetization is isolated as the last and most stable NRM component. However, it is not always possible to prove that this stable component is a primary magnetization, even if its direction differs greatly from that of the present magnetic field at the sampling site or when direct methods suggest its prefolding nature.

The study of NRM stability under different experimental conditions is the first stage of any investigation. Such a study uses so-called pilot specimens which account for 2–5% of the collection. It demands thorough plotting of gradual demagnetization curves and the observation of changes in magnetic direction due to increasing laboratory treatment. If the pilot specimens show similar results, thus implying that the largest precision parameter for the J_n distribution will be obtained within a certain field or a narrow temperature interval, and the higher values of the fields or temperatures in this interval do not, in fact, change the NRM directions, then this field or temperature is used for cleaning the entire collection. However, the NRM values often continue to decay, but changes of their directions may cease at different values of the field, or temperature interval or different times of maintaining in acid. This means that a so-called stable end point is attained at different instants, i.e. the magnetization becomes a single component. In these cases, demagnetization curves need be plotted for all specimens. Hence, the directions resulting from the different steps of demagnetization fall within a distribution showing the best precision parameter.

It is not unusual that a single-component pattern of isolated magnetization appears. However, spectra of coercivities $d(J_{nh}/J_n)/dh$, blocking temperatures $d(J_{nT}/J_n)/dT$ and solubilities in acid $d(J_{nl}/J_n)/dt$ of different NRM components may overlap, thus preventing their separation. The complete isolation of magnetizations incorporated in the NRM, along with the determination of their values and directions, is only possible when the above-mentioned spectra are clearly discrete (see Sect. 1.4.8).

Paleomagnetic studies commonly use a traditional approach in choosing the type of cleaning for igneous and sedimentary rocks. The following factors are taken into consideration: the typical composition of ferrimagnetic-NRM carriers, the nature of primary and secondary magnetizations and practical factors, e.g. technical difficulties resulting from the design of the equipment. Alternating field (A.F.) cleaning, complemented in some cases by thermal cleaning, is applied to igneous, magnetite-bearing rocks. Thermal and chemical cleaning, or their combination, are employed to investigate sedimentary rocks in which a single-domain hematite is the main carrier of the primary and, very often, of the secondary magnetization (A.F.-cleaning for such a highly coercive material would necessitate the presence of a very strong, hardly achievable fields). Temporal cleaning can be used for all rocks, and is sufficient even when the secondary magnetization is of a viscous nature.

Before we proceed to the discussion of different cleaning methods, we shall refer to a system necessary for the compensation of geomagnetic field and magnetic distortions in the laboratory. It is necessary to use such systems for A.F.-cleaning because any steady magnetic field present during the demagnetization by the a.c. field (the amplitude of which decreases from a chosen original value to zero) produces in the specimen an anhysteretic magnetization directed along the steady field, sometimes exceeding the natural magnetic component which it is intended to isolate. This field can result from any alternating field asymmetry for instance due to the presence of even harmonics (see, e.g. As 1967) etc., all of which may produce a magnetic bias along the solenoid axis. The diurnal variations of the geomagnetic field, along with laboratory distortions, usually do not exceed 4×10^{-2} A/m. Experiments are not carried out when magnetic storms are recorded by laboratory magnetometer (for instance, proton magnetometer). The formation of new thermoremanent or thermochemical components, often exceeding magnetization J_n (mainly in sedimentary rocks), may result from imperfect steady field compensation in the case of thermal cleaning. The destruction of viscous magnetization greatly decreases due to imperfect field compensation during thermal and temporal cleaning.

Field compensation can be carried out using two or three mutually perpendicular pairs of square loops or coils (Helmholtz coils). Sometimes, four-circuit systems, compensating each of two geomagnetic field components (vertical and horizontal), are used to increase the volume of "nonmagnetic space" (McElhinny et al. 1971). The compensating current is first calculated theoretically and then experimentally by a null magnetometer. The adjustment is made before each heating-cooling cycle or A.F.-demagnetization. With this method geomagnetic field variations are neglected during the experiment (2-3 h for thermal-cleaning), however, they are small, as mentioned above. A system of self-compensation coils is the most reliable, namely amplified antiphase signals from the sensitive magnetic units are input into the set of coils. Many laboratories, in order to produce nonmagnetic space, employ magnetic screens (for instance, those produced from permalloy). Screens can be added to coils and compensate the field to a level of $(8-72) \times 10^{-4}$ A/m. They are also used for the storage of cleaned specimens before measurement and during their transportation, after cleaning, to the measuring device.

2.4.2 Temporal Cleaning and the Compensation Method

Any paleomagnetic study of the collection should be preceded by the evaluation of viscous magnetization, which sometimes is the only secondary component. In fact, temporal cleaning involves the positioning of specimens inside non-magnetic space for a certain period of time in order to reduce or remove viscous magnetization. However, it is a technically difficult procedure to establish a space "without a field" large enough for all samples to be placed inside and to retain such conditions for a long period. The compensation of viscous magnetization by the Earth's field is much simpler. The technique was elaborated by Khramov (1965). The collection is placed in an in situ position for 2 weeks (or a month) and the NRM is measured. The specimens are then rotated through 180° about EW-axis and then maintained for 3 months. In practice, if the slope of the orientation planes exceeds 10°, specimens are placed symmetrically in a position around the magnetic meridian and, only then, the specimens are swung by 180° around the y-axis. The second measurement of J_n (after being kept "against the field") gives the NRM direction and value free of a natural viscous component. The vectorial difference in the results between the first and second measurements gives the J_{nv} component.

Experiments have shown that there are cases when viscous magnetization, usually removed under the influences of weak alternating fields and upon heating to low temperatures, is resistant to these factors. Such tests only lead to the partial destruction of the laboratory viscous magnetization. The compensation method is thus more effective in these cases (Fig. 2.10). Since viscous magnetization is parallel to the present magnetic field at the sampling site, a positive result of the compensation method indicates an increase in the angle between the mean direction of the magnetization vector and the present field direction. If the collection consists of normally and reversed magnetized samples, then the directions are subdivided into two groups which, ideally, differ by 180° (Fig. 2.11).

2.4.3 Alternating Magnetic Field Cleaning

This type of cleaning is based on the ability of ferromagnetics to follow decreasing hysteretic cycles under the influence of an alternating field of decreasing amplitude. This causes a reduction of the remanence which is the greater the higher the initial demagnetization field. In isothermal remanent magnetization the effect of an a.c. field, with an initial amplitude which is equal to the amplitude of the field

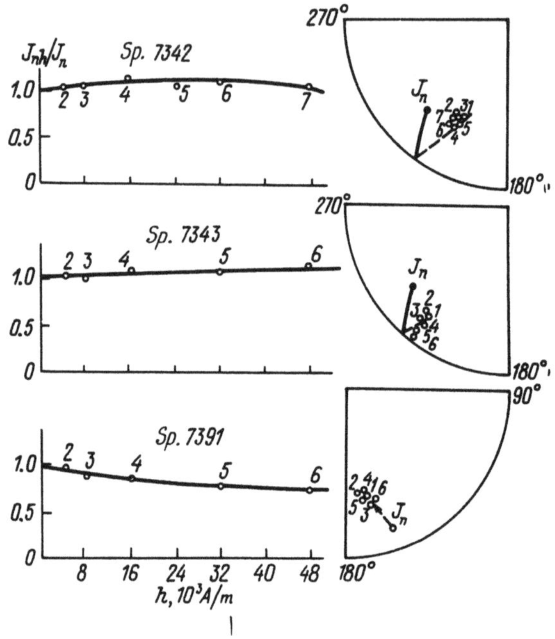

Fig. 2.10. Results of compensation and A.F. cleaning applied to the Upper Devonian rocks from the Main Devonian field. *1* After 90 days of compensation; *2–7* after A.F. cleaning in a.c. fields presented on left plots. The stereoplots show A.F. cleaning, used after the compensation, incompletely eliminate the laboratory viscous magnetization (the points *2–7* are closer to J_n than point *1*)

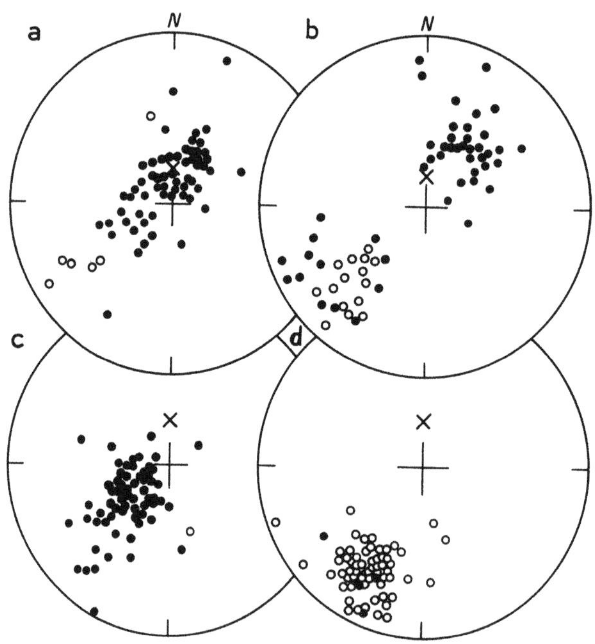

Fig. 2.11a-d. Distributions of NRM directions for specimens of sedimentary rocks before **(a,c)** and after **(b,d)** compensation. **a,b** Red beds, the Tatarian stage, Sukhona River; **c,d** limestones, the Ludlov stage, Dniester River

in which it was acquired, results in its complete destruction. The demagnetizing alternating field for other types of magnetization should be considerably larger than their original magnetizing fields.

Different types of magnetization are not equally resistant to the alternating field. Thus, the magnetizations of igneous rocks, in decreasing order of their stability with respect to alternating fields, are usually as follows: complete thermoremanent[4], chemical, thermoviscous, viscous, dynamic and isothermal magnetization. Thermoremanent and chemical magnetizations of single-domain particles have the same stability. The following series are characteristic of sedimentary rocks, namely chemical magnetization of single-domain particles, detrital and chemical magnetization of multidomain particles, thermoviscous and viscous magnetization, etc. However, a rock is a multicomponent system, not only in the way it acquires its components of magnetization, but also in the composition of the ferromagnetics and the size of their grains. Fine-grained hematite and hydroxides are referred to as magnetically hard constituents, while coarse-grained magnetite is considered magnetically soft. There are hemoilmenites, titanomagnetites, maghemites, titanomaghemites and fine-grained magnetites among the above-mentioned extreme members. Their coercive spectra overlap. Moreover, fine-grain phases are magnetically harder than the coarse-grained ones in similar minerals (Fig. 2.12).

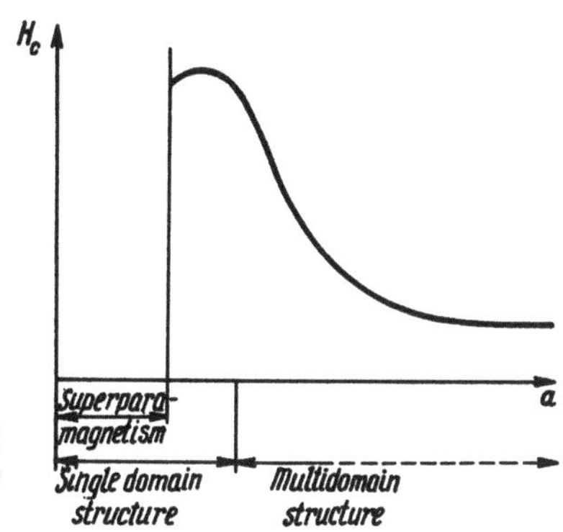

Fig. 2.12. Schematic representation of coercive force and diameter dependence for grain assemblage. (Slightly modified after Nagata 1961)

Thus, when magnetic cleaning is applied to igneous rocks, it results in the isolation of secondary chemical magnetization, the carrier of which is by single-domain hematite (hematite is more common in rocks than single-domain magnetite; the hematite of igneous rocks is usually secondary, except for high-temperature oxidation during autometamorphism). In other cases, the efficiency of A.F.-cleaning depends on complex combinations of the magnetization type, its carriers and the grain size of the magnetic minerals, i.e. the degree of overlap of the coercive spectra of NRM components. The existence of single-domain grains in rocks possibly enables the NRM to be resistant to fields over 72×10^3 A/m (McElhinny 1973). Considerable magnetic hardness of such grains indicates a long

[4]Partial thermoremanent magnetization occurs in the succession according to the temperature interval of its origin.

relaxation time, i.e. the possibility of retaining their NRM during the entire period of the rock's existence.

All the above mentioned are also true for sediments. Chemical magnetization of single-domain hematite and hydroxides can be both secondary and primary, often coexisting. Red beds, which are often used in paleomagnetic studies, due to large NRM values, have a high content of diffused hematite, whereas hydroxides are responsible for the rock colour and frequently forming part of its matrix. The clastic (often single-domain) hematite fraction of a rock amounts for only a few percent. Thus, the primary and secondary NRM components in red beds are associated, in most cases, with high coercive materials. Thus, the isolation of such components necessitates the presence of fields over 16×10^4 A/m.

The solenoid is a main component of the alternating field demagnetization equipment. The test specimen is placed in the centre of the solenoid, which corresponds to the centre of the compensating coils or to the centre of the cylindrical magnetic screen. Some parameters, which define the solenoid construction, are prearranged, namely internal diameter, depending on the size of the specimen holder, line voltage and maximum alternating field, necessary for demagnetization. One solenoid design has been described in detail (Roy et al. 1973). All the solenoids, used as rock demagnetizers, operate in a resonance mode and have filters to suppress even harmonics.

A gradual decreasing of the alternating field from the chosen initial amplitude to zero is necessary in order to obtain the demagnetizing effect. This can be obtained either by the slow removal of the solenoid from the centre in which the specimen was placed, or by decreasing the current in the solenoid. The second method is preferable, because the mechanical movement of a heavy coil cannot be uniform. Furthermore, high currents, used to produce strong fields, and maintained during the slow movement of the solenoid may cause its overheating.

There are two methods for demagnetization in the alternating magnetic field, namely (1) the specimen is stationary relative to the field (demagnetization is performed successively along the three J_n components); (2) the specimen revolves in the field (or the field rotates about the specimen). The second method was developed in order to reduce the experimental time and to prevent the possible formation of anhysteretic magnetization in case of inadequate steady field compensation. The complex (simultaneous) rotation of specimens about two (or three) mutually perpendicular axes provides a simultaneous demagnetization of all the NRM components. The more complex the rotation of the specimen, the faster and more complete it is demagnetized. Simultaneous rotation of the field and specimen is often used to hasten demagnetization.

Specimen tumblers are usually made of nonmagnetic material, such as chipboard, nylon or fibreglass. The rate of rotation around the different axes is chosen so that the specimen can take as many positions, relative to the demagnetizing field direction, as possible. Usually, speeds are of the same order and their ratios do not constant integers. The speed of specimen rotation should be sufficiently high compared to the rate of decrease of the alternating field amplitude from maximum to zero. When there are two axes of rotation, they fall within the plane perpendicular to the solenoidal axis. When there are three axes, two of them (not coinciding with the solenoidal axis) lie within the plane of the solenoidal longitudinal section, whereas the third axis is perpendicular to this plane.

Magnetizations, distorting the regular J_{nh} change, were observed in specimens rotating in alternating fields. The first thorough investigation of the phenomenon (Wilson and Lomax 1972) revealed that rotational remanent magnetization (RRM) occurs even at slow rotation of the ferrimagnetic specimen in a decreasing, alternating field and is not a variety of anhysteretic remanent magnetization (ARM). This RRM considerably exceeds the ARM in value (the noncompensated steady field was not above 4×10^{-2} A/m). The RRM direction was antiparallel to the rotation vector when the rotational axis was perpendicular to the alternating field axis. RRM increases with the initial amplitude of the alternating field and has a tendency to saturation. When the angle between the alternating field axis and that of the rotational axis changes from 0° to 90°, the RRM increases from zero to maximum.

In later works some conclusions, made by the authors mentioned above, were considered debatable. However, in general, they have been confirmed and complemented. Besides, comparative RRM and ARM studies have revealed a disparity in the principle of their behaviour (Stephenson 1976; Edwards 1980) and, hence, their different nature. Stephenson obtained very important additional data on RRM (Stephenson 1976), namely the behaviour of soft magnetic materials and conductors, rotating in the steady field, is absolutely consistent with theory. Magnetic effects have no assymmetry relative to the rotational axis, as in the case of RRM. When the speed of specimen rotation is close to the a.c. frequency, RRM has a peak value. RRM = 0, when the rotation speed and current frequency are equal. RRM changes its sign and becomes extremely large at high rotation speed. RRM versus rotation velocity (within a range of 0.001 to 1 Herz) curves are, in fact, identical both for a specimen rotating in a demagnetizing field and for the field rotating around the specimen. Different methods have been proposed to eliminate the RRM effect. This can be exemplified by two, rather simple methods, namely the use of a two-axis tumbler, having a rotation speed of 1 and 2 Herz with a sharp change in rotation every 2 s (Wilson and Lomax 1972); two demagnetization experiments were carried out under similar conditions with opposite directions of rotation and the results were averaged (Hillhouse 1977).

However, there is no theory to explain the origin of RRM, although many scientists have tried to explain its nature, using such phenomena as rotational hysteresis (Owens 1978) and temperature fluctuations under the influence of the precession torque moment, associated with the alternating field (Smith and Merrill 1980). The equipment, used in the a.c. demagnetization of rocks, is being constantly improved and, in different laboratories, the various elements of their construction are being refined with respect to the basic problems in which scientists are interested.

An a.c. demagnetizer, equipped with a generator of a fine sinusoidal a.c. spectrum, and a multilayer screen for protection from outer, steady fields were designed in the USSR Academy of Sciences, Borok Observatory. An apparatus was also designed here, which in the automatic mode, allows a graph plotter during a 20–30 s period to construct a continuous demagnetization curve in fields up to 1.6×10^5 A/m with sensitivity to the magnetic moment of 1×10^{-3} Am^2 (Tsirel' et al. 1983).

An apparatus based on a two-component spinner magnetometer for obtaining demagnetization curves for different types of remanence in an alternating magnetic field was designed and constructed in Kazan' University. An alternating field is produced by ferrite-cored coils and the spinner magnetometer measuring coils are set in the interpole gap of the ferrite core. A continuous record of the two remanence components during a.c. amplitude growth is recorded, which results in a vector diagram in one of the orthogonal planes. Zijderveld diagrams (see below, Sect. 2.4.8) are constructed, in combination with a diagram for another orthogonal plane which includes the third magnetization component measured for the duplicate specimen from the same sample (Nurgaliev 1983).

2.4.4 Thermal Cleaning

Thermal cleaning, as well as alternating field cleaning, must separate NRM components and, under favourable conditions, isolate the primary magnetization J_n^o when it has the highest stability. Thermal activation processes become more intense in specimens on heating, the higher the temperature, the easier to overcome the high energy barriers. For example, isothermal and viscous magnetizations, associated with magnetic transitions due to low energy barriers, are the least resistant to heating, whereas thermal and chemical magnetizations are the most stable. A high blocking temperature, like high coercivities found on A.F.-cleaning, implies that magnetic grains have a long relaxation time at 20°C.

The multicomponent nature of NRM is shown in thermal demagnetization curves by intercepts showing a different rates of changes (rapid or low) and inflection points. However, this means that one and the same magnetization is carried by various ferrimagnetics with different blocking temperatures. Control over the J_n directions enables this problem to be solved by seeing whether directions change with a change in slope, or whether the directions remain unchanged. Heating allows magnetic components to be removed, not only in those carriers in which the blocking temperature will be exceeded, but also in magnetic components associated with the magnetic minerals which transform into other varieties in the course of thermal demagnetization. For instance, pyrrhotite and titanomaghemite, together with most products of low temperature oxidation such as maghemite, and ferric hydroxides, are the most common carriers of secondary chemical magnetization. Usually, these are transformed into hematite on heating. Maghemite changes into hematite in the range between 250° and 450°C. The decay and transformation of hydroxides occurs in the range 150°–300°C. Therefore, in some cases, heating to 300°C may remove secondary magnetization. If hydroxides are carriers of primary magnetization (which is rather common in sedimentary rocks), the primary magnetization is also removed and vector difference analyses must be employed to determine it (see Sect. 2.4.8).

Transformation of paramagnetics into ferrimagnetics occur on heating (e.g. siderite to maghemite). If this occurs in a laboratory instrument in which the geomagnetic field is not compensated, then it can cause the formation of parasitic magnetization of such an amplitude that the remaining part of the NRM becomes undetectable. However, even by good field compensation, a specimen still needs to be shielded when it is brought to the magnetometer because newly formed minerals, during their short stay in the geomagnetic field, can acquire strong viscous magnetization. The transformation of rock during thermal cleaning should be monitored, for instance, by magnetic susceptibility measurements – the increased growth of $\varkappa$ implies the formation of a stronger magnetic phase. Despite such complications, thermal cleaning is widely used for both igneous and sedimentary rocks, as it is often impossible to remove highly coercive secondary components by means of A.F. cleaning.

The choice of a suitable thermal cleaning mode for the pilot specimens differs slightly from that used in A.F. demagnetization, namely the heating step becomes shorter with increasing temperature (as the Curie point of major minerals, i.e. magnetite and hematite, is approached), whereas in A.F. demagnetization, the size of the steps increases simultaneously with increase in the applied field. The disadvantage of thermal cleaning is that the heating-cooling cycle of 2–3 h is time-consuming. However, it is possible to compensate this disadvantage by using a uniform field, or shielding the geomagnetic field, over a large volume. Except for the field compensating system, thermal cleaning equipment also includes an electrically heated, non-inductive furnace. Despite some differences, the following conditions are usually met in the construction of furnaces, namely the materials from which the furnace is made (including the spiral) should be non-magnetic; the materials should have low external heat radiation, whereby each specimen should be uniformly heated, i.e. temperature deviations should not exceed 10°C; measurement and maintenance of temperature should be accomplished within an accuracy of a few degrees. The furnace can be removed in order to provide quicker cooling of specimens in the non-magnetic space. When shielded furnaces are used, such a removal is possible when the shield is of a sufficient length (usually cylindrical), thus leaving the specimens cooling in a very weakened field, i.e. the specimens must be placed far away from the open end of the cylindrical shields.

These procedures imply a progressive method of thermal demagnetization, i.e. stepwise heating of specimens to gradually increasing temperatures, cooling of specimens to 20°C in the field free space and the measurement of their magnetization. Another method incorporates NRM measurements at the heating temperature (at specific points of a given temperature interval or continuously) and is used to obtain a detailed thermal demagnetization curve for every specimen.

A device, based on a two-component, spinner magnetometer and equipped with an electric furnace, temperature control and plotter, was designed at the Kazan' State University. It is used to plot $J_n(T)$ vector diagrams in different coordinate planes during continuous heating of the specimens. Comparison of these diagrams with data from thermomagnetic differential analyses, which enables the determination of the composition of ferromagnetic fractions in the rocks studied, suggests that this continuous method is more effective than stepwise thermal demagnetization (Yasonov 1981).

2.4.5 Chemical Cleaning

Collinson was the first to introduce chemical cleaning (Collinson 1967), which is now used extensively in paleomagnetic studies of red beds in which detrital magnetization very often coexists with primary and secondary chemical magnetization.

Ferrimagnetic compounds form coatings on nonmagnetic minerals (frequently on quartz) and on detrital grains of magnetite and hematite. Such fine grains, which joints or enter the cement and pigment, are dissolved by 8–10 N hydrochloric acid (more rarely, oxalic acid) when they are transformed into paramagnetic salts. Thus, the NRM can be divided into components associated with high and low soluble magnetic phases, ranging from coatings to the detrital grains of specularite which is the most resistant to acid and is the carrier of detrital magnetization.

Undoubtedly, chemical cleaning is not always effective, since a rock can have low porosity (clay rock) or it can have a carbonate matrix which dissolves in acid much more quickly than the other constituents of the rock. A rock may also contain only a fine-grain fraction, carrying different components, which is destroyed by acid equally quickly. Coarse and medium-grained red beds, including quarzite cement, are ideal for chemical cleaning. However, chemical cleaning can be effectively used in rocks of low porosity when the acid is under pressure (Burek 1971). Another way to achieve greater acid penetration is chemical cleaning in vacuum, e.g. of 10 kPa (Dankers 1982). At first, to avoid the destruction of specimens containing carbonate (specimens literally explode due to violent carbon dioxide liberation), it is necessary to treat them with a weak hydrochloric acid, which facilitates the gradual escape of the gas and does not destroy the rock. Specimens, including cement irresistant to acid, are placed into cells, the shape of which is similar to that of a specimen; their material is acid-resistant (Collinson 1967).

The chemical cleaning procedure is simple, but laborious. Specimens are first prepared for cleaning. They are cut into thin plates (3–5 mm), or holes are drilled in them in order to widen the surface in contact with the acid. The magnetization is then measured and the specimens are placed into individual cells and filled with 10 N acid (sometimes, during the first 1000–1300 h, they are kept in an 8 N HCl solution). At regular intervals (first short, and then longer), the specimens are removed from acid, washed with distilled water and placed into a nonmagnetic space for drying (2–3 days). The use of a ventilator may help to shorten this time. Then the J_{nc} is measured again and the specimens are immersed into fresh acid. This procedure continues until the J_{nc} intensity and direction cease to change or until magnetization is so negligible that the accuracy of the measurement is very low.

The J_{nc} measurement is carried out as a specimen loses its colour because the secondary chemical magnetization, or the late phase of the primary chemical magnetization, can be due to the pigment. Such decolourization becomes discernible, if we make a cut or break the plate. The cleaning process is often terminated due to the increasing fragility of the specimen or even its complete destruction. In the classical work of Roy and Park (1974), chemical magnetizations CRM_A (the older component is associated with hematite, the growth of which is caused by diagenesis) and CRM_B (the younger one, associated with pigment) were separated by means of chemical cleaning, from the DRM, whose carrier was specularite. Heating to 674°C removed CRM_B only. Thus, it appears that only a single component of magnetization was left after heating because the spectra of the CRM_A and DRM blocking temperatures overlapped.

This example is characteristic of red beds, namely that the multiphase composition of the chemical magnetization makes it impossible, in many cases, to separate CRM and DRM, and results in an apparent single-component behaviour of remanence, during thermal cleaning. Thus, there is a need to combine different types of cleanings.

2.4.6 Low-Temperature Cleaning

This cleaning is based on the property of magnetite to lose its magnetization (the first constant of the crystallographic magnetic anisotropy changes its sign) at a low temperature, called the isotropic point (which is −143°C for stoichiometric magnetite). Then, on heating back to room temperature, 80–90% of component J_{rt} is restored and less than 20% of J_r is lost (Nagata 1965). This cleaning procedure is very simple, as the specimen is immersed into a Dewar flask with liquid nitrogen and cooled to −143°C, then it is naturally heated, in the field-free space, back to room temperature.

Ozima et al. (1964) discussed low temperature cleaning, which was initially tested for basalts having an NRM, with an isothermal magnetization improvised in the laboratory. Later, it was tested for serpentinite. In the first case, the isothermal component was completely removed. In the second case, the precision parameter for the NRM distribution increased from 13 (before cleaning) to 25 (after cleaning). However, this cleaning has not found wide application, because it is effective only for specimens containing coarse-grained magnetite, which is not typical of rocks that retain their primary magnetization.

2.4.7 Combined Cleanings

The compulsory application of temporal cleaning, or the compensation method, prior to A.F.-, thermal- and chemical-cleanings, has proved, by long-term studies, to be a prerequisite for combined cleaning. Chemical cleaning in combination with heating (before and after cleaning) appears to be very effective. For instance, the influence of 10 N hydrochloric acid on Silurian-Devonian red beds from Tuva which are resistant to A.F.-cleaning in the fields up to 64×10^3 A/m and to thermal-cleaning at 600°C, has revealed the following features. After acid treatment (37 to 1375 h), the NRM decreased to 0.74–0.18 of the initial value. The magnetization vector changed direction by several degrees to as much as 108° for different specimens. The subsequent thermal-cleaning continued to show NRM directions which deviated from the direction of remagnetization in the present field (Fig. 2.13). As a result, the magnetization vectors formed two groups of approximately opposite directions (Fig. 2.14). Only thermal cleaning did not produce such an effect.

The application of combined cleaning allowed scientists from the Kazan' State University to estimate the contribution of the CRM and DRM primary components to the NRM, and to isolate these components. For this purpose specimens were treated by the A.F.-cleaning in fields of 8×10^3 A/m and then by thermal-cleaning at 150° to 200°C (Boronin 1979). A similar combined cleaning (two-stage demagnetization) is widely used for igneous rocks. It is efficient when the secondary

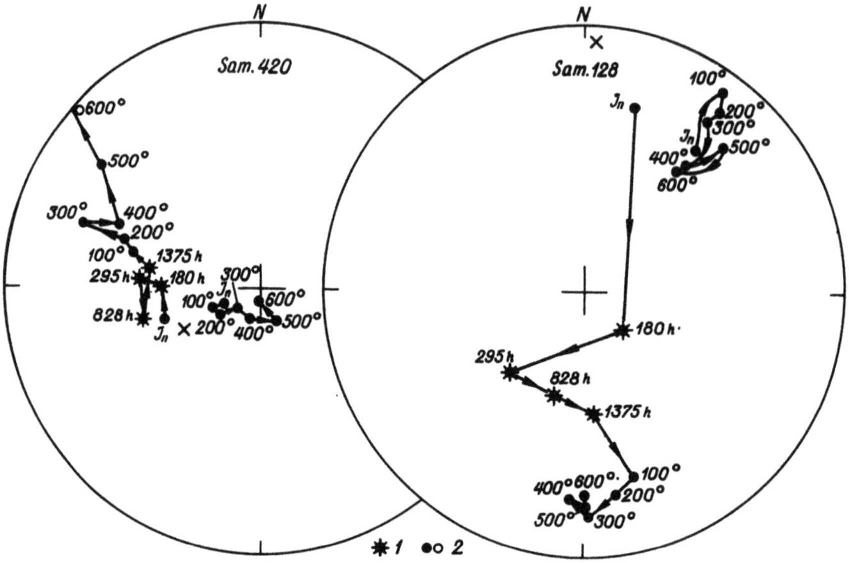

Fig. 2.13. NRM directional change during cleaning (the Silurian-Devonian red beds from the Tuva region). *1* Chemical cleaning; *2* thermal cleaning

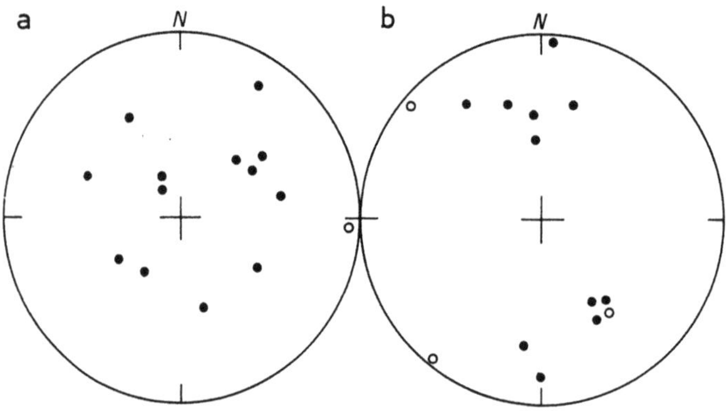

Fig. 2.14a,b. Distributions of the NRM **(a)** and ChRM **(b)** for specimens of the Silurian-Devonian red beds (Tuva region). ChRM is obtained after combined (chemical + thermal) cleaning

(dominating) and primary (negligible) ancient components have practically equal coercivities (their coercive spectra show almost complete overlap). The A.F.-cleaning in the field of 8 to 12×10^3 A/m removes all of the unstable components. The stable part of the NRM in the field up to a few tens of kA/m then behaves like a single-component magnetization whose direction almost coincides with the dominant component. Further heating first destroys the secondary magnetization and leaves the ancient primary magnetization intact as this is resistant to high temperatures. This accounted for less than 10% of the original NRM. As a result, three magnetization phases could be isolated.

2.4.8 Methods for the Presentation of Magnetic Cleaning Data and Their Interpretation

The remanent magnetization of rocks is a vector, so the presentation of magnetization changes due to cleanings should reflect changes of both magnitude and direction. The angles I, D and the intensity of J_n versus temperature or demagnetization field intensity curves will not be quite so informative, because in each case, the vector is characterized by scalars only. Data are generally represented by stereograms, showing the distribution of the NRM directions at different stages of demagnetization, along with the plots, showing the magnitude of the total magnetization vector (usually normalized) versus laboratory treatment. Stereograms illustrate the vector deviation from a remagnetization direction and reflect the increase in the precision parameter of distributions (i.e. when one component becomes dominant). The demagnetization curves represent the vector components, the relationship between cleaned and remaining components, and the coercivity and blocking temperature spectra. This method provides a direct representation of the behaviour of the vector resultant of several J_n components, which remain at each steps of cleaning and, indirectly, show the behaviour of destroyed components.

For the purpose of estimating the contribution of each NRM component and tracing their behaviour during cleaning, the total vector is separated into vectors, corresponding to given directions, namely the directions of (i) the present field at the sampling site, (ii) the assumed ancient remagnetization (one or several, if known from the geologic or tectonic history of the region) and, finally, (iii) the primary magnetization. In general, at the different steps of specimen cleaning, stable (to certain a degree) clustered distributions of vectors occur, the mean directions of which correspond to particular remagnetization fields. In such cases the resultant vector can be resolved into components along these directions, taken from the stereograms, which include the direction of the present field and the final stable direction. Changes in magnetization are plotted for every component and stereograms are constructed for the distribution of directions in different specimens. The analysis of vectors, directly within the remagnetization planes, provides the best representation of changes of the J_n components on cleaning (see Sect. 7.1). For example, the analysis of the J_n^o and J_{nv} vectors in the plane of remagnetization caused by the present field is widely used. This plane is defined by the primary magnetization direction (when known) and the direction of the present field (Boronin 1979). The total vector is defined by its magnitude and angle Δ (see Sect. 2.3.3) at different steps of cleaning.

Another method is the use of Zijderveld vector diagrams (Zijderveld 1967). The Zijderveld diagram is a plot combined from two orthogonal projections of the end point of the total magnetization vector trajectory in at each cleaning step. One projection is in the vertical plane (x vs. z or y vs. z). The other projection is in the horizontal plane (x vs. y). Both projections have a common NS (x-axis) or EW (y-axis). The destruction of one of the NRM components (all the other components being constant) manifests itself on the diagram as a linear segment over some specific range of temperatures, alternating fields and time in acid. When there is a scatter of points on the diagram, averaging lines are drawn. In this case the least squares method is usually used. There are cases when, at the end of the experiment (large fields or high temperatures), a stable NRM component is still retained. On the diagram it is represented by a straight line drawn to the origin of the coordinates. This is the characteristic magnetization and corresponds to the so-called stable end point on the diagram. The direction of the NRM components, removed at certain intervals, and of the characteristic magnetization are calculated from the diagrams. Applying a graphic technique, one can plot the NRM component demagnetization curves projected on one of the planes (Fig. 2.15).

The procedure used to isolate the NRM components by the Zijderveld diagram, as well as the advantages and limitations of the method, have been discussed by Dunlop (Dunlop 1979). The main advantage, apart from its simplicity and descriptiveness is that the "windows" of non-overlap in the coercive or the blocking temperature spectra of NRM-components "automatically indicated by linear segments ..." (Dunlop 1979, p. 14).

The method of remagnetization circles, proposed by Khramov (Khramov 1958) to isolate prefolding magnetization, was used by Halls in the interpretation of data obtained after cleaning. He called it the method of converging remagnetization circles (Halls 1978). This method can be used when the rocks studied have two magnetization components differing in time and also in the scatter of directions (before cleaning) due to geologic processes as well as in the coercive spectra (or blocking temperatures). Then, in the course of cleaning, the directions of the total J_n vector for every specimen lie along a great circle of remagnetization, whereas the circles converge to a point which corresponds to the least scattered, component direction. Halls (Halls 1978) applied this method to determine the direction of the secondary NRM component in pre-Cambrian rocks from the Lake Superior area.

Thus, the present state of laboratory studies and their interpretation enables the isolation of the secondary components of the ancient magnetization not only to remove them, but also to reconstruct the history of rocks (chemical, tectonic). This can be of great help when the geological evolution of the entire region is studied.

2.5 Analysis of the Origin of Ancient Magnetizations

Cleaning methods allow the isolation of the ancient component from multi-component rock magnetizations and the determination of its value and direction. It is also possible to estimate its resistance to destructive laboratory treatment effects (alternating field, heating, leaching) and, in some cases, one can establish whether or not the ancient magnetization is synchronous with the age of the rock. However, these methods cannot determine the origin of the ancient magnetization.

To obtain an estimation of the origin of the ancient magnetization, it is necessary to answer the following questions.

1. Which magnetization carriers exist in rock and what is their origin and concentration?
2. What is the contribution of a particular magnetic mineral to the NRM?
3. How have the magnetic moments of the mineral grains acquired their orientation in the ancient geomagnetic field, i.e. what type of magnetization do they possess?

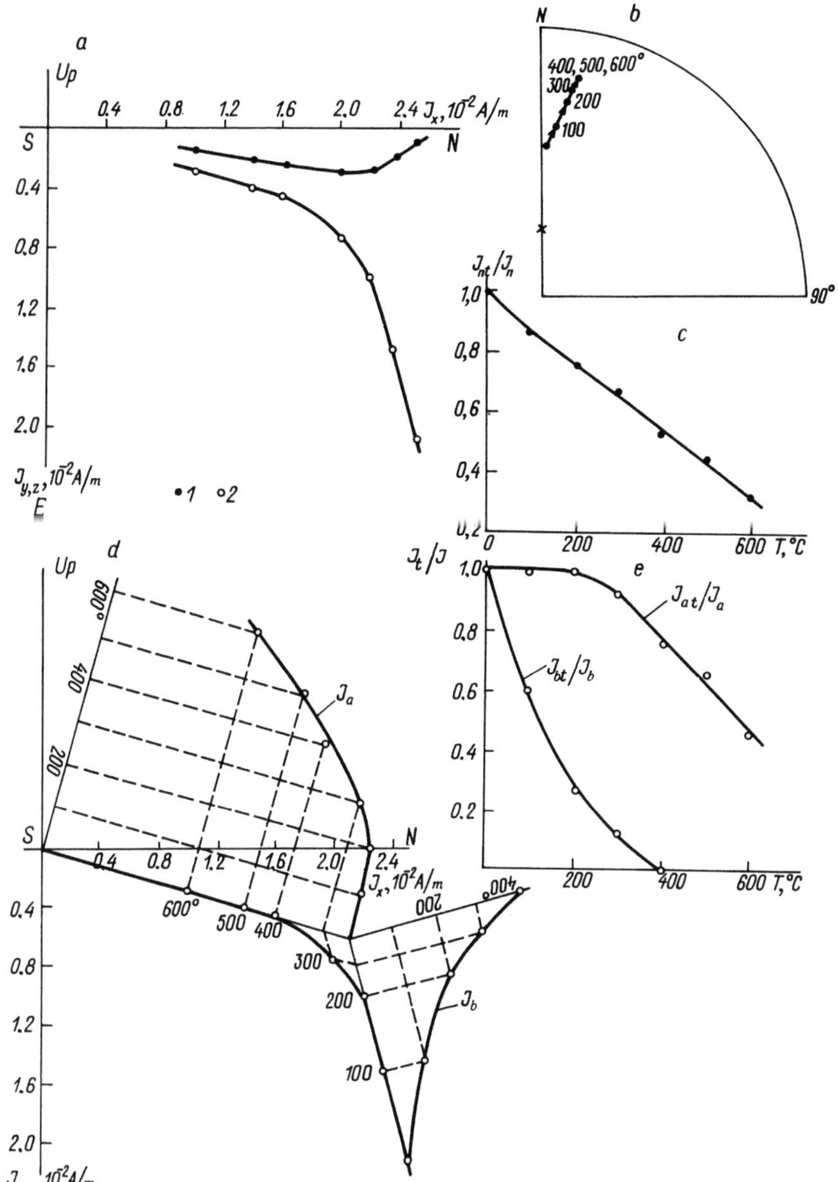

Fig. 2.15a-e. Results of the thermal demagnetization of the Devonian siltstone specimen (Tuva region). **a** Zijderveld vector diagram (projections of NRM vector end points: *1* on the horizontal plane; *2* on the vertical meridional plane); **b** NRM direction change, plotted on the unit sphere (polar equidistant projection); **c** changes of J_{nt}/J_n value; **d** changes of the NRM components, characteristic J_a and unstable J_b, separated by means of a Zijderveld diagram on the vertical plane (see **a**); **e** the same for normalized intensities (magnitudes) of J_a and J_b

Methods, used in magnetic mineralogy provide an answer to the first question, when, for example, the thermal relationships of the saturation magnetization $J_s(T)$, or its remanent part $J_{rs}(T)$ as well as the partial thermoremanent magnetization J_{rpt} spectra, are studied. The analysis of sections and polished sections of the magnetic, monomineral fractions under the light microscope, accomplished by electromagnetic separation allow the determination of the origin (allothigenous or authigenous) of the particular magnetic minerals in the rock.

The answer to the second question can be given by means of mathematical statistics, defining, for instance, the statistical relationship between the concentration of a particular magnetic mineral distributed in the rock and the J_n value. The results obtained show that the contribution of any particular mineral to the NRM has a wide range due to facies and genetic differences in rocks (Boronin 1979). There are rocks, in which the magnetic moments of some minerals are oriented along the resultant vector and some which have an opposite orientation.

The comparison of J_{rs} and J_n during heating makes it possible to establish which of the ferromagnetics are stable and which are unstable (Petrova 1977). The analysis of the J_n variation at temperatures at which magnetic minerals, according to the $J_{rs}(T)$ curve, pass the Curie point or undergo the chemical, structural or crystallographic changes, also shows which part of the magnetization is associated with a given mineral (based on the amplitude relation of the changes of J_{rs} and J_n curves on passing this temperature range). Proceeding from this analysis, one can also determine whether the minerals acquired their magnetization synchronously or at a different times from study of the change in the NRM direction (Petrova 1977). The third problem, concerning the type of magnetization connected with the assemblage of mineral grains in the ancient geomagnetic field, cannot be completely solved, if only the contribution of the magnetic mineral to the NRM generation is known.

Redeposition experiments are very important in the determination of the NRM type in sedimentary rocks (Bagina 1970). Methods and techniques used in these experiments are well known. Thus, some problems concerning data interpretation need to be discussed. Comparing the magnetization J_n of the original rocks and the J_{ro} value of the redeposited sediments, we can established whether the NRM is detrital or chemical. If the redeposition coefficient $K = J_n / J_{ro}$ is close to or equals unity, then the NRM is detrital in origin. If $K \gg 1$, the NRM is either chemical or thermoremanent (Bagina 1970). Such an interpretation is based on the assumption that the precision parameter for a particle magnetic moment direction in the case of a chemical origin of the NRM in natural sediments is considerably higher than that in redeposited rocks. This is only partially true (Khramov and Sholpo 1967).

First, the exsolution of magnetic particles is a long-term process. If a geomagnetic field reversal takes place at some instant, then the orientation of newly formed magnetic particles follows this new direction and, hence, the NRM decreases. The redeposition coefficient of such rocks will be less than unity or, in an extreme case, equal to zero. Second, in all the foregoing discussions it was presupposed that the formation and orientation of all of the magnetic particles were independent. However, newly formed particles can originate in the vicinity of old ones and their magnetic moments may be oriented by the demagnetizing field of those pre-existing particles. Thus the total magnetization of rock can decrease and under certain conditions become even opposite to the Earth's magnetic field during sedimentation (Khramov and Sholpo 1967).

Despite such limitations, experiments on redeposition are of great importance in the establishment of the origin of the ancient magnetization. To obtain a reliable interpretation of the redeposition experiments, it is necessary to establish independently the distribution pattern of the particle magnetic moments relative to the ancient geomagnetic field direction and, in particular, to estimate the contribution of the moments oriented opposite the field during redeposition. A joint study of all the results would solve the problem of the ancient magnetization origin (Boronin 1979). Unfortunately, this method cannot be applied to the lithified rocks.

To determine the type of natural magnetization, the widely known Thellier method is applied to igneous rocks and baked sediments. Thermodemagnetization $J_n(T)$ and thermomagnetization $J_{rt}(T)$ curves of the same rock are used for this purpose. Coincidence of the $J_n(T)$ curve and the reversed $J_{rt}(T)$ curve proves the thermoremanent nature of the NRM. However, this method can be applied only to rocks containing multidomain magnetite and hematite grains, since the TRM and CRM characteristics in single-domain grains are the same (Petrova 1977).

Comparison of the J_n and J_{ri} magnetization stability characteristics is a versatile method, providing the determination of the magnetization type. It is based on the fact that the J_{ri} is higher in value and stability for all types of magnetization in which the general pattern of potential barriers in a ferromagnetic are not affected (Petrova 1977). If the NRM is detrital then its value is one-third to one-fifth of the ARM induced in the same steady field and its stability is higher than for ARM. If the NRM is of thermoremanent or chemical origin (for a single-domain state of ferrimagnetics), then its magnitude as well as its stability exceeds J_{ri}. The stability of CRM carried by multidomain grains is less than their J_{ri} stability.

Though the problem of the origin of ancient magnetizations is complex, it can be solved on a compound basis. To increase the reliability of the paleomagnetic reconstructions, it is very important to know the nature of the ancient magnetization J_n^a and the time in its geological history when its various components were formed. A method of detection and removal of intermediate magnetization has been proposed for its study (Danukalov et al. 1977) (see Sect. 7.1).

2.6 Statistical Analysis of Paleomagnetic Directions and Poles

2.6.1 The Best Estimates of Mean Direction and Precision Parameters

In Sec. 1.5.2 it was shown that the sum R_o of N projections for the j_{ni} unit vectors in the direction of the magnetizing field H is $R_o = N(\text{cth } K - 1/K)$. A distribution of vectors having precision parameter $K<3$ is not considered as significant in paleomagnetology and are used in special problems only. At $K>3$, with good approximation, $\text{cth } K \approx 1$, and

$$R_o/N = (K-1)/K. \tag{2.5}$$

The left side of the equation, in accordance with the mean value theorem, is the mean integral value of $\cos\psi$, namely

$$\overline{\cos\psi}=(K-1)/K, \tag{2.6}$$

which gives the relation between the precision parameter K of the j_{ni} vectors and mean angle $\vartheta=\arccos\overline{\psi}$ of their deviation from the magnetizing field H.

As the direction of the ancient geomagnetic field H is not predetermined, it can be estimated from the direction of the geometric sum R for N of the j_{ni} vector components along the coordinate axes,

$$X=\sum_{i=1}^{N}\cos I_i \cos D_i;\quad Y=\sum_{i=1}^{N}\cos I_i \sin D_i;\quad Z=\sum_{i=1}^{N}\sin I_i. \tag{2.7}$$

The vector R direction is defined by angles of declination D_R and inclination I_R:

$$D_R=\arctan(Y/X);\quad I_R=\arcsin(Z/R), \tag{2.8}$$

where $R=(X^2+Y^2+Z^2)^{1/2}$ is vector R length.

In accordance with statistical laws, when true values are exchanged for their estimates, then Eq. (2.5) should be substituted by

$$(R-1)/(N-1)=(K-1)/K,$$

which gives the estimate of the precision parameter for vectors j_{ni}

$$K=(N-1)/(N-R) \tag{2.9}$$

and for the mean value of $\cos\psi$, due to relation (2.6)

$$\overline{\cos\psi}=(R-1)/(N-1). \tag{2.10}$$

Value $\vartheta=\arccos\psi$ serves to estimate the scatter of vectors and presents the standard deviation of the unit vector j_{ni} from direction R, i.e. it denotes the root mean square error (i.e. this is its analogue in Fisher's distribution).

2.6.2 Estimates of the Accuracy of the Determination for Unit and Mean Directions of NRM

A number of vectors, which have an angle of deviation ψ exceeding that of ϑ, is

$$N(\psi>\vartheta)=\int_{\vartheta}^{\pi} dN.$$

Value $p(\psi>\vartheta)=N(\psi>\vartheta)/N$ may be considered as the probability of the event that an angle between any j_{ni} vector, of a given population and field H, shall be more than ϑ. The substitution of dN from Eq. (1.42) gives

$$p(\psi>\vartheta)=\frac{K}{2\operatorname{sh}K}\int_{\vartheta}^{\pi}\exp(K\cos\psi)\cdot\sin\psi\, d\psi;$$

after integration

$$p(\psi>\vartheta)=[\exp(K\cos\vartheta)-\exp(-K)]/[\exp K-\exp(-K)]. \tag{2.11}$$

This suggests, neglecting exp(−K), as we consider a case when $K > 3$, that

$$\cos\vartheta = 1 - \ln(1/p)\cdot K^{-1}. \qquad (2.12)$$

It is evident that $\cos\vartheta = \cos\psi$ in Eq. (2.6), if one takes $1/p = e$. Hence, the probability $p'(\psi > \vartheta) = 1 - p = 0.63$. This probability corresponds to the standard deviation in Fisher's statistics.

Let us now find the estimation error of the ancient field direction by the vector R direction. The probability of any angle between H and R being more than a preset angle α can be considered as the probability of a complex event when each of the j_{ni} vectors has a deviation from H through an angle exceeding α. This probability, by the product rule, equals the product of probabilities of Eq. (2.11), i.e.

$$p(\psi > \alpha) = \{[\exp(K\cos\alpha) - \exp(-K)]/[\exp K - \exp(-K)]\}^N; \qquad (2.13)$$

from which, neglecting exp(−K), we obtain

$$\cos\alpha = 1 - \ln(1/p)\cdot(NK)^{-1}. \qquad (2.14)$$

Angle α, defined as the confidence angle, is one-half of the circular cone angle, described around the vector R. The true direction of the ancient field H lies inside this cone with a probability $p' = 1 - p$. The circle with a radius α (confidence circle), described around the point representing the projection of the resultant vector R corresponds to angle α on a unit sphere.

Taking into account all members of Eq. (2.11), one can obtain the following formula for the confidence angle α (Shipunov, personal communication):

$$\cos\alpha = 1 - (RK)^{-1}\ln[1/p - (1-p)\exp(2RK)]. \qquad (2.15)$$

When estimating the error in the R-direction, $p' = 0.95$ is usually taken, i.e. $1/p = 20$. Substituting this value in Eq. (2.14), changing $1 - \cos\alpha = 2\sin^2(\alpha/2)$, and taking into account that at for small angles $\sin\alpha \approx \alpha$[if angle α is in radians or $\sin\alpha = (\pi/180)\alpha$ if angle α is in degrees], one obtains (for the second case):

$$\alpha = 140(NK)^{-1/2}. \qquad (2.16)$$

Another often used approximation is expressed by the following formulae for the confidence angle:

$$\cos\alpha = 1 - [(N-R)/R]\,[(1/p)^{1/(N-1)} - 1],$$

which for $1/p = 20$ gives

$$\cos\alpha = 1 - [(N-R)/R]\,[20^{1/(N-1)} - 1], \qquad (2.17)$$

or

$$\alpha = 81[20^{1/(N-1)} - 1]^{1/2}\,[(N-R)/R]^{1/2}. \qquad (2.18)$$

Values of $81[20^{1/(N-1)} - 1]^{1/2}$ for $N = 3$–290 have been tabulated in Khramov and Sholpo (1967).

As mentioned above Eqs. (2.5, 2.6, 2.9, 2.10, 2.12–2.18) show good approximations only if $K > 3$. In the case of more dispersed vectors, such as in the study of moraines or conglomerates when using the conglomerate test (see Sect. 2.3.2), it is necessary to use the exact formulae (1.44) and (2.15), solve them by iteration to obtain the K and α values.

2.6.3 Choice of the Set of the NRM Directions

Any statistical analysis presupposes that if the number N of independent vectors belonging to the population used in the analysis increases, the confidence circle radius α decreases. In general, the necessary accuracy for the mean direction of the vector, can be obtained even for sets having a low precision parameter K if the number of vectors are increased.

This seems to be natural, since the precision parameter of any set of the NRM vectors is defined primarily by the ancient geomagnetic field changes over a time interval. This time interval is incorporated in the set by the freezing process of the field and, only to some extent, by the sampling and measurement methods, i.e. it is practically independent of the investigator. Therefore, we can see from Eqs. (2.16)-(2.18) and Fig. 2.16 that the value of α falls rapidly until $N = 30$–40, and any further moderate increase in the number of vectors only decreases the confidence circle radius to a small extent. Thus, the basic method of eliminating the error in the paleomagnetic determinations is the rational choice of the objects to be studied (geologic bodies, lithologic types of rocks), which should be as "clean" as possible, and a more thorough determination of the primary magnetization. A reasonable choice of the set of vectors for statistical analysis is necessary to obtain an unbiased estimate.

Statistical analyses can be applied at different stages of processing of the paleomagnetic data. For instance, the analysis can be applied on the following sets of vectors, (i) repeated measurements of the properties of the same specimen; (ii) the NRM directions in specimens from one sample or bed; (iii) mean directions of the NRM in every studied bed of the given formation.

Let there be two sets of NRM mean direction determinations for a given suite (Khramov and Sholpo 1967). The first set includes three determinations of the NRM direction for each of 20 samples (or specimens) taken from four beds of the suite (five samples from each bed). Another set includes one determination of the

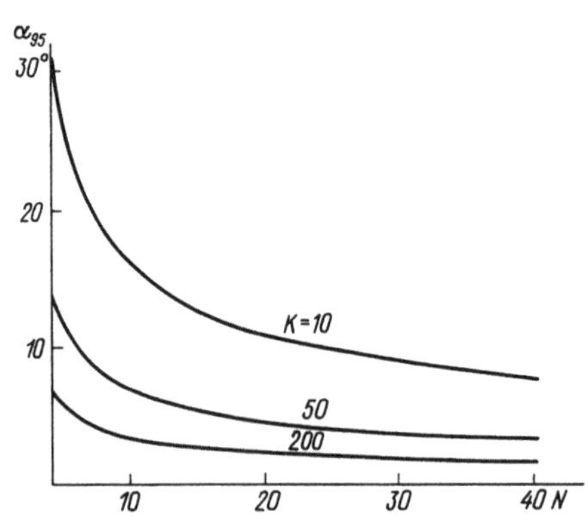

Fig. 2.16. Values of angles of confidence α_{95} versus number of measurements for different values of the precision parameter K

NRM direction for each of 60 samples, taken from 60 beds of the same suite. If, in both cases, we analyze all 60 directions obtained as a common set, it should be possible to obtain less distinct values of α_{95} to characterizing the error in the J_n mean direction determination. It is not difficult to show that the procedure, used in the first case, is erroneous, and only in the second case, is the error estimated correctly. In fact, in the first case each of 20 sets consisting of three vectors, referring to the NRM measurements in the samples, only shows the accuracy of the individual measurements. The mean directions of NRM for each sample comprise four sets of five vectors, in analyzing each of these sets, so it is possible to obtain the J_n vector precision parameter within each of the four beds and the accuracy, with which the mean directions are determined in them. Only the set of four mean directions of NRM for each bed can be used to calculate the NRM mean direction of the suite as well as the K and α_{95} values. In the second case, all 60 vectors can be analyzed simultaneously, because they refer to homogeneous objects, i.e. beds. Similar examples have been discussed by Irving (1964).

Taking into account the fact that the precision parameters of vectors are different at different levels of groupings then the impossibility of uniting these sets into one becomes more evident. Thus, the precision parameter in the course of repeated NRM measurements of one and the same specimen as well as the precision parameter for the NRM directions in specimens, prepared from one sample, strongly depend on the ratio between the threshold sensitivity of the method and the J_n magnitude, which ranges from 100 to 3000. The precision parameter of the NRM direction for samples selected from the same bed is usually 300–6000 for rocks having a stable single-component magnetization, whereas the between-bed precision parameter is 30–200 (Table 2.1).

The best way of sampling a suite is to take a sample from as many beds as possible. The number of samples from each bed should be increased only when the number of beds in a suite is limited (or when the primary magnetization is isolated from a few beds only) or a high within-bed scatter of directions is observed. It is evident that a single sampling from as many beds as possible is profitable, when the goal is the determination of the NRM mean direction in the suite, but not that of the precise NRM direction in each bed or flow. Generally, it is better to take two or three samples that allow the control and recognition of occasional mistakes, due to incorrect sampling or broken bedding. The necessity of such sampling is evident, because one can then obtain pilot collections for the determination of type and parameters of the magnetic cleaning and for laboratory study as well as for standard collections for permanent storage.

2.6.4 Fit of the Observed NRM to Fisher's Distribution

The fit to Fisher's distribution is a feature of the single component magnetization distinguished by the random deviations of its direction from the mean direction. Checking this correspondence with the observed distribution of paleomagnetic directions (j_{ni} vectors of the characteristic magnetization) is made, first of all, to establish the efficiency of the cleaning applied, e.g. to establish the clearness of the isolated NRM component. Besides, in the course of the temporal succession

Table 2.1. Mean directions and precision parameters for NRM populations at different statistical levels. (After Khramov and Sholpo 1967 with additions)

Declinations D and inclinations I of resultant vectors (degrees)								Precision parameter	Sampling site and statistical level
D	I	D	I	D	I	D	I	K	
225	−37								
226	−36								
226	−39							2553	Specimens from
		226	−37						sample 6423-3
226	−37								
229	−36								
226	−36								
		224	−37	224	−36			3318	Samples from bed
		223	−36						6423
		224	−36						
		223	−36						
		224	−34						
				214	−25				
				217	−24	217	−31	177	Site Gzhel – I
				222	−35				All beds
				220	−30				6417 – 6424
				213	−37				
				216	−33				
				218	−29				
						213	−32		Gzhel – II
						220	−32		Gzhel – III
						217	−32		Noginsk – I
						217	−29		Noginsk – II
						219	−30		Voskresensk – I
						218	−31		Voskresensk – II
Mean value for Upper Carboniferous clays of the Moscow region						217	−31	1320	$\alpha = 1.6°$

analyses of the j_{ni} vectors, such a comparison of distributions helps to determine the existence of any regular components, which is necessary in order to define a synchronous magnetization.

This control of the correspondence, as one sees from Eq. (1.43) for Fisher's distribution density, may be carried out separately for the radial deviation from the mean and for the azimuthal distribution. It is convenient to analyze distributions by nets, which divide the unit sphere projection into segments, which equalizes the probability of vectors being found within them. (Khramov and Sholpo 1967).

The nets are plotted using precision parameter K values. Setting the values of probability successively at $p_1 = (m-1)/m$, $p_2 = (m-2)/m, \ldots, p_n = 1/m$, one obtains $m-1$ angles of α from Eq. (2.10), which subdivide the unit sphere into m rings whose common centre coincide with the projection of the resultant vector R. $p = 1/m$ is then the probability of the j_{ni} vector lying inside any of these rings. Figure 2.17 shows the α angles versus the precision parameter K for $m = 10$. When the precision parameter K for a given distribution is known, we may define, by the curves of Fig. 2.17, nine angles $\alpha_1, \ldots, \alpha_9$ and plot the corresponding circles, i.e. the boundaries of the rings against the sphere's projection.

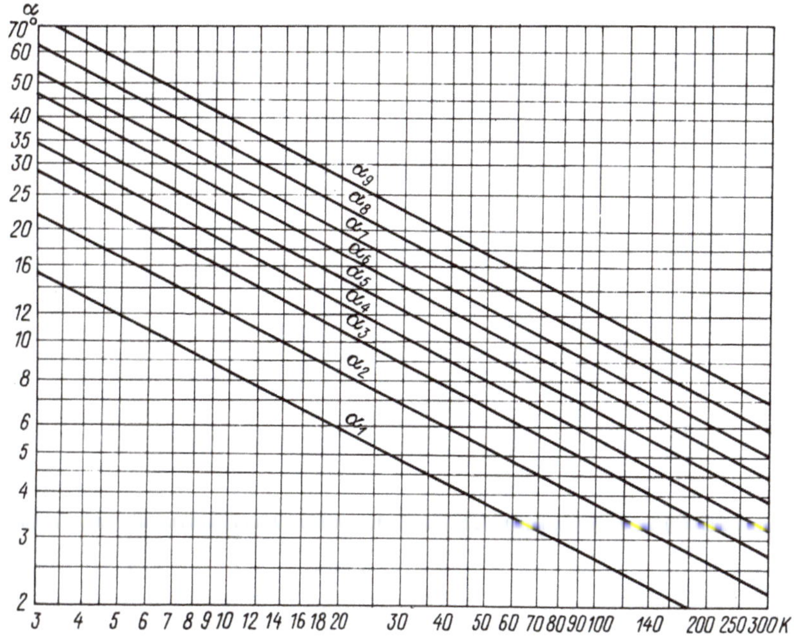

Fig. 2.17. Values of angles $\alpha_1 \ldots \alpha_9$, which divide the unit sphere into rings of equal probability, versus precision parameter K. The probability of getting a unit vector inside each of these rings equals 0.1. (After Khramov and Sholpo 1967)

Assuming N number of j_{ni} vectors for a given set with a precision parameter K and $n_1, n_2, \ldots, n_i, \ldots, n_m$ vectors are in each of the areas. As a measure of the observed and theoretical distribution discrepancy, the following value is obtained

$$\chi^2_\nu = (\sum_1^m n_i^2)/(pN) - N. \tag{2.19}$$

The χ^2_ν criterium values define the probability p_f correspondence, or, strictly speaking, the significance level of the starting hypothesis, assuming the analyzed distribution is a Fisher's distribution. Values of p_f depend on χ^2_ν and the number of degrees of freedom:

$$\nu = m - 1 - t,$$

where t is the number of independently determined parameters. Here, $t = 3$: the precision parameter and two coordinates of the distribution centre, i.e. of vector R (Bazhenov and Rjabushkin 1978).

The probability p_f value at $\nu = 4$–7 is taken from the graphs on Fig. 2.18.

The uniformity of the azimuthal distribution is controlled in the same manner, namely the sphere projection is divided into l equal sectors and the number $s_1, s_2, \ldots, s_i, \ldots, s_l$ of vectors in each sector is counted and the χ^2_ν values are calculated. Then, the probability p_a of the fit to uniform distribution of vectors by the azimuth is defined. In this case, the number of degrees of freedom is taken at $\nu = l - 3$ (here, $t = 2$, i.e. two coordinates of vector R).

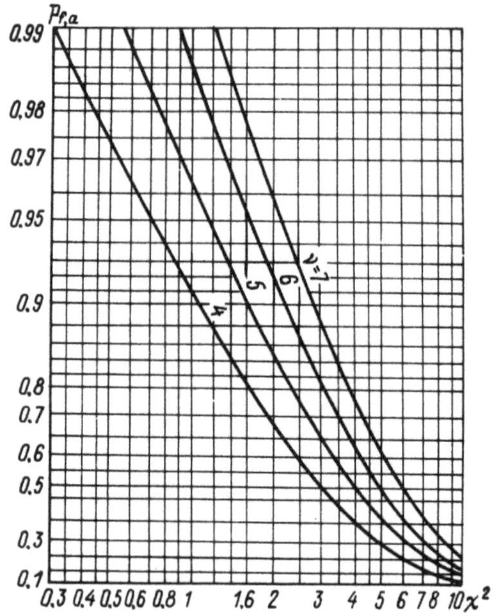

Fig. 2.18. Values of probability $p_{f.a}$ for the theoretical distribution versus χ^2 criterion for different degrees of freedom ν. (Khramov and Sholpo 1967)

Before the analyses, the j_{ni} vector distributions should already be centered, i.e. the vector R is placed at a sphere pole, which lies on the projection centre. The j_{ni} vector projections are then centered by the same methods used to recalculate the vectors from the ancient system into the present. Let D_R, I_R be coordinates of the distribution centre, i.e. the resultant vector R, defined by formula (2.6). D_i, I_i (i = 1.2 . . . N) are coordinates of N vectors of j_{ni}. Thus, new coordinates of D_i^x, I_i^x of the j_{ni} vectors can be defined by Eq. (2.1), having changed $I_c \rightarrow I_i^x$, $I_a \rightarrow I_i$, $D_c \rightarrow D_i^x$, $(D_a - A) \rightarrow (D_i - D_R)$, $B \rightarrow (90 - I_R)$. Centering can be also done graphically (Khramov and Sholpo 1967). All the j_{ni} and resultant vector directions are plotted on the polar grid (see Fig. 1.26). Then the grid is placed on an equatorial grid (See Fig. 1.27) placing the centres of grids together and the projection of R on the equator of the equatorial grid. All projections of the j_{ni} are then displaced along the parallels by the same longitudinal degrees which is equal to angular distance between the R and the grid centre.

Examples can be found of two distributions of the j_{ni} vectors. One of the examples belongs to Fisher's distribution, the other does not. Figure 2.19 a depicts the j_{ni} vector projections of the rock samples from the Kuk-Karauk suite, the Ashinsk series in the southern Urals. The resultant vector direction is placed in the centre of the projections. Circles of equal probability are drawn at $p_f = 0.2$ intervals. Azimuthal layout is performed through 45°. Analysis has shown that the probability level values are as follows: $p_a = 0.99$; $p_f = 0.85$. These values are close to unity. Hence, the distribution of the J_n directions correspond to Fisher's distribution, i.e. the NRM has one component only. Checking by the fold test (see Sect. 2.3.3) has shown that the magnetization is prefolding.

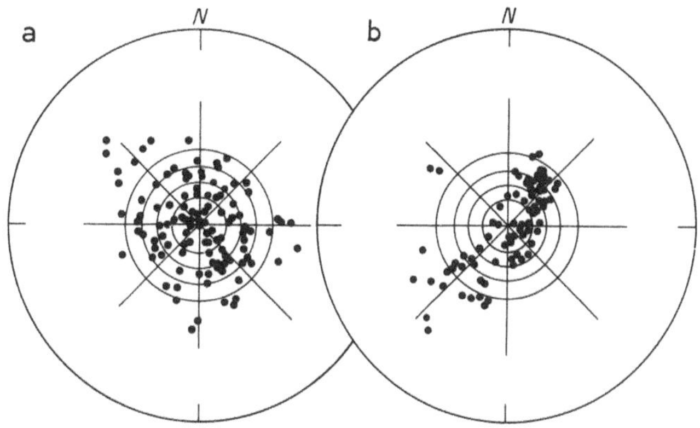

Fig. 2.19a,b. Analysis of the fit of the NRM direction distribution to Fisher's distribution. (Khramov and Sholpo 1967). Distributions are centred and the projections of the unit sphere are divided into ring segments and sectors of equal probability. Examples of NRM distributions: **a** Kuk-Karauk suite of the Ashinsk series, Vendian, Southern Urals; **b** Upper Carboniferous red clays from the Avilov Formation, Donets Basin

Let us now consider the second example where the j_{ni} distribution does not correspond to Fisher's distribution (Fig. 2.19b). The samples are from Upper Carboniferous red clays from the Avilov Formation in the Donets Basin. The probabilities $p_f = 0.85$ and $p_a = 0.001$ were obtained, i.e. the azimuthal distribution was nonuniform. Such a band like distribution is characteristic of two-component magnetization when there is a different relationship between components in different samples. The remagnetization circles method makes it possible to prove that one of the components corresponds to a prefolding magnetization and the other to a recent one. Temporal cleaning has revealed that recent magnetization is viscous and further analyses of the j_{ni} distributions (characteristic magnetization after cleaning) has established that prefolding magnetization has one component ($p_a = 0.90$).

In this case agreement between the distribution of the j_{ni} radial deviation and Fisher's distribution indicates that due to monotonous lithology of rocks (red clay), only one probable relation between ancient and recent magnetization can be observed. Deviations from this relation are random.

The sphere's surface, as can be seen from these examples and other paleomagnetic studies, was divided into five ring zones and eight sectors. It has been shown (Bazhenov and Ryabushkin 1978) that the application of criterion χ^2 on such a small number of intervals is not reasonable. It was found further that p_f and p_a estimates depend greatly on the number and position of the intervals, especially when the collection of samples (set of the j_{ni} vectors) is small. Therefore, it is advisable to increase the number of intervals to 18–20 and to use the χ^2 method repeatedly at different numbers and positions of the subdivision interval. Thus, the total number of vectors should be not less than 100.

Bazhenov and Ryabushkin (1978), in order to analyze sets of paleomagnetic directions with $N = 10–100$, proposed the use of Cuiper's criterion, which does not compare the distribution densities, but instead, their integral characteristics (distribution functions). As mentioned above, the resultant vector R is coincident with the projection centre, i.e. the entire population of vectors is centred. Therefore, azimuths φ of the vector directions can be found from any point in the increasing order of the azimuth as well as the radial deviations ψ from the vector R (projection centre). The tier functions of the observed distribution (height of the tier is 1/N) and the functions of the theoretical distribution are plotted. The azimuthal distribution is represented by the function as a straight line $F(\varphi)=[1/(2\pi)]\varphi$ and the function of $p(\psi)$ is given by Eq. (1.43). The sum of the maximum deviations of the tiers curve for the analyzed distribution from the theoretical curve is taken as a measure of correspondence. Empirical data are considered in poor agreement with the supposition of Fisher's distribution, if the value of

$$V_n=[(p_N-p)_{max}+(p-p_N)_{max}]N^{1/2} \tag{2.20}$$

is more than the certain critical value, defined by the choosen level of p_f and p_a significance. If $V_n \leqslant 0.9$, then the values of p_f, $p_a \geqslant 0.8$ at $N=10$ and p_f, $p_a \geqslant 0.9$ at $N \geqslant 100$; if $V_n \geqslant 1.5$, then p_f, $p_a \leqslant 0.1$ at $N=10$ and p_f, $p_a \leqslant 0.15$ at $N=100$. A higher accuracy of the p_f and p_a values can be found from the plot presented by Bazhenov and Ryabushkin (1978).

The example presented in Fig. 2.19 was analyzed in the same manner. The comparison of the distribution functions is shown in Fig. 2.20. Estimates of the significance level were obtained ($p_f=0.80$; $p_a=0.98$), which practically coincided with those obtained by the χ^2 method ($p_f=0.85$; $p_a=0.99$).

2.6.5 Two-Tier Statistical Analysis

Two-tier statistical analysis (Watson and Irving 1957) is used to establish the ancient magnetization synchroneities. This method allows one to define the significance of between-bed variation and between-bed and within-bed (K_b; K_w) precision parameters. In the case of synchronous magnetization the standard angular deviation $\theta_{63}=81\ K_b^{1/2}$ characterizes the secular variation amplitude.

According to Watson and Irwing, the J_n^a measurements, obtained for a group of samples taken from b number of beds with n_i number of samples in each, can be tabulated (Table 2.2) and the ratio of the mean squares can be determined:

$$F_e=\frac{\sum_1^b r_i-R}{b-1}\cdot\frac{\sum_1^b (n_i-1)}{\sum_1^b (r_i-n_i)}$$

Hence, the calculated F_e is compared to its theoretical value F (F-test). If the between-bed dispersion is significant ($F_e \geqslant F$), then the K_b and K_w precision parameters are determined by equating the mean squares to their mathematical expectations. The significance of the between-bed dispersion with a probability above 95% indicates that the magnetization process was regulated by a succession of bedding.

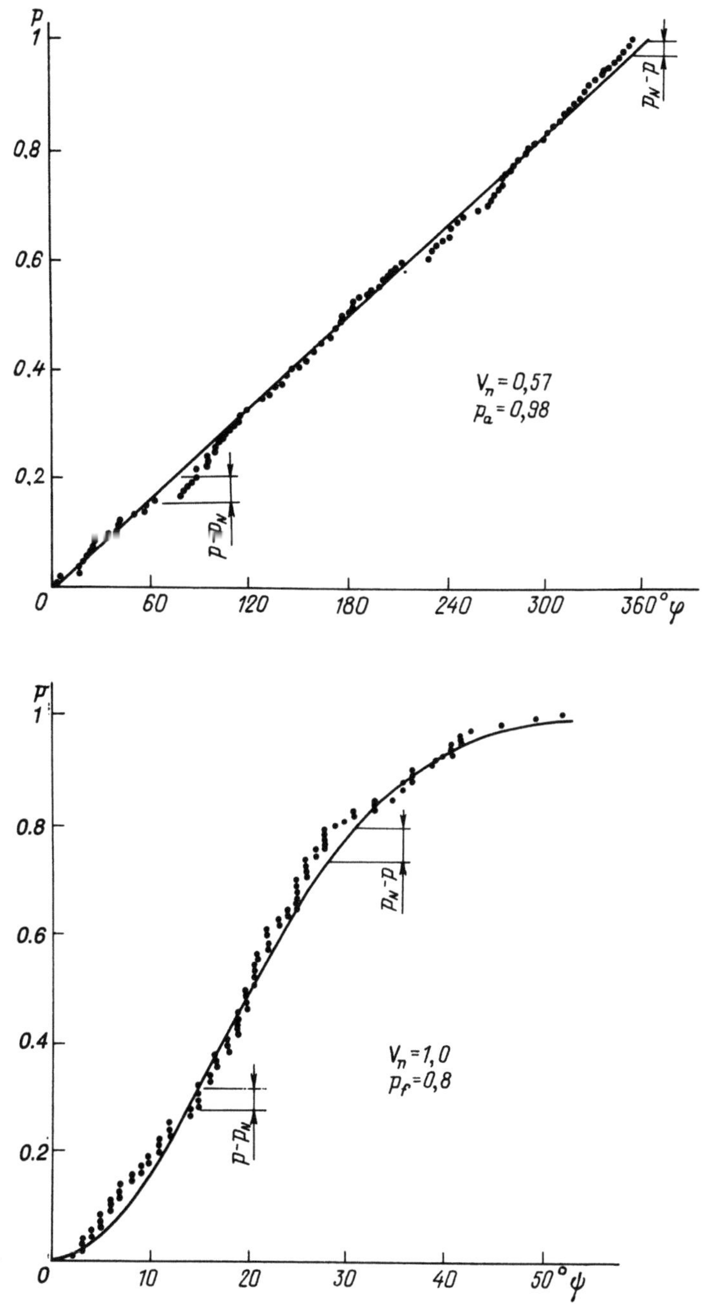

Fig. 2.20. Analysis of the fit of the NRM direction distribution to the theoretical ones, using integral (cumulative) curves. Azimuthal distribution (*above*), radial distribution (*down*). φ is the azimuth of NRM directions plotted on the centred projection, and ψ is the radial deviation from the resultant vector (projection centre)

Table 2.2. Two-tier analysis[a] (Watson and Irving 1957; Irving 1964)

Population	Degrees of freedom	Sum of squares	Mean square	Expectations of mean squares
Between beds	$2(b-1)$	$\sum_1^b r_i - R$	$\frac{\sum_1^b r_i - R}{2(6-1)}$	$\frac{1}{2}\left(\frac{1}{K_w} + \frac{\bar{n}}{K_b}\right)$
Within beds	$2\sum_1^b (n_i - 1)$	$\sum_1^b (n_i - r_i)$	$\frac{\sum_1^b (n_i - r_i)}{2\sum_1^b (n_i - 1)}$	$\frac{1}{2K_w}$

[a] r_i, geometrial sum of n_i unit vectors for i bed; R, vector sum of N unit vectors, $N = \sum_1^b n_i$; $\bar{n}$, weighted mean number of samples from one bed $\bar{n} = [N - (\sum_1^b n_i^2)(N^{-1})]/(b-1)$.

2.6.6 Virtual Geomagnetic and Paleomagnetic Pole Distributions and Their Confidence Ovals

Equations (1.3) and (1.4) establish a one-to-one correspondence between the geomagnetic field direction at a given point and time, and a specific virtual geomagnetic pole. Thus, N vectors of the j_{ni} set, obtained at a point with φ and λ geographic coordinates, correspond to N poles, which can also be considered as v_i unit vectors with constituents along the x, y, z axes:

$$v_{ix} = \cos\Phi_i \cos\Lambda_i; \; v_{iy} = \cos\Phi_i \sin\Lambda_i; \; v_{iz} = \sin\Phi_i \tag{2.21}$$

and analyzed in the same way as the j_{ni} vectors, i.e. by computing Φ, Λ coordinates of the mean virtual (paleomagnetic) pole by summing the constituents:

$$\Lambda = \mathrm{arctg}(\Sigma v_{iy}/\Sigma v_{ix}); \qquad \Phi = \arcsin[(\Sigma v_{iz})/R], \tag{2.22}$$

where

$$R = \sqrt{(\Sigma v_{ix})^2 + (\Sigma v_{iy})^2 + (\Sigma v_{iz})^2}.$$

The precision parameter (K) and the confidence circle radius (A_{95}) are calculated by formulae (2.9) and (2.16)–(2.18) and the distribution of the poles is analyzed as described in Sect. 2.6.4. The same method is used to determine the mean paleomagnetic pole and the errors from paleomagnetic poles are calculated from observations at different points of the Earth's surface.

There is another method, which can be applied to observations at any given point. Computations of the latitude Φ and longitude Λ of the paleomagnetic pole by the mean vector R (resultant vector of the j_{ni} set), using formulae (1.3) and (1.4), are performed. Here, the oval around the paleomagnetic pole, with the semi-axes θ_1 and θ_2 situated perpendicular to the paleomagnetic meridian at the observation site (semi-axis θ_1) and directed along the meridian (semi-axis θ_2), correspond to the confidence circle, having a radius α on a unit sphere, described around the vector R (estimated paleomagnetic field direction) (Irving 1964; Khramov and Sholpo 1967).

Taking into consideration that angle $\theta_1 = \cos\varphi_m dD$ is defined by the error in declination $dD = \alpha/\cos I$, we have

$$\theta_1 = \alpha(\cos\varphi_m/\cos I). \quad (2.23)$$

Differentiating the equation $tg\varphi_m = (1/2)tgI$ at $dI = \alpha$; $d\varphi_m = \theta_2$, we obtain

$$\theta_2 = (\alpha/2)(\cos\varphi_m/\cos I)^2. \quad (2.24)$$

Relations (2.23) and (2.24) show that the confidence oval semi-axes, for a given α value, depend only on the magnetic inclination I and increase with it. The exact conversion of a confidence circle, with a radius α, into an oval with semi-axes $\alpha/2$ and α, takes place at $I=0°$ or into a circle of 2α radius at $I=90°$. In this case the area of the oval, described around the pole, increases eight times. Thus, the paleomagnetic pole position (at one and the same value of α) has a precise definition when the point of observation is on the paleomagnetic equator.

It is evident that the first method is wholly correct when the distribution of the poles corresponds to Fisher's distribution. Similarly, the second method is only correct if the j_{ni} directions have Fisher's distribution. In so far as both distributions cannot simultaneously have Fisher's distribution one should use both methods interchangeably. Practice has shown that in this case the Φ and Λ coordinates of the mean pole differ slightly and the $\theta_1, \theta_2, \alpha$ and A_{95} values are in approximate relation

$$A_{95} \approx \sqrt{\theta_1\theta_2} = (\alpha/\sqrt{2})(\cos\varphi_m/\cos I)^{3/2}. \quad (2.25)$$

2.7 Methods Used in the Determination of the Geomagnetic Field Intensity in the Geological Past

2.7.1 The Basic Principles of the Ancient Geomagnetic Field Intensity Determination

The paleointensity of the geomagnetic field has been poorly studied, which can be explained by the uncertainties imposed by the natural remanence J_n of rock using known methods. The determination of the paleointensity H_a can be carried out on rocks which have preserved their primary magnetization direction as well as its intensity. Thus, the primary magnetization of igneous and sedimentary rocks contains information on the intensity of the geomagnetic field at the time when the primary magnetization was acquired, but only when their magnetization (or one of its components) has not changed.

All methods used in the determination of the intensity of the magnetizing field, synchronous with rock formation, are based on the proportionality of the magnetization to weak magnetizing fields. Only magnetic anisotropic constants (see Sect. 2.7.3) and stepwise remagnetization methods (see Sect. 2.7.6) use other effects. The first method uses structural anisotropy, the second is based on the asymmetry of the magnetic particle assemblages.

The process of the primary magnetization, based on heating or redeposition of rocks, can be simulated in laboratory studies, when the absolute values of the geomagnetic field intensity are determined from the NRM of detrital or thermoremanent origin. These methods are now often applied. Many questions arise from the relative methods of field H_a determination when the properties of anhysteretic magnetization are used.

The values of detrital, thermoremanence or anhysteretic magnetizations obtained in the course of the experiment are compared to those of the natural remanent magnetization. Proceeding from the known intensity of the laboratory field H_l and the proportionality of magnetization to the field, the value of the ancient field can be estimated from the following equation:

$$H_a = H_l \, (J_n / J_l). \tag{2.26}$$

To compare all H_a values, they are transformed to the equator value (the equator's recent field is $B_e = 0.035$ mT; $M_e = 27.9$ A/m) or the ratio of the Earth's ancient magnetic moment to the present one is calculated. If the Earth's radius R and its field is constant and dipolar, then comparison of magnetic moments can be calculated by the formula of the magnetic potential for dipole U (Yanowsky 1978):

$$U = M \cos\varphi / R^2, \tag{2.27}$$

where M is the dipole magnetic moment and φ is the latitude.

Taking into consideration that the constituents of the geomagnetic field are

$$X = -\frac{1}{R}\frac{dU}{d\varphi}; \quad Y = O; \quad Z = -\frac{dU}{dR}; \quad H_a = \sqrt{X^2 + Z^2},$$

we obtain the law of the geomagnetic field intensity change for a central axial dipole, as a function of latitude φ, of

$$H_a = H_e \sqrt{1 + 3\sin^2\varphi} \tag{2.28}$$

and the expression for the Earth's magnetic moment:

$$M = H_e R^3. \tag{2:29}$$

At present, there are only a few estimates of H_a values for different geological epochs. It is, therefore, necessary to broaden the observations in order not only to estimate the field intensity in the geological past, but also to trace the changes of the Earth's magnetic moment and its relationship with geological events. This would also aid in the evaluation of the central axial dipole hypothesis for different epochs and in the estimation of changes in the Earth's radius. With numerous applications, one can apply the paleointensity of the geomagnetic field as a paleomagnetic scale parameter in geochronology and stratigraphy.

2.7.2 Estimation of the Ancient Field H_a by the Redeposition Method

The redeposition method is mainly used in paleointensity determinations for sedimentary rocks having a detrital nature of NRM. The type of rocks for this method is limited because, in their redeposition, they should be macerated and pass through a suspension phase, all without destruction of terrigenous particles. Clays and clayey silts represent such weakly lithified rocks. Coarse-grained sediments (sands) cannot be used for redeposition because of the complex mechanism of their detrital magnetization which, in practice, cannot be reconstructed in the laboratory

Multiple experiments in the redeposition of sediments showed that most of the detrital magnetization J_{ro} is formed within a sediment when the water content decreases form 70% to 30% (Khramov 1968). The orientation of particles in a half-liquid sediment also leads to a smaller dispersion of inclinations. Comparing NRM and DRM values, one obtains a relation between the ancient field and the laboratory field intensities.

Earlier, the NRM and DRM were corrected for natural and artificial viscous magnetization.

2.7.3 Method of Magnetic Anisotropic Parameters

Ivkin and Aparin proposed the use of the redeposition method for sediments with detrital magnetization in the estimates of the ancient field H_a using magnetic anisotropic parameters (Ivkin and Aparin 1980). The magnetic anisotropic constants, before and after redeposition, are compared. The magnetic structures of the original sediments and those redeposited from them are assumed to be similar. This method is based on the rock magnetic structure, acquired during sedimentation. This structure is more stable to external physical effects than remanent magnetization. The geomagnetic field is defined by the linear relation between anisotropic constants and the actual field during redeposition.

In this case one can use the phenomena and technique of rotational magnetic hysteresis. The amplitudes of the first (T_1) and second (T_2) harmonics of the torsional moment versus field intensity are plotted for sediments redeposited in different fields. They can increase the "hematite" structure or decrease the "magnetite" structure linearly. The same amplitudes are calculated for samples of the original sediments, and their values, taken from graphs $T_1 = f(H)$ and $T_2 = f(H)$ for the redeposited sediments, are used to obtain the intensity of the field H_a, corresponding to the time when the rocks acquired their natural magnetization. This method can also be applied when the primary magnetization is not preserved, but any alteration of the rock has not changed the primary magnetic structure. The most reliable results were obtained by the authors from Late Carboniferous Gzhel' clays (see Table 3.5).

2.7.4 Thellier's Method

This method was first proposed and applied to check the thermoremanent magnetization nature of archeomagnetic artefacts (Thellier and Thellier 1959). This method and its different modifications are now widely used in the estimation of paleointensity. The experiment reproduces the process of the rock thermal magnetization. The method can be applied to igneous and baked sedimentary rocks whose thermoremanent origin of NRM has been proved. Rocks in which the minerals have been subjected to changes on heating, cannot be used in the estimates of H_a using this method.

The similarity of curves of the thermal decay of NRM and of artificial TRM is checked by preheating the samples chosen up to the Curie point. Simultaneously, the absence of mineral changes is checked. If the similarity between the J_n and J_{rt} curves is proved and the control heating has not shown changes in the minerals, then we apply the two-heating method of a sample in the Earth's magnetic field for each temperature interval. During the first heating, the NRM orientation of a specimen is parallel to the Earth's magnetic field in the laboratory, while during the second heating, its orientation is reversed. In practice, each step of heating is 50°C. In general, the initial part of the curve (100°–200°C) is not used due to the influence of viscous magnetization. The comparison of that part of the NRM disappearing on heating (J_{npt}) with that of the partial thermoremanent magnetization J_{rpt} appearing within the same temperature interval on cooling is carried out for each temperature interval. During the first heating (orientation along the Earth's magnetic field), the partial thermoremanence at a given temperature interval is destroyed and, hence, it is subtracted from the J_n value, whereas the partial magnetization in the laboratory field is added.

The magnetization, measured after heating, is

$$J_{n1} = J_n - J_{npt} + J_{rpt}. \tag{2.30}$$

During the second heating of the specimen, over the same temperature interval but in the opposite position to the first one, then

$$J_{n2} = J_n - J_{npt} - J_{rpt}. \tag{2.31}$$

Solving these equations for each temperature interval, we obtain two partial magnetizations, namely the ancient, arising in an unknown field, and that arising from the known laboratory field. The coefficient k, showing a magnetic field intensity ratio of $k = H_a/H_l$, is found from these partial magnetizations for each temperature interval.

The results of heating can be shown diagrammatically with J_n and J_{rt} values plotted on the ordinate and temperature plotted on the abcissa. The coefficient k, in this case, is determined as the J_n/J_{rt} ratio in the middle of the same temperature intervals. The latter can be determined using the so-called Arai-Nagata diagram. The increments of J_n and J_{rt} for each temperature interval are plotted on the axes and the H_a value is then defined by the slope of the curve passing through the experimental points.

The two-heating method, within each temperature interval, has experimental errors resulting from mistakes in the orientation of a specimen between the two heatings, and temperature differences between the first and second heatings. Methods using one heating do not cause such mistakes, and also decrease the time of the experiment.

Domen (1977) proposed to cut the specimen into two parts and to treat both of them in an antiparallel manner. Here, the similarity of the two halves should correspond to the NRM intensity and the direction as well as to the magnetization carriers. Then, taking a double measurement (halves are firstly parallel and afterwards antiparallel to each other), one can obtain the J_n and J_{rt} values. To calculate H_a, any of the above methods can be used.

Kono and Ueno (1977) proposed a very helpful single-heating method. In this method a specimen is heated in the magnetic field perpendicular to the total NRM vector. Like all other methods of H_a determination, the use of this method is based on the assumption that the J_n direction remains stable during t- and A.F.-cleaning. On heating of a specimen, oriented so that its total vector J_n is normal to the Earth's magnetic field, the following process takes place: the PTRM component arises in the direction of the applied magnetic field; a part of the J_n magnetization disappears, leading to changes, after heating, of the total vector, thus turning it towards the PTRM component. Successive heating at a higher temperature makes the vector J_n move further towards J_{rt}, and the geometrical sum of the NRM left after heating and the PTRM component are measured.

The vectors observed can be shown on a diagram with J_n and J_{rt} axes. Taking into account the magnitude of the resultant vector for all temperature intervals are plotted according to magnitude and direction. The vectors approach the J_{rt} axis as the temperature increases. A line drawn through the vector end points, the slope of which is on the abcissa, defines the ratio of the ancient and laboratory fields. If necessary, the resultant vectors can be projected onto the axes and $J_n(T)$ and $J_{rt}(T)$ determined.

There are several modifications of Thellier's method. Of great interest is the Van Zijl method (Van Zijl et al. 1962), which enables the calculation of the H_a value, using the NRM/TRM ratio during A.F.-cleaning. Using this procedure, one can remove unstable, secondary components having low blocking temperatures and can compare those components whose ratio is closer to those of the present and ancient field intensities than to the J_n/J_{rt} ratio.

Van Zijl used a 16 kA/m alternating field, because fields of such an amplitude removed unstable, secondary components in the rocks studied. The magnetic ratio was interpreted as the ancient to the applied field intensity ratio. Mean values of the geomagnetic field intensity within a normal polarity zone (0.02 mT) and within a transition zone (0.004 mT) were obtained from lavas at the Triassic-Jurassic boundary.

2.7.5 Method of Anhysteretic Magnetization

This method, like Thellier's method, utilizes the gradual demagnetization and magnetization of rock samples, but heating is replaced by an alternating field. The results, obtained using Thellier's and anhysteretic magnetization methods, show a good agreement when the J_n and J_{ri} coercive spectra maxima differ by no more than 4 kA/m.

The H_e method is one of estimating the ancient field intensity H_a (Petrova and Bagina 1976), where H_e is the value of the field, thus forming magnetization $J_{ri} = J_n$. When this method is considered as absolute, i.e. the H_a and H_e fields are identified, then the ARM should be formed in the alternating field equal in amplitude to that of saturation. This method cannot be applied to hematite-bearing rocks, with a saturation field of a few tesla, – in this case the H_a values will be overestimated.

The experiment was carried out as follows. The magnetization J_n of a specimen is destroyed by alternating fields having a maximum amplitude h. The remaining part of J_n is measured. The magnetization J_{ri} is then found in the field whose H and h_{max} values are known. The steady field H is increased gradually and the J_{ri} values are measured. The components of the NRM destroyed in the alternating fields up to 8 kA/m are considered to be of a viscous nature and are ignored. J_{ri} is plotted against H. If the NRM is completely destroyed by the alternating field, its curve should extend to the origin of the coordinates. If a part of J_n remains, the abcissa needs to be elevated by this value (i.e. the remainder of J_n should be subtracted both from J_{ri} and the initial NRM value). Then, using the total magnetization J_n, in the first case, or the part, destroyed by the alternating field, in the second case, a specific value of the field H_a is taken from abcissa.

The method of coercive spectra (Petrova and Bagina 1976) is based on the comparison of the NRM and ARM coercive spectra for one and the same rock. If the coercive spectra are similar, then the ratio of $k = \Delta J_n(h)/\Delta J_{ri}(h)$ within different field intervals can be determined from the demagnetization curves of $J_n(h)$ and $J_{ri}(h)$. The mean coefficient is the H_a/H_l ratio. The method applied is as follows. The magnetization of the rock J_n is destroyed by a stepwise alternating field, each step being 2×10^3 A/m up to a maximum amplitude. The magnetization J_{ri} is then induced in a specimen and destroyed by a field changing in the same way. The initial intervals up to (4–8) kA/m, within which the viscous magnetization J_{nv} is destroyed, are omitted. The relationship of the destroyed parts of J_n and J_{ri} is determined for each interval. The lines are plotted from the mean values of ΔJ_n and ΔJ_{ri} for each specimen. The ratio of the ancient to recent field is determined as the tangent of a slope a, namely:

$$H_a = H_l \text{ tga}. \qquad (2.32)$$

2.7.6 Method of Stepwise Remagnetization

Theoretically, rocks can be considered as assemblages of Neel's "effective" particles (see Sect. 1.3). Any magnetization can be expressed as a certain asymmetry within the assemblage. The asymmetry resulting from thermomagnetization is governed by diffusion processes, whereas that of detrital magnetization is governed by the statistical alignment of the magnetic moments of detrital grains. Experimental and theoretical studies of the magnetization types show that the asymmetry of the assemblage depends on the steady magnetic field in which magnetization was acquired. Stepwise remagnetization method is based on a single-valued bond between the magnetic state and the intensity of the active magnetic field (Shashkanov et al. 1975). An asymmetry of the particle assemblage in a rock results in the formation of an induced, magnetic anisotropy, thus giving rise to constricted hysteretic loops (Krupicka 1973). In laboratory anhysteretic magnetization, this effect is revealed by the distortion of the proportionality between I_{ri} and the field established in the Rayleigh area with an interval of distortion close to the field H_a of the primary magnetization. The technique, used to determine this effect, and proposed in a number of publications (Shashkanov and Metallova 1982; Metallova et al. 1984a) allows the determination of the distorted interval.

However, it has been shown experimentally that the determination of the paleointensity, using the ARM curve, is unreliable. So, a special procedure, known as the stepwise remagnetization method, was proposed (Shashkanov and Metallova 1982; Metallova et al. 1984b):

1. An anhysteretic magnetization I_{ri} is given to a rock specimen and its coercive spectrum is measured. The amplitude of the alternating field h_1, corresponding to the coercive spectrum maximum is then determined.
2. Along the NRM direction of a specimen, a so-called quasi-anhysteretic magnetization is formed with an initial alternating field amplitude of $h_m = 2h_1$ at the changing steady field H_1. The H_1 value changes with a decrease of the alternating field h amplitude according to the law

$$H_1 = H_m(1 - h/2h_1), \tag{2.33}$$

where H_m is a steady field of 40–120 A/m.
3. The specimen is then remagnetized and, at the same time, is affected by a steady field, varying according to the relationship $H_2 = H_m\ h/2h_1$ by a stepwise increase of the alternating field from 0 to $2h_1$. The number of steps is uncertain, but usually equals 10.

According to the curve of $\delta I_r(h)$, plotted as a result of stepwise remagnetization, H_a can then be calculated from one of the following formulae:

$$H_a = H_m\ (h_m - 2h_1)/(2h_m); \tag{2.34}$$
$$H_a = H_m\ (h_m + 2h_2)/(2h_m). \tag{2.35}$$

An example of stepwise remagnetization is shown in Fig. 2.21. The H_a values from different rocks obtained by this method are summarized in Table 3.5.

2.7.7 Estimation of the Ancient Field Intensity H_a by the Q_o Factor

As the primary magnetization of rocks is proportional to the geomagnetic field at the time of their formation and also depends on the conditions of rock formation

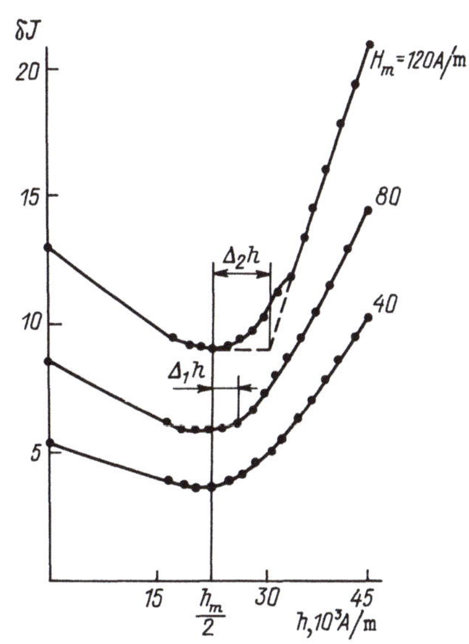

Fig. 2.21. Stepwise remagnetization curves for different values of H_m, h_m is the initial A.F. amplitude; H_m is the steady field chosen within a range of 40–120 A/m; *J* is the resulting magnetization; $\Delta_1 h$ and $\Delta_2 h$ are values used for the calculation of the ancient field H_a according to the formulae (2.34) and (2.35)

as well as their alteration (i.e. the type of magnetization, its preservation, composition, size and concentration of grain-carriers of J_n), we can determine H_a with respect to or by the elimination of the influence of all these factors. The choice of a fairly homogeneous series of rocks to be measured, with a similar NRM carrier and primary magnetization type and which were formed in similar environments, can be considered as one of the methods which enables the solution of this problem. We should only take into account the concentration of the carriers. This can be done approximately, using relative values, such as $J_n^o/(\varkappa H)$, J_n^o/J_{rv}, J_n^o/J_r^s, J_n^o/J_{ri}, J_n^o/J_{rt}.

Two of these relationships have already been discussed. The factor $Q_o = J_n^o/(\varkappa H)$, i.e. the relationship between the primary remanence and the induced components, is one of the most widely used.

The field H_a is derived from the study of red beds, and has a chemical magnetization, associated with finely dispersed, hematite grains (approx. single-domain size). These are distinguished by a high NRM stability and homogeneity of magnetic susceptibility $\varkappa$ across the section and can be estimated from the Q_o factor. The latter is proportional to the ancient/applied field ratio (Boronin 1979). This method enables the determination of only a relative geomagnetic field intensity. Values of $Q_o = 0.7$–0.9 were obtained for the Upper Permian red beds from the River Volga area. The transitional intervals, separating the polarity zones, show a decrease of the Q_o value down to 0.1–0.2 (Boronin 1979).

This method can be used when the samples are selected from a homogeneous succession of rocks with a constant NRM carrier and ferrimagnetic grains of constant size, thus maintaining the persistance of $\varkappa$ across the section. The Permo-Triassic rocks were studied in a similar way (Slautsitais 1973). H_a values were found similar to those obtained using the redeposition method (see Sect. 3.6 and Table 3.5).

CHAPTER 3

The Geomagnetic Field in the Phanerozoic and the Magnetostratigraphic Scale

3.1 Geomagnetic Reversals and Geomagnetic Polarity Scales

3.1.1 Evidence of Geomagnetic Reversals

The first observations of Brunhes and David (1903) and Brunhes (1906) on the magnetization of lava flows and of clays baked by this lava led to the conclusion that the geomagnetic field in the past was directed in a sense opposite to that of the present. Later, it was established that a reversed magnetization, as well as a normal one, is registered in rocks of various age. This was the basis for the argument that the geomagnetic field recurrently changed its polarity during geologic history. This hypothesis was seriously debated when it was theoretically and experimentally revealed that magnetization possesses self-reversal properties, i.e. the rocks might be magnetized opposite to the magnetic field or they might subsequently, be spontaneously remagnetized (see Sect. 1.5.5).

The following factors were most important in testing the geomagnetic reversals (Khramov and Sholpo 1967):

1. Coeval rocks ubiquitously exhibit the primary magnetization corresponding to the same polarity of the geomagnetic field.
2. Rocks baked by intrusions and lavas, irrespective of their composition and initial direction of vector J_n, are identically magnetized (in normal or reversed directions) as these intrusions and lavas.
3. Exceptions are rare and usually these are the cases of self-reversals of magnetization J_n confirmed experimentally and theoretically.
4. Regularities in the changes of primary magnetization in transitional beds, separating the zones of normal and reversed magnetization, cannot be explained by self-reversal phenomena.

It was necessary to collect and examine voluminous factual material in order to test all these factors. The results of this study appear promising. One of the first and most striking results was obtained upon examination of the relationships of J_n directions in dykes and their baked rocks in volcanogenic-sedimentary strata of the Cenozoic in Iceland (Table 3.1). Later, similar data were obtained from many localities and Wilson (1962) had the opportunity to examine 87 cases of relationships of the vectors J_n in volcanogenic rocks and their baked rocks. This results, supplemented with new evidence, are given in Table 3.2. It is shown that only 3 of 176 cases might be interpreted as a result of self-reversal of magnetization J_n.

The transitional beds, in which vector J_n rotates, were recognized soon after the alternation of the zones of normal and reversed magnetization became known in

Table 3.1. NRM polarities in the volcanic sequence of Iceland

Units	Different cases of NRM polarity			
	1	2	3	4
Dyke	N	N	R	R
Total country rock	N	R	N	R
Baked part of rock				
a) Observed polarities (after Hospers 1954; Einarsson 1957b)	N	N	R	R
b) Theoretical polarities for				
field reversals	N	N	R	R
self-reversals (reversible mechanism in rock)	N	R	N	R
self-reversals (irreversible mechanism in rock or in dyke) (Khramov and Sholpo 1967)	N	N	N	N

Table 3.2 NRM polarities for igneous and baked rocks. (After McElhinny 1973 with additional data from Bol'shakov and Solodovnikov 1980a,b)

Types of interrelations[a]	NRM polarity		Number of cases
	Igneous rock	Contact baked rock	
I	N	N	57
II	I	I	7
III	R	R	109
IV	N	R	3
V	R	N	0

[a]I denotes intermediate NRM directions.

these sections. However, the existence of such zones of vector rotation do not account for the self-reversal of magnetization J_n and, thus, the hypothesis of geomagnetic reversals was accepted as an alternative. Detailed studies of transitional beds are being intensely conducted to investigate the morphology of transitions and consequently the regularities in the process of geomagnetic reversals (see Sect. 3.3.2).

The most evident argument for geomagnetic reversals is the synchroneity of unipolar zones (of normal or reversed magnetization) in sedimentary and volcanogenic strata that cover large areas. In certain cases the synchroneity of these zones, i.e. the isochroneity of their boundaries, is shown on a world-wide scale (see Sect. 3.2).

3.1.2 Types of Geomagnetic Polarity Scales

One of the most important problems of paleomagnetology is the construction of a time sequence of geomagnetic reversals for the entire geologic time range when these reversals were recorded. The construction of this time sequence may be performed by several methods; the method chosen depends on the details needed, on the completeness of the sequence and also on the possibilities of dating the reversals.

Geomagnetic reversals are known to be an event on a global scale and, thus, a most precise, stratigraphic and chronologic correlation of the formations magnetized in normal and reversed directions all over the world is needed. Thus, the scale of geomagnetic reversals might be constructed, in general, as a pure chronologic scale if the rock samples with an established magnetic polarity fit the dating by physical methods. This is then called the magnetochronological scale. This approach in the study of the history of geomagnetic reversal might be true only on the condition that the duration of the epochs when the geomagnetic polarity was preserved, exceeds the errors of the dating method. For instance, classical paleontological methods hardly seem suitable in this case. The construction of a magnetochronological scale became feasible only with development of K-Ar analysis, which allowed a high precision in the dating of volcanogenic rocks, mainly using basic lavas, the magnetic polarity of which is determined.

The second mode in tracing the history of geomagnetic reversals is the examination of linear anomalies in oceans and includes the construction of profiles perpendicular to the strike of these anomalies. Interpretation of these profiles gives the sequence of anomaly-forming bodies magnetized in normal and reversed directions which, according to the concept of seafloor spreading, reflect the chronologic record of geomagnetic reversals. When the rate of seafloor spreading and the datings of certain anomalies obtained on bedrocks or on the lower bed of oceanic sediments are available, the construction of the so-called anomaly scale of geomagnetic reversals can be realized.

The third method of studying the history of geomagnetic reversal is the examination of sequences of bottom sediments in drilcores and in columns obtained with the help of corers. As the seafloor is young, this method and the previous one are useful in the construction of a scale of reversals since the Jurassic period only.

And finally, the fourth method of constructing a scale of geomagnetic reversals is the application of a standard stratigraphic procedure: paleomagnetic examination of the most complete sequences of sedimentary and volcanogenic-sedimentary strata in stratotypical areas, correlation of the established sequences of normal and reversed magnetization zones and compilation of generalized magnetostratigraphic sequences, at first for separate regions and then their unification on a global scale. The time equivalent of the magnetostratigraphic scale constructed in such a manner is the sequence of intervals of normal and reversed geomagnetic polarity set in a standard geochronologic scale, i.e. into periods, epochs and ages (see Sect. 3.2).

3.1.3 Dating of Polarity Epochs on Lava Flows; the Magnetochronologic Scale

Cox et al. (1963) constructed the first scale of geomagnetic reversals based on K-Ar datings. It embraced the last 3.5 m.y. and consisted of three zones: two zones of normal polarity and one zone of reversed polarity, each embracing approximately 1 m.y. An increased number of determinations revealed that many of them do not fit this simple scale. And thus, certain shorter intervals of the order of 10^5 years were recognized within the major intervals of polarity (10^6 years). The shorter ones were called episodes or events and were given the names of the localities in which they were established (Jaramillo, Gilsa, Olduvai, etc.); the longer ones, called epochs, were given the names of prominent geomagnetologists of the past: Brunhes, Matuyama, Gauss and Gilbert.

The most widely used is the scale of the geomagnetic polarity for the last 4.5 m.y. based on 150 determinations of K-Ar age and of the polarity of basic lavas from different parts of the world. It is to be noted that data obtained on examination of linear anomalies and deep-sea sediments in oceans were used to refine the age boundaries of the units of this scale (Cox 1969). This scale (Fig. 3.1) is the classical scale of Cox with certain refinements and supplements introduced later (Denham et al. 1977; McDougall et al. 1976.)

Covering of the older epochs by the magnetochronologic scale is problematic due to the increase in absolute errors in dating by the K-Ar method. An eror of 5% for a rock with an age of 5×10^6 years gives an absolute error of 2.5×10^5 years which exceeds the duration of many events. Still, certain satisfactory attempts were made to extend the scale of Cox into the time of 12.5 m.y. (Harrison et al. 1979; McDougall et al. 1976) and even of 16 m.y. ago. This study, however, was conducted under stratigraphic control; sequences of lava series in Iceland and the Trans-Carpathians were examined where the sequence of flows is known and the datings can be checked (Mikhailova et al. 1974).

Thus, an age boundary at 5–7 m.y., i.e. the beginning of Pliocene, serves at the moment as the lower limit of applicability of a purely geochronologic method in constructing a scale of geomagnetic reversals based on K-Ar datings. In the pre-Pliocene reversal studies, the stratigraphic approach is the only feasible one.

3.1.4 Zones of Geomagnetic Polarity in Bottom Sediments of the World Ocean

Studies of series of lava flows and sheets usually do not provide any continuous or regular temporal data on geomagnetic reversals, as the volcanic activity is characterized by an essential temporal irregularity. In pursuing this goal, more comprehensive information is available from paleomagnetic studies of bottom sediments in seas and oceans, particularly in their abyssal parts which do not suffer from the action of turbidity currents. Typical oceanic sediments have a deposition rate of 0.1–1 cm/10^3 years, thus the Brunhes-Matuyama boundary is at a depth of 0.7–7 m, and the Gauss-Gilbert boundary, between 3.3 and 33 m. Thus, the construction of a sequence of reversals for the period of the same duration and of even longer duration by the examination of columns obtained with the help of cores is possible.

The geomagnetic reversal sequence in a column of bottom sediments was first recorded in 1964 (Harrison 1966). Now the examinations of deep-sea bottom sediments in the Atlantic Ocean (Sclater and Cox 1970), and Indian and Pacific Oceans (Lin'kova 1976, 1984) are being conducted on a large scale. The longest columns (up to 30 m) first permitted the extension of the Cox scale into the time of 9 m.y., to the middle of the late Miocene and then throughout the whole Neogene (Theyer and Hammond 1974), (Fig. 3.2). Construction of geomagnetic

Fig. 3.1. Time scale of the geomagnetic reversals for the last 7 m.y. (After Cox 1969, with additions). Events: *1* Laschamp; *2* Blake; *3* V-zone; *4* Jaramillo; *5* Gilsa; *6* Olduvai; *7* Reunion; *8* Kaena; *9* Mammoth; *10* Cochiti; *11* Nunivak; *12* Sidufial; *13* Tvera. Symbols used for Figs. 3.1-3.13 see p. XV

Fig. 3.2. Guaternary and Neogene magnetic polarity time scale according to data from deep-sea sediments. (After Theyer and Hammond 1974)

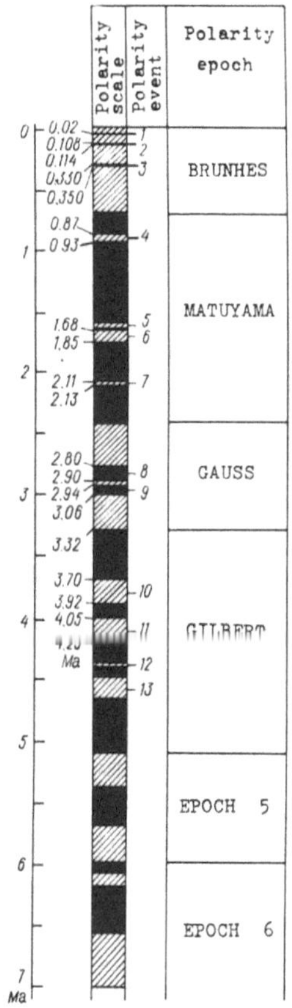

Fig. 3.1.

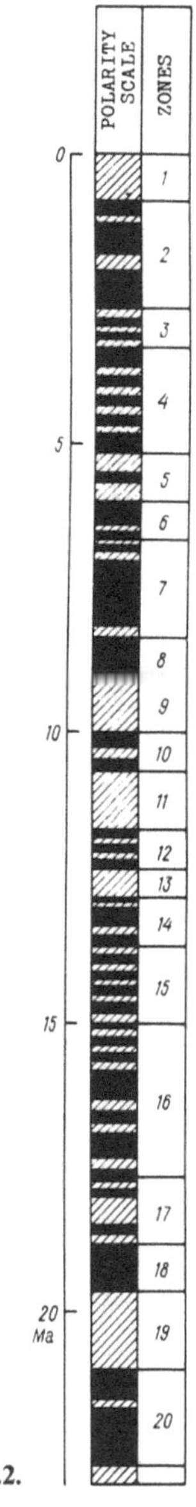

Fig. 3.2.

reversal scales for the more ancient epochs using data from the oceanic deposits started with the realization of the DSDP legs of "Glomar Challenger". These scales, however, were constructed as magnetostratigraphic scales with reference to the faunal zones of the magnetic zones (for instance, Poore et al. 1982). Therefore, some of the results of these studies are examined in Sect. 3.2.

Another approach to the examination of bottom sediments is the detailed study of columns in the areas of high-rate sedimentation, their purpose is to recognize the events (episodes) and anomalous deviations (excursions) of a geomagnetic field in the Brunhes epoch (see Sect. 3.4.1).

3.1.5 Scale of Marine Magnetic Anomalies

The hypothesis of seafloor spreading provided a simple explanation for the peculiarities of the anomalous magnetic field in oceans: linearity, alternate signs, symmetry relative to the mid-oceanic ridge axis, similarity in anomaly sequences counted from this axis around the Earth (Heirtzler et al. 1968; Karasik 1978). This seafloor spreading hypothesis involves the supply of matter from the Earth's interior along fissures cutting the mid-oceanic ridges, the transportation of this matter in different directions from the ridge axis under the effect of convection in the mantle and consequently, a continuous spreading of the seafloor, parts of which are older, the further they are from the ridge axis.

This process occurs during regular geomagnetic reversals. The basaltic matter melted from the mantle acquires a thermoremanent magnetization while cooling in the uplift while at a certain geomagnetic polarity and preserves this direction later when moving away from the ridge axis. After several geomagnetic reversals, the so-called reversible magnetically active layer is formed, consisting of a sequence of N- and R- (normally and reversed) magnetized bodies, which are symmetrical relative to the ridge axis; this sequence might be regarded as a double record of chronologic succession of geomagnetic reversals.

This formation mechanism of a magnetically active layer of the oceanic crust, proposed by Vine and Matthews (1963), is thus confirmed, since it provides an explanation for all the major properties of the geomagnetic field in oceans. Furthermore, it is verified from correlations of the bottom relief and the shape of positive and negative magnetic anomalies and from drilling data, which revealed a correlation of polarity of vector J_n in core samples and with the sign of the anomaly. The most convincing argument for a sequence of geomagnetic reversals, originating from the interpretation of linear anomalies, is the concurrence of the upper part of the anomaly scale and the Cox scale and also the consistence in anomaly age and basement age according to drilling data (Karasik 1978; La Breque et al. 1977; Tarling and Mitchell 1976). The age of anomalies is estimated by the identification of the upper part of the anomaly scale and the Cox scale. One of the last versions of the anomaly scale of geomagnetic reversals for the Cenozoic is illustrated in Fig. 3.3.

3.1.6 Construction of a Magnetostratigraphic Polarity Scale: Principles and Problems

A stratigraphic equivalent of a geomagnetic polarity scale is a magnetostratigraphic scale. Geomagnetic reversals are the most striking phenomena in paleomagnetism, their history is sufficiently well known and global nature is fairly evident. Thus, the magnetostratigraphic scale of polarity is the only one among the paleomagnetic scales that is being successfully improved and applied. Actually, the magnetostratigraphic or paleomagnetic scale implies the magnetostratigraphic scale of polarity.

The geomagnetic reversals examined from the viewpoint of their distribution in time during long intervals (eras and longer periods) exhibit a complex rhyth-

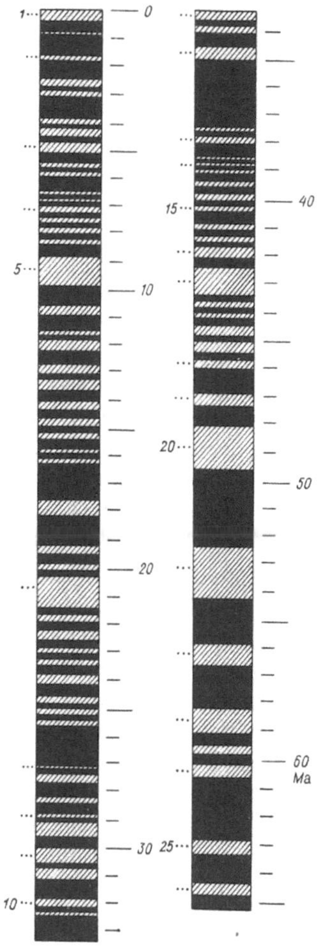

Fig. 3.3. Cenozoic magnetic polarity time scale according to data from marine magnetic anomalies. (After Larson and Pitman 1972; LaBregue et al. 1977)

micity (see Sect. 5.2). Moreover, their distribution in the time scale appears to be irregular, displaying clustering and scattering, typical grouping of reversals, etc. Thus, it may be possible to recognize in magnetostratigraphic and magnetochronologic scales, units of different rank, i.e. taxonomic units.

At the present time there are no generally accepted names for the different units of magnetostratigraphic and magnetochronologic scales. Thus, in discussing the project of the magnetostratigraphic code of the USSR, the following major units of a magnetostratigraphic scale were proposed (in descending order): megazone, hyperzone, superzone, orthozone and subzone and their time analogues: megachron, hyperchron, superchron, orthochron and subchron, respectively. It was suggested to call the time analogue of a paleomagnetic anomaly an "excursion". The Subcommission on magnetic stratigraphy of the International Commission on stratigraphy recommended different magnetostratigraphic and magnetochronologic units (Table 3.3).

Table 3.3 Summary of magnetostratigraphic polarity-unit terminology. (After Hedberg (ed.) 1976 with additions according to the Subcommission on a Magnetic Polarity Time Scale, SMPTS)

Magnetostratigraphic polarity units	Chrononostratigraphic equivalent	Geochronologic equivalent	Approximate duration, years
Polarity megazone	Megachronozone	Megachron	10^7–10^8
Polarity superzone	Superchronozone	Superchron	10^6–10^7
Polarity zone	Chronozone	Chron	10^5–10^6
Polarity subzone	Subchronozone	Subchron	10^4–10^5
Polarity microzone	Microchronozone	Microchron	$< 10^4$

In this work we use the terms for the magnetostratigraphic units proposed at the discussion on the magnetostratigraphic code of the USSR, thus the terms, epoch and event, are used for the chronologic analogues of orthozone and subzone, respectively. An analogue of an excursion of magnetic polarity is considered a marker horizon and is not included in the scale. The term "magnetic polarity" is usually omitted for brevity, for instance, the Matuyama epoch, the Jaramillo event.

The unit rank is defined according to the duration and importance of the stage corresponding to it in the general history of the geomagnetic field.

The volume of megazones corresponds approximately to erathems of a general stratigraphic scale; hyperzones, to systems; superzones, to series or several stages; orthozones, to stages or zones. However, we have no well-defined criteria to distinguish an event and an excursion. This seriously hinders the correlation of data obtained by different workers. It has been suggested to distinguish these events only according to their duration: a short one is an excursion, a longer one is an event; others have suggested their characterization according to the situation of the virtual geomagnetic pole at the moment of the event. Thus, for Watkins (1972) an event is a relatively short-time, complete reversal of a geomagnetic field. He contended that the duration of an event might even amount to 10^4-10^5 years. An excursion of magnetic polarity is considered to be a short-time event (less than 10^4 years) when the geomagnetic pole declines from the normal (reversed) position not less than 60° and not more than 120° and then it returns to the initial position.

A paleomagnetic scale includes a sequence and a distribution of reversals in time and serves as a basis for studies of the geomagnetic field history. The general level of paleomagnetic studies and the details of regional paleomagnetic scales indicate the completeness and comprehensiveness of this scale.

A paleomagnetic scale is compiled as a stratigraphic scale by successive paleomagnetic studies of the most complete sections of formations and groups in their stratotypical areas, and by the correlation of sections and the construction of standard paleomagnetic sections of stages, series and systems. When comparing regional paleomagnetic scales, special attention must be paid to the level of details of the paleomagnetic examinations conducted. Thus, under the same level of details in regions with different sedimentation regimes, the major principle of correlation, i.e. comparing units of the same rank, might be subverted (Khramov et al. 1981). Upon continuous sampling in areas of a platform regime of sedimentation, it is feasible to distinguish the peculiarities of the geomagnetic field behaviour corresponding to excursions and events, but this is impossible with less detailed sampling. Only the general regularities of the geomagnetic field behaviour embracing long-time intervals would be revealed by such sampling. In regions of geosynclinal sediments, fairly fine peculiarities of the geomagnetic field behaviour are revealed even at a low level of details in sampling, and, thus, minor magnetostratigraphic units are recognized.

An important problem in paleomagnetic stratigraphy, is thus the search for additional diagnostic features for the individual units of a paleomagnetic scale. In addition to polarity and paleomagnetic poles, which permit the identification of only large units in a paleomagnetic scale, e.g. hyperzone and superzone, typical peculiarities of the field behaviour, for instance, might be additional features.

A different problem arises when constructing a scale for the early Paleozoic, i.e. the problem of polarity. Which of the two poles is to be considered the northern pole, e.g. for the Siberian plate. The scale given here corresponds to the minimal length of the APWP (apparent polar wander path) during the middle Paleozoic. However, this is not the only solution to the problem (Gorodnitsky and Zonenshain 1979; Khramov 1978). The problem of polarity also concerns other plates. Therefore, the answer must be sought in the compilation and correlation of detailed paleomagnetic scales of the early and middle Paleozoic for all continents.

3.2 The Phanerozoic Magnetostratigraphic Scale

3.2.1 Structure of the Magnetostratigraphic Scale for the USSR

Early paleomagnetic studies in the USSR can be clearly characterized as stratigraphic works with the goal of solving special stratigraphic problems, such as dismembering and age correlation of sedimentary and volcanic formations, age determination of geologic bodies, mineralization stages, etc. These paleomagnetic studies, directed mainly to the solution of special problems, here provided many data of varying reliability and completeness. Systematic use of the available data became easier due to paleomagnetic studies which compiled standard regional paleomagnetic scales for nearly all the intervals in the Phanerozoic.

First attempts to summarize data for the Paleozoic section of the scale were made in 1965 (Molostovsky 1970; Khramov and Sholpo 1967). Later, reviews on Mesozoic magnetostratigraphy were published (Pergament et al. 1971). In 1976 the first Phanerozoic magnetostratigraphic scale was proposed (Molostovsky et al. 1976). Subsequent reviews on the magnetostratigraphy of the Paleozoic, Mesozoic and Upper Cenozoic, together with new data (Molostovsky et al. 1976; Khramov et al. 1981), offer a new, more detailed and reliable magnetostratigraphic scale for the Phanerozoic of the USSR (Fig. 3.4).

In the USSR the pattern of the paleomagnetic scale is most evident for the Cambrian, Ordovician, Upper Paleozoic, Lower Triassic and Upper Cenozoic (Khramov 1963; Rodionov and Sidorova 1963; Boronin 1973, 1979; Kravchinsky and Davydov 1972; Khramov et al. 1974; Slautsitais 1975; Molostovsky 1983b; Ananiashvili and Pevzner 1977; Zubakov and Kochegura 1976).

At the same time, many intervals of the Phanerozoic paleomagnetic scale (Upper Cretaceous-Lower Neogene, Upper Triassic-Lower Jurassic, Lower Permian, Silurian-Devonian) have still been insufficiently studied. Construction of the

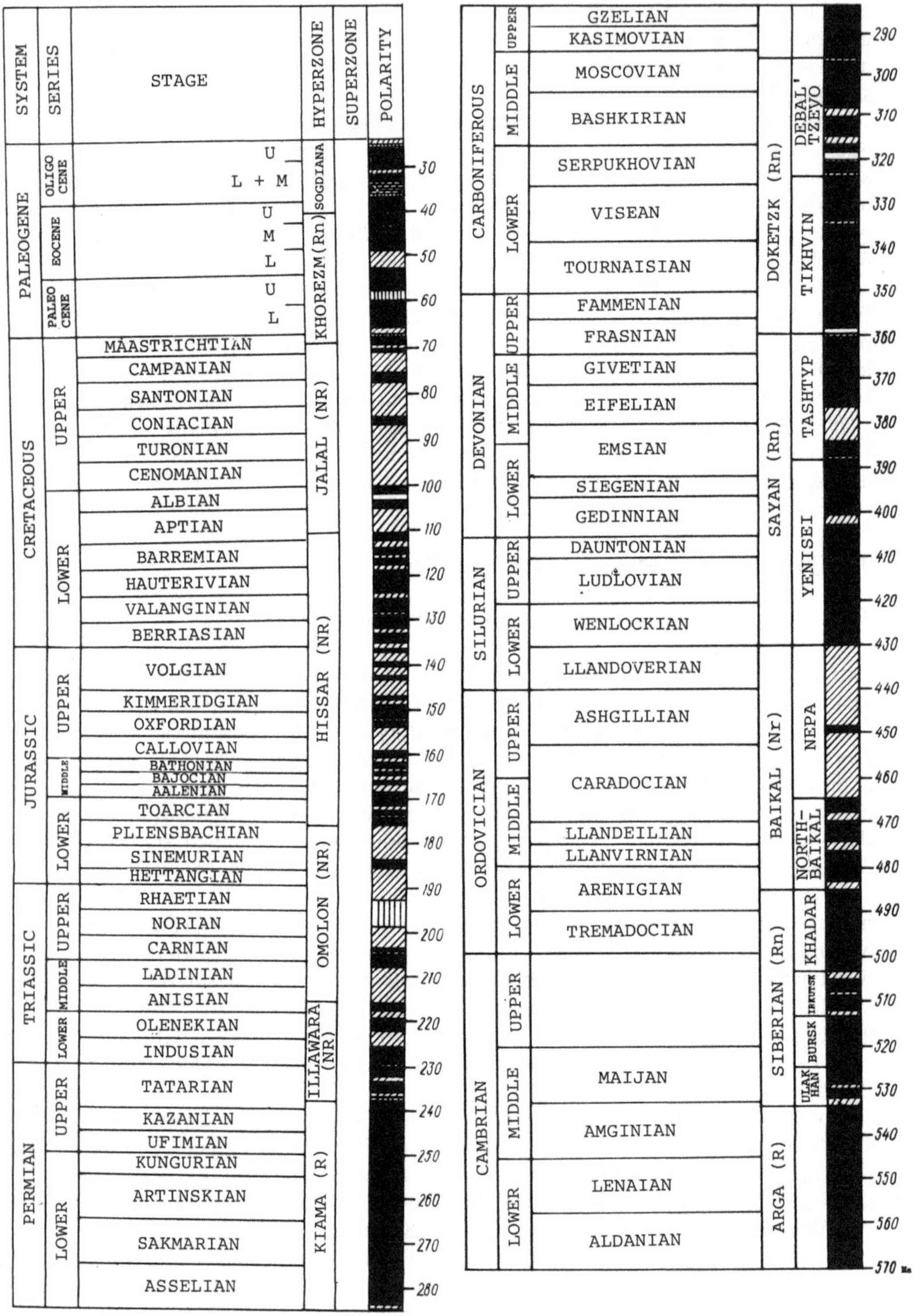
SYSTEM
SERIES
STAGE
HYPERZONE
SUPERZONE
POLARITY
PALEOGENE
OLIGOCENE
EOCENE
PALEOCENE
U
L + M
U
M
L
U
L
SOGDIANA
KHOREZM (Rn)
CRETACEOUS
UPPER
LOWER
MAASTRICHTIAN
CAMPANIAN
SANTONIAN
CONIACIAN
TURONIAN
CENOMANIAN
ALBIAN
APTIAN
BARREMIAN
HAUTERIVIAN
VALANGINIAN
BERRIASIAN
JALAL (NR)
HISSAR (NR)
JURASSIC
UPPER
MIDDLE
LOWER
VOLGIAN
KIMMERIDGIAN
OXFORDIAN
CALLOVIAN
BATHONIAN
BAJOCIAN
AALENIAN
TOARCIAN
PLIENSBACHIAN
SINEMURIAN
HETTANGIAN
OMOLON (NR)
TRIASSIC
UPPER
MIDDLE
LOWER
RHAETIAN
NORIAN
CARNIAN
LADINIAN
ANISIAN
OLENEKIAN
INDUSIAN
ILLAWARA (NR)
PERMIAN
UPPER
LOWER
TATARIAN
KAZANIAN
UFIMIAN
KUNGURIAN
ARTINSKIAN
SAKMARIAN
ASSELIAN
KIAMA (R)
30
40
50
60
70
80
90
100
110
120
130
140
150
160
170
180
190
200
210
220
230
240
250
260
270
280
CARBONIFEROUS
UPPER
MIDDLE
LOWER
GZELIAN
KASIMOVIAN
MOSCOVIAN
BASHKIRIAN
SERPUKHOVIAN
VISEAN
TOURNAISIAN
DEVONIAN
UPPER
MIDDLE
LOWER
FAMMENIAN
FRASNIAN
GIVETIAN
EIFELIAN
EMSIAN
SIEGENIAN
GEDINNIAN
SILURIAN
UPPER
LOWER
DAUNTONIAN
LUDLOVIAN
WENLOCKIAN
LLANDOVERIAN
ORDOVICIAN
UPPER
MIDDLE
LOWER
ASHGILLIAN
CARADOCIAN
LLANDEILIAN
LLANVIRNIAN
ARENIGIAN
TREMADOCIAN
CAMBRIAN
UPPER
MIDDLE
LOWER
MAIJAN
AMGINIAN
LENAIAN
ALDANIAN
DOKETZK (Rn)
DEBAL'TZEVO
TIKHVIN
SAYAN (Rn)
TASHTYP
YENISEI
BAIKAL (Nr)
NEPA
NORTH-BAIKAL
SIBERIAN (Rn)
KHADAR
IRKUTSK
BURSK
ULAKHAN
ARGA (R)
290
300
310
320
330
340
350
360
370
380
390
400
410
420
430
440
450
460
470
480
490
500
510
520
530
540
550
560
570 Ma

Proterozoic paleomagnetic scale is at its initial stage and presents one of the main problems in magnetostratigraphy (see Sect. 7.3).

The present state of knowledge permits reliable identification of paleomagnetic units large enough to correspond to a stratigraphic system rank, i.e. hyperzone (10^7 to 10^8 years). For the best-known intervals of a scale, especially within the N- or R-hyperzone, a subdivision into minor units, i.e. superzones, is possible. The paleomagnetic units are given geographic names, which are not employed in stratigraphy, according to the priority of their subdivision (Molostovsky et al. 1976; Khramov et al. 1981.)

3.2.2 Paleozoic

The Lower Paleozoic section (Cambrian-Lower Silurian) of the scale was compiled mainly using material from the Siberian platform sections (Rodionov and Sidorova 1963; Kravchinsky and Davydov 1972; Rodionov and Osipova 1973; Khramov et al. 1974); the Middle-Upper Paleozoic from material on the Russian platform and its margin (Khramov 1963; Molostovsky 1970, 1983b; Boronin 1973, 1979; Khramov et al. 1974; Slautsitais 1975). Data from the Altai-Sayan area (Aparin and Vlasov 1965; Aparin et al. 1970) and the Tunguska syneclise have also been included (Khramov et al. 1974).

Reversed polarity was dominant in the Paleozoic (70% of the normal stratigraphic column), and more than 30 normal polarity zones of 0.5 to 20 m.y. duration contrast with this background. The Paleozoic section of the scale contains six hyperzones with reversed (R), predominantly reversed (Rn) and predominantly normal (Nr) polarity, with a time span of 50 to 70 m.y. each.

The Arga hyperzone (R) starts the Paleozoic paleomagnetic section covering a wide interval of the Upper Vendian, Lower Cambrian and Lower Middle Cambrian (610–535 m.y.). Two N-zones of 0.5 m.y. duration are recognizable in the upper part of the Arga hyperzone.

The Siberian hyperzone (Rn) comprises a 48 m.y. interval from the Middle Cambrian through the Lower Ordovician, and is subdivided into four superzones each of 10-15 m.y. duration: the Ulakhan (mixed polarity NR), the Bursk (Rn), the Irkutsk (NR-polarity with a similar time span of 1.5–2 m.y. for each polarity) and the Khadar R-superzone.

The Baikalian hyperzone (Nr) includes the Middle and Upper Ordovician and lower Lower Silurian, covering a time span of 55 m.y. This hyperzone differs from the other Paleozoic hyperzones by the predominance of N-polarity. Two superzones are recognized within this hyperzone: North Baikalian (with three N-zones of nearly equal duration, which contrast with an R-polarity background) and the Nepa N-superzone with an intermediate minor R-subzone of 2 m.y. duration.

The Sayan hyperzone (Rn) of 70 m.y. duration corresponds to the stratigraphic interval from the lowermost Silurian to the uppermost Devonian; its lower limit corresponds to the Llandovery-Wenlock boundary in the Silurian and its upper limit divides the Frasnian stage of the Upper Devonian into two parts. Two superzones are recognized within the Sayan hyperzone: the Yenisei Rn-superzone and the Tashtyp NR-superzone with the alternation of N- and R-intervals of differing duration.

The Donetsk hyperzone (Rn), with a duration similar to the Sayan, is generally characterized by a predominance of R-polarity. This hyperzone was identified in the interval between the Frasnian of the Upper Devonian and the uppermost Moscovian stages of the Middle Carboniferous. The Donetsk hyperzone is subdivided into the Tikhvin R-superzone with a set of thin N- and R-subzones in its

Fig. 3.4. Paleogene, Mesozoic and Paleozoic magnetic polarity time scale for the USSR. (After Khramov et al. 1981)

lowermost part and the Debal'tsevo superzone, which represents an interval of frequent field reversals with NR-polarity zones of different duration.

The Kiaman hyperzone (R) covers a time span of about 70 m.y. and corresponds to a stratigraphic interval from the uppermost part of the Moscovian stage (Carboniferous) to the Upper Tatarian substage (Permian). This long interval of reversed polarity is traced through all continents, and its upper limit serves as a marker horizon in many Permian sections.

3.2.3 Mesozoic

The Mesozoic part of the scale was compiled using material from the Lower-Middle Triassic sections of the Russian platform (Molostovsky 1970; Boronin 1973) and the Mesozoic sections of the Caucasus (Vagabova et al. 1973; Gasanova et al. 1973; Grishanov and Molostovsky 1981; Vorob'eva et al. 1981), Central Russian Asia (Gamov and Pen'kov 1970; Nazarov 1973), northeastern Asia (Pechersky 1969; Lozhkina 1982; Bretshtein et al. 1982) and Siberia (Pospelova et al. 1967; Pospelova 1976a).

The Mesozoic section is dominated by zones of normal polarity (70% of the column) and more than 30 zones of reversed polarity, with about a 1 m.y. and longer duration, are also registered. The distribution of the R-zones in the time scale is of an extremely irregular pattern which permits the division of the Mesozoic paleomagnetic scale into four hyperzones of 20 to 65 m.y. Data for certain time intervals are incomplete and yield poor correlation, thus, the minor units are not recognized within the hyperzones.

The Illawarra hyperzone (NR), of about 22 m.y. duration, overlies the Kiaman hyperzone. The lower boundary of Illawarra is placed in the middle of the Tatarian stage (Permian) and the upper, in the lowermost Anisian (Triassic). The Illawarra hyperzone starts with a narrow NR-zone of frequent reversals (nr) which was recognized in the sections of the Middle Volga and in the north of the Russian platform (Khramov et al. 1974; Boronin 1979). Five R-zones have been traced and recognized in many sections. A thin subzone of N-polarity, placed within the second R-zone, is identified in the sections of the northeastern parts of the Russian platform. Molostovsky (Molostovsky et al. 1976; Molostovsky 1983b) subdivided the Illawarra hyperzone into three superzones: the Nagorsk (mixed polarity, with N- and R-zones of different duration), the Povolgan (R-zone which occupies the main part of the Indus stage) and the Perovsk (N-, R- and N-zones in the upper Lower Triassic with a dominance of N-polarity).

The Omolon hyperzone (Nr) embraces a stratigraphic interval from the lowermost Anisian (Middle Triassic) to the top of the Pliensbachian (Lower Jurassic), which corresponds to a 40 m.y. time span. Three R-zones of different duration with a dominance of N-polarity and a large interval of frequent reversals in the middle part are characteristic of this hyperzone.

The Hissar hyperzone (NR), recognized as the longest (65 m.y.) interval of mixed polarity, ranges from the Pliensbachian (Lower Jurassic) to the lowermost Aptian (Lower Cretaceous). Two N-superzones (Callovian-Lower Oxfordian and Upper Kimmeridgian-Volgian) of about 10–15 m.y. are identified in the middle part of the hyperzone against a background of NR-alternation.

The Jalal hyperzone (Nr) comprises the Cretaceous stages from the Aptian to the lowermost Maastrichtian, corresponding to a 40 m.y. time span. Four R-zones (2 to 5 m.y. each) are identified within this long hyperzone.

3.2.4 Cenozoic

The Cenozoic scale given in Figs. 3.4 and 3.5 is the first generalized paleomagnetic section of the Cenozoic of the southern part of the USSR (Molostovsky 1983a). This scale was compiled according to recent data of detailed paleomagnetic studies of some reference sections of the Paleogene and Neogene marine strata of the

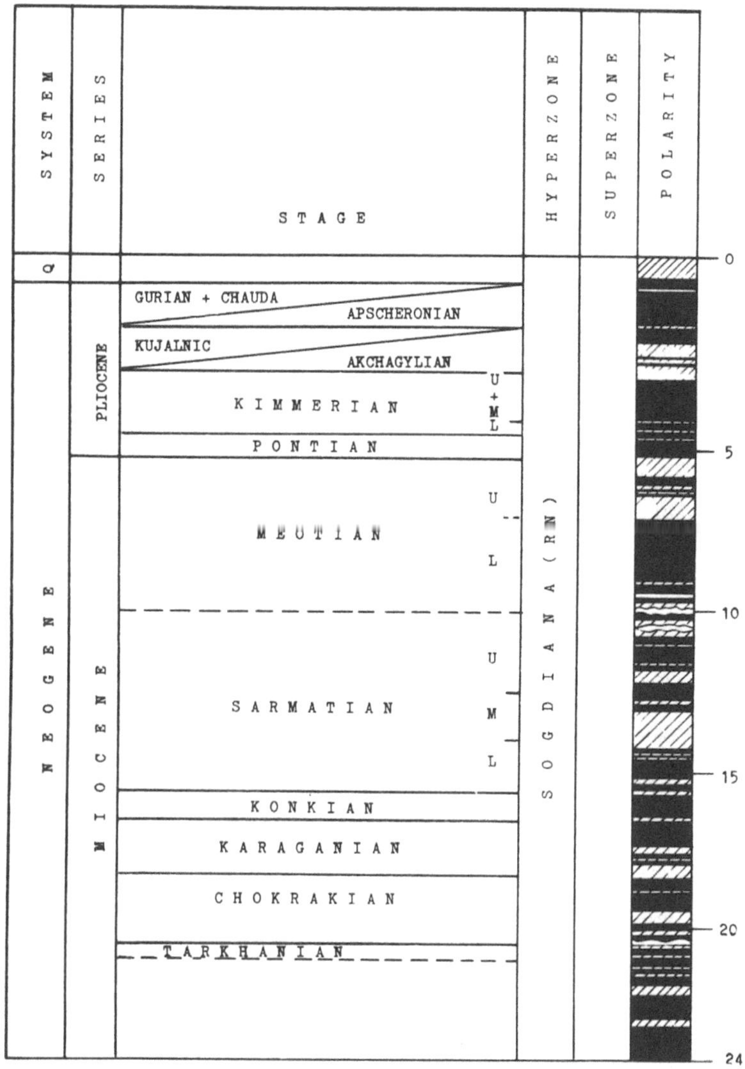

Fig. 3.5. Late Cenozoic magnetic polarity time scale for the USSR. (After Molostovsky 1983a)

Crimea-Caucasian region (Gonnov et al. 1981; Karakhanyan 1981; Imnadze et al. 1982; Gurary and Trubikhin 1981) and according to data from previous studies (Ananiashvili and Pevzner 1977; Ali-Zade et al. 1973; Pevzner and Semenko 1979; Tretyak et al. 1980). Two sharply distinctive parts are seen clearly on this scale according to the volume of data used. The Lower Cenozoic section (Paleogene) of the scale was mainly compiled according to data on a few sections in Turkmenia (Nazarov and Davtyan 1970), in Tajikistan (Pen'kov 1970) and Nakhichevan' (Ismail-Zade and Gasanov 1973). The Upper Cenozoic section (Neogene-Quaternary system) of the scale was compiled using abundant data from the Dnieper area (Pevzner 1970), the Trans-Carpathians and the southern Ukraine (Tretyak 1973; Agadzhanyan et al. 1981; Mikhailova et al. 1974), the Crimea and Northern

Caucasus (Pevzner and Semenko 1979; Tretyak et al. 1980; Pevzner and Chikovani 1978), Georgia (Ananiashvili and Pevzner 1977; Imnadze et al. 1982), Armenia (Karakhanyan 1981), Azerbaijan (Ali-Zade et al. 1973) and data from Uzbekistan (Yeroshkin 1980), Turkmenia (Trubikhin 1977), Tajikistan (Pen'kov 1970; Pen'kov and Gamov 1980), Siberia (Pospelova and Gnibidenko 1977; Pospelova et al. 1982), Sakhalin (Remizovsky 1982b) and Kamchatka (Remizovsky 1982a) were also taken into account.

The Cenozoic geomagnetic field is typified by an exceptional instability, which is seen from numerous zones (about 70) in the scale for a time span of 65 m.y. In the uppermost Cenozoic (Miocene-Quaternary) there are several zones where the field polarity changes so frequently that it can be shown on a scale only as a single zone of frequent polarity alternations. The Cenozoic paleomagnetic zones are distributed in such a way that the fields of N- and R-polarity each cover 50% of the scale. Nevertheless, the lower interval embracing Paleocene and the major portion of Eocene are characterized by the dominance of R-polarity. The upper part of the Cenozoic scale, since the Oligocene, reveals a frequent alternation of narrow zones of normal and reversed polarity and the most frequent change occurred during the last 20 m.y. The difference observed in the lower and upper parts of the Cenozoic scale permitted the recognition of two hyperzones (Molostovsky 1983a).

The Khorezm hyperzone (Rn), with a duration of 35 m.y. (75–40 m.y.), embraces the stratigraphic range between the middle Maastrichtian and presumably the uppermost Eocene. This hyperzone starts with frequent field reversals that mark the Cretaceous-Paleogene boundary. The R-polarity is typical of the Paleocene and the major portion of the Eocene. The structure of the Upper Eocene section of the hyperzone is not yet certain.

The Sogdiana hyperzone (Rn), with a time span of 40 m.y. (40–0 m.y.), embraces the Oligocene and the entire Neogene and Pleistocene and is characterized by frequent reversals of the field with a certain dominance of R-polarity in the Oligocene and Lower Miocene and of N-polarity in the Upper Miocene.

3.2.5 Correlation of the Phanerozoic Paleomagnetic Scale for the USSR and the World Paleomagnetic Data

The Phanerozoic paleomagnetic scale for the USSR is compiled as a stratigraphic scale. This is the main difference between this scale and the widely known scales of Cox, Larson, and others, which were compiled as time scales based on radiometric age determinations (Cox 1969; Larson and Pitman 1972). The magnetostratigraphic method, which is preferred in studies of the more ancient parts of the paleomagnetic scale, has been widely applied in many countries (Helsley 1969; Peterson and Nairn 1971; Luck 1972; Larson and Hilde 1975; Alvarez et al. 1977; Sinito et al. 1979). It widens the range of correlation of various paleomagnetic scales and offers an additional control to on the paleomagnetic scale of the USSR.

The comparison of paleomagnetic columns on a global scale, however, is difficult due to the following reasons: (1) as a rule, the stratigraphic control is poor and the magnetic zones are incomplete, thus their boundaries have no certain allocation; (2) there is no reliable, generally accepted time scale for the Phanerozoic, for the stage boundaries, in particular; (3) various stratigraphic and geochronologic scales have been used. Therefore, the correlation of paleomagnetic scales is rather tentative, especially for the intervals of frequent polarity reversals.

As the magnetostratigraphic data for the early and middle Paleozoic of other continents are quite fragmentary, a correlation is only possible starting from the Debal'tsevo superzone. Its upper part and boundary with Kiaman hyperzone were established in sections of North America in the uppermost Pennsylvanian (McMahon and Strangway 1968) and in the Lower Paganzo Formation of South America (Thompson and Mitchell 1972; Valencio et al. 1977b) (Fig. 3.6). The Kiaman hyperzone, traceable through all continents, is the most complete within these sections. Its upper boundary with the Illawarra hyperzone was reported from the uppermost part of the Permian Guadalupian Stage of North America (Peterson and Nairn 1971), i.e. at the same level as on the Russian platform. The Illawarra hyperzone was clearly traced in the Upper Permian and Triassic sections in many regions of the world. The number of N- and R-zones and their stratigraphic range can be correlated with the USSR scale. The Illawarra hyperzone and its upper boundary are best known in the sections of the Moenkopi and Chugwater Formations in North America. According to the generalized section of Helsley (Larson

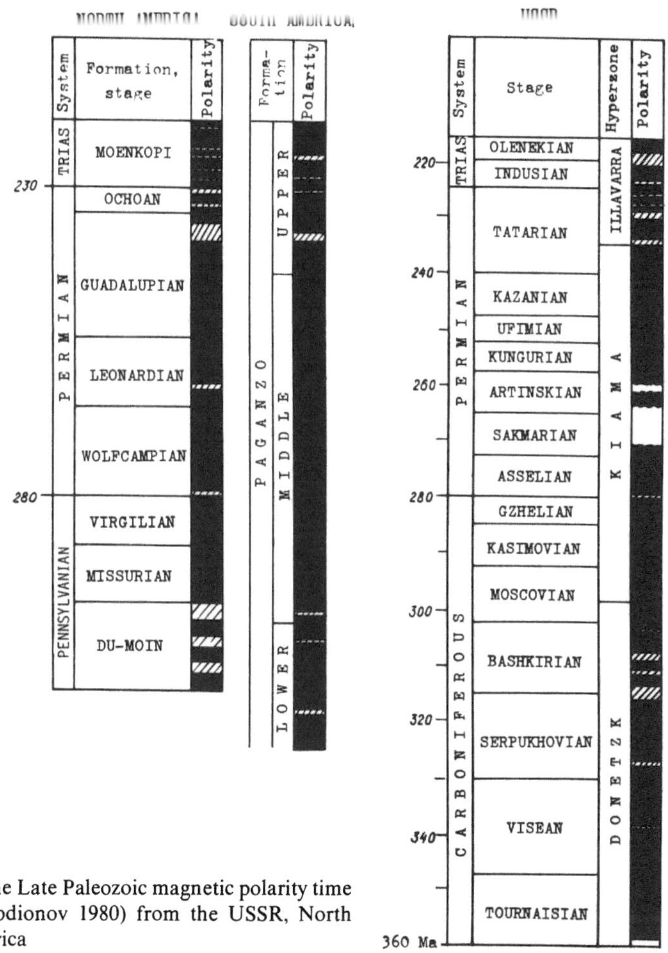

Fig. 3.6. Correlation of the Late Paleozoic magnetic polarity time scales (Khramov and Rodionov 1980) from the USSR, North America and South America

and Helsley 1975), this boundary lies very close to that of the Middle and Upper Triassic (in the USSR it is drawn at the base of the Anisian stage.)

This generalized paleomagnetic section of Helsley, compiled using data from North America, Greenland and Africa, comprises the lower half of the Omolon hyperzone and appears to be its correlative: two R-zones within the Carnian (we, however, place them slightly lower), a large number of narrow R-zones and subzones of the Norian-Rhaetian, which correspond to the Nr-zone within the same interval of the USSR scale.

The material for the Hissar hyperzone is known only from the sections of narrow stratigraphic span (Steiner and Helsley 1975). In general, they show a similarity, but differ greatly in detail, which might be due to the incomplete study of the sections (Fig. 3.7).

The Jalal hyperzone has been reported from many sections of the world (see Fig. 3.7). It is best known in Western Europe (Lowrie and Alvarez 1977; Alvarez and Lowrie 1978; Petersen et al. 1981; Mörner 1981; Channell and Medizza 1981), North America (Lerbekmo et al. 1979; Butler and Lindsay 1981) and from the oceanic floor sediment cores (Heirtzler 1981). The lower boundary is drawn below the Cenomanian (in the lowermost Aptian). The upper boundary of this stage is traced by the R-zone in the USSR, North America and Italy in the lowermost Maastrichtian. In some sections of the USSR the Klyuev R-zone (Coniacian) is recognized within the Jalal hyperzone: in North America and Italy it has not been identified. The uppermost Santonian-lowermost Campanian is traced by an excellent marker horizon, the Kuldzha R-zone (Gubbio A).

The upper part of the Sogdiana hyperzone (0–4.5 m.y.) is traced in detail in the Neogene-Quaternary sections of France, Iceland, Japan and North America as well as in the deep-sea sediments (Theyer and Hammond 1974; Tarling and Mitchell 1976; La Breque et al. 1977; Heirtzler 1981). This hyperzone is also recognized in many sections in different regions of the USSR. The magnetostratigraphic scale of this interval is comparable with the Cox scale (1969), which is based on the radiometric ages of paleomagnetically studied basalts. Four large polarity zones occur in this scale: the Brunhes (normal) 0–0.69 m.y., the Matuyama (reversed) 0.6–2.43 m.y., the Gauss (normal) 2.43–3.32 m.y. and the Gilbert (reversed) 3.32–4.45 m.y. are clearly traced in the Pliocene-Pleistocene sections throughout the world.

3.2.6 Comparison of the Mesozoic Paleomagnetic Scale of the USSR and the Scale of Marine Magnetic Anomalies

The comparison of the paleomagnetic scale with that of the marine magnetic anomalies is impossible by means of stratigraphic units, but it can be accomplished when a time scale is used. However, difficulties arise due to the absence of a generally accepted time scale for the Mesozoic stage boundaries. Furthermore, the marine anomaly scale has been repeatedly modified. The magnetostratigraphic and marine anomaly scales generally reveal a common pattern in polarity distribution (Fig. 3.8). However, special discrepancies should be noted (Khramov et al. 1981):

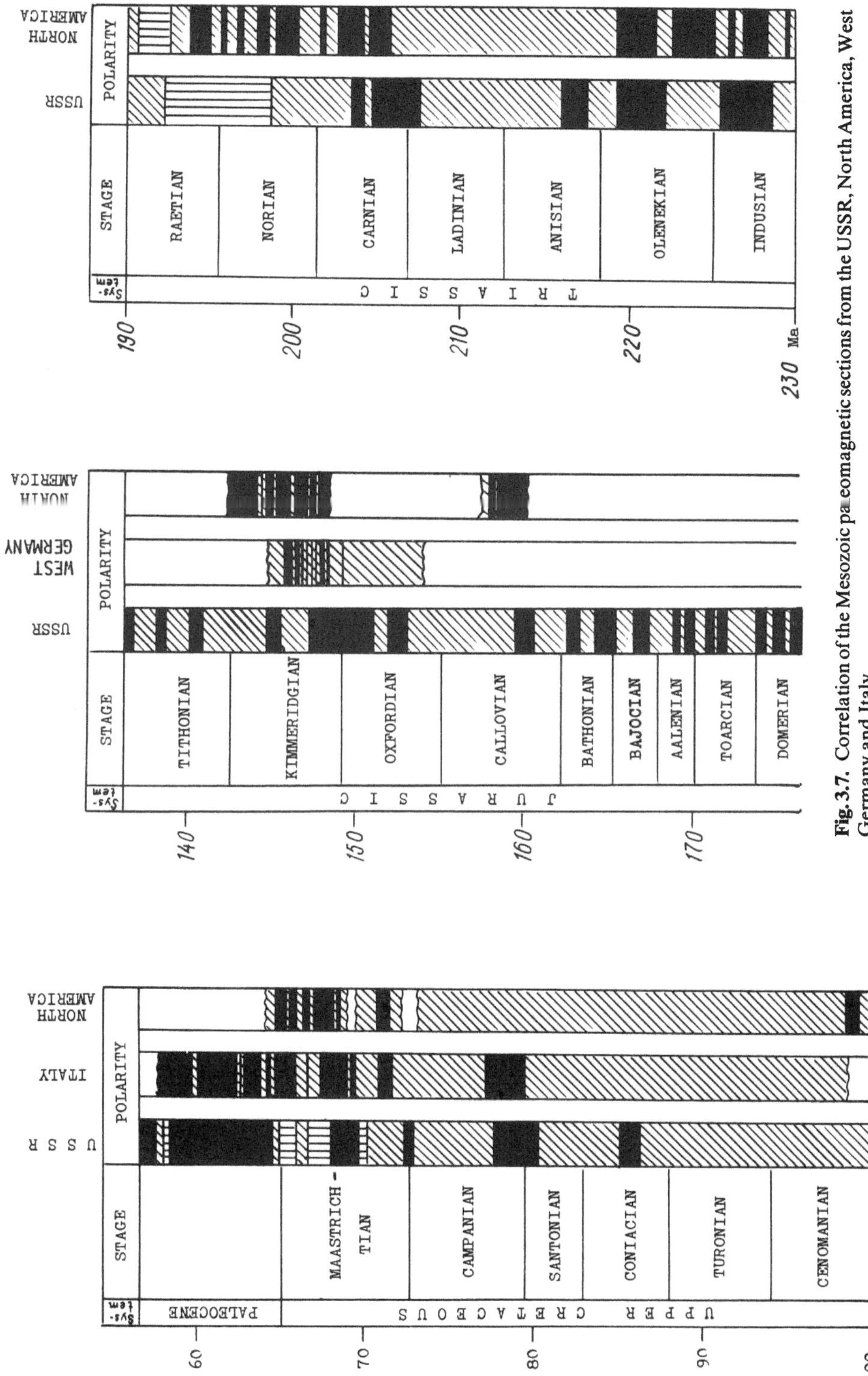

Fig. 3.7. Correlation of the Mesozoic palaeomagnetic sections from the USSR, North America, West Germany and Italy

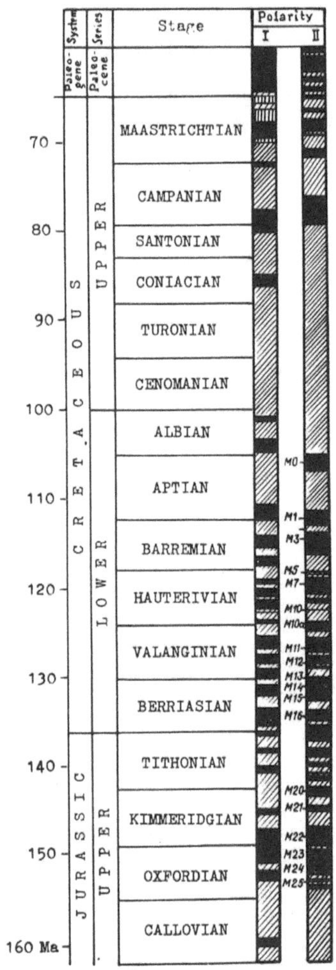

Fig. 3.8. Correlation of the Mesozoic magnetic polarity time scale for the USSR (I) with the scale of the linear magnetic anomalies (II). (Khramov et al. 1981)

1. In the marine anomaly scale of Larson and Hilde (1975), the Klyuev R-zone (Coniacian) is not recognized within the Jalal hyperzone.
2. The Gatan R-zone (Albian) of the paleomagnetic scale is absent in the marine magnetic anomaly scale, but there we have the MO anomaly corresponding to the uppermost Aptian. It is probably one and the same zone. However, the Aptian is not well known in the USSR in a paleomagnetic sense.
3. An incomplete fit of the number of R-zones in the Berriasian-Barremian in the scales is mainly due to numerous reversals during this time interval (13 and 19 R-zones in our scale and the marine magnetic scale, respectively). It is quite possible that some zones of the paleomagnetic scale have merged or that anomalies of the marine scale were unduly subdivided.
4. Thus, it is impossible to explain the small number of R-zones in the Tithonian and their absence in the Siberian sections of the paleomagnetic scale, whereas the marine anomaly scale contains more than four R-zones.

5. The Kimmeridgian-Oxfordian sections studied in the USSR are evidently incomplete (mainly the volcanic sequences of the Caucasus). This might account for the small number of R-zones in the paleomagnetic scale and the great span of the middle R-zone. However, there are over ten R-zones in Oxfordian-Kimmeridgian marine anomaly scale. The Kimmeridgian sections studied in the USA (Steiner and Helsley 1975) and FRG (Heller 1977) are similar to those of the marine anomaly scale, in particular within the interval of the anomalies M22-M25; but in this scale the anomalies mentioned are mainly Oxfordian in age.

The examples presented show the necessity of introducing corrections to both scales. These corrections, it concern mainly the Tithonian and Oxfordian with a difference of about half of a stage.

3.3 Mechanisms of Geomagnetic Reversals and Their Morphology

3.3.1 Possible Mechanisms of Geomagnetic Reversals

The principle difficulty in the theory of geomagnetic reversals is in determining whether the reversals are inherent in the mechanism of the geomagnetic dynamo proper or whether they occur as a result of distortion in the function of this dynamo under the effect of extraneous mechanisms. Bullard and Gellman (1954) showed that a single-disk dynamo cannot spontaneously change the direction of the induced field to an opposite one under any conditions of normal behaviour. In the model of the double-disk dynamo, proposed by Rikitake (1966), field reversals may occur. A double-disk dynamo, consisting of two mutually related disks, might yield self-excitation and even instability (under certain conditions), resulting in a change of intensity as well as in the direction of the induced magnetic field. The principle of this process is thus reduced to the solution of equations of disk rotation and their electromagnetic state.

The solution of these equations, calculated by numerical integration, shows that the parameters of the dynamo (resistance, inductivity, mutual induction between contours and disks, inertia moments of disks and frequency of their rotation) and also the initial conditions of the dynamo can give different variants of the current function I with time. However, all these solutions are of quasi-periodic nature, i.e. the magnetic field induced by a double-disk dynamo changes its magnitude and direction recurrently and in different time intervals, whereas the system rotation is supported by an external supply of energy; one of the solutions is shown in Fig. 3.9.

Thus, the theory of the geomagnetic dynamo allows spontaneous reversals. The reality of this mechanism, however, has not been confirmed as the model of a double-disk dynamo is a strong simplification of the interaction of movements and currents existing in the Earth's core. Moreover, Cox proposed a different mechanism of reversal, i.e. of an external origin relative to the geomagnetic dynamo. This mechanism implies a change in polarity as a result of the interaction between spontaneous, slow oscillations of a geomagnetic dynamo, as in the Rikitake model,

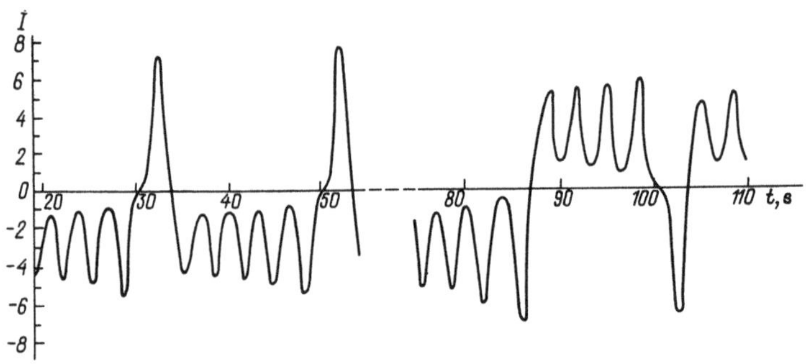

Fig. 3.9. The oscillation of the resultant current in the system of the double-disk dynamo. (After Rikitake 1966)

and certain, more rapid random process (Cox 1969). In this process, the slow oscillator is the field (Cox 1968, 1969) of a central dipole and the random variations are the nondipole fields which, as in the model of Alldredge and Hurwitz (1964), are approximated by the effect of a system of small radial, eccentric dipoles. If M is the magnetic moment of a central dipole and M′ is the vector sum of the moments of the eccentric dipoles, then a prerequisite for polarity reversal would be the change of sign of the sum of the moment when projected onto the Earth's rotational axis. This can, in fact, occur when a minimal M coincides with a strong M′ fluctuation in the opposite direction (Irving 1964). Then, the weak R-field created, in accordance with the fundamental property of a geomagnetic dynamo, would be amplified and, furthermore, oscillations of the moment M would occur in the negative area.

Analysis of the model shows that with an actual relationship between the dipole and the nondipole part of the field, and with the values of the amplitude and period of variations of the dipole moment M established from the archeomagnetic observations, the probability of a polarity reversal during one period ($\sim 10^4$ years) is 0.05. This gives the mean time of existence for a single polarity as $\tau = 2 \times 10^5$ years, which is similar to that observed during the last 4.5×10^6 years.

In the Cox model the dipole part of the geomagnetic field disappears during the reversals and only the nondipole components are preserved. This model, as well as any other in which the nondipole field remains during a reversal, might be tested within the limits of model A: a geomagnetic field is regarded as the sum of the central axial dipole field with an intensity H_e at the equator and the sum of a field H_N of nondipole fluctuations with a constant amplitude, but with a randomly distributed direction in space (see Sect. 3.4.2). For the random distribution of the angles of these directions, the following relationship (Irving 1964) is given as:

$$f_o = H_N / H_e \sim 1.73 K^{-1/2} (1 + 3 \sin^2 \varphi)^{1/2},$$

where K is the precision parameter of the geomagnetic field directions at a latitude φ.

Analysis of the geomagnetic field for the 1945 epoch showed the relationship $f_0 = 0.40 \pm 0.01$ and its mean value according to all paleomagnetic data $f_0 = 0.38 \pm 0.13$. If the model is accurate, then the relationship of the field intensity during the reversal to its previous mean value and afterward (recalculated for the equator) is close to these values of f_0. A test of two reversals, i.e. for the Jurassic lavas of South Africa and for the Neogene basalts of Japan, gave $f_0 = 0.39$ and 0.41 respectively (McElhinny 1973). This is not sufficient, however, to validate this model. An analysis of the behaviour of geomagnetic field directions during such reversals is also important for its verification. If these directions are converted into a sequence of virtual geomagnetic poles (VGP), then with the dipole field absent during a reversal, the trajectories of VGP, drawn according to data from different parts of the Earth's surface, would differ for the same reversal. Obviously, it should not be expected that the trajectories of VGP for various reversals are identical in the same region. However, the first analysis of seven Cenozoic reversals known in detail, showed that their VGP paths are similar (Creer and Ispir 1970). This similarity is not easily explained when a nondipole field is supposed to dominate during a reversal. Later analyses, including many more reversals, produced a more complicated pattern in the distributions of the VGP (Gurary 1977b; Petrova et al. 1972). It appeared that a concentration of VGP within several meridional belts occurred during the major part of the reversals studied, the reversals that follow each other have significantly different VGP trajectories. Reversals with a latitudinal component of VGP movement were also registered (see Sect. 3.3.5).

More models of reversals are known, but they have not been so extensively elaborated, so as to allow the determination of distinct features which would test the validity of any of these from a limited number of observations. Thus, it is necessary, together with further elaboration of existing and new models, to conduct a detailed morphological analysis of a great number of synchronous geomagnetic reversals, but examined in different geographic sites, and also some that are asynchronous in a time sequence, thus embracing the entire Phanerozoic. The following characteristics of a field are required in particular:

1. Change in the geomagnetic field intensity during a reversal and during the corresponding time intervals.
2. Morphology of a reversal, i.e. variations of the angular components of a field and of the VGP coordinates.
3. Duration of reversals.
4. Character of reversal commencement and termination (which elements of a field: intensity or angular components begin changing earlier); the presence or absence of "precursors".
5. The sequence of the reversal process with time: whether it is symmetric or asymmetric.
6. Dependence of all characteristics of a reversal on the geographic situation of a site.

Extremely labor-consuming and voluminous work is awaiting us in pursuing this aim, as we are now only at the initial stage. Some previously obtained results are given below.

3.3.2 Morphology of Geomagnetic Reversals

The principal problems and aims in studying reversal morphology concern the theoretical aspects of the geomagnetic field (the dynamo theory, the Earth's core structure, processes occuring in the core). Besides, (1) a comparison of the typical features of geomagnetic reversals of different age within a single region, (2) a combined analysis of geomagnetic field behaviour during a reversal irrespective of time and place and (3) a correlation of these data with those with coeval reversals in different regions, would be of great assistance in understanding two major questions: (a) whether the field was dipolar during a reversal; (b) whether every given reversal possesses its own peculiarities. The answers to these questions will help to solve certain geologic problems, the stratigraphic ones in particular.

Paleomagnetic stratigraphy is known to be based on the division of sections into paleomagnetic zones according to the primary magnetization of the rocks from which these sections are composed. A complex of up-to-date laboratory procedures permits the recognition of the direction of primary magnetization in the rocks under study with satisfactory reliability and the identification of adjacent beds of different polarity in every region with high precision. Lithologic, faunal and much more data must be used in comparing the paleomagnetic zones in different regions, because paleomagnetic characteristics alone are inadequate for this purpose. The most problematic are the time intervals, which are called "critical": the reversals of the geomagnetic field within these intervals are very frequent and the paleomagnetic characteristics of successive zones of the same polarity are very similar. If the pattern of the field reversal appears to change in detail from zone to zone, then the typical features of the geomagnetic reversals would probably provide additional aspects for synchronization and the distinction of reversals following each other and for the description of reversals of a certain time (period, epoch).

Studies of geomagnetic reversals include the following aspects.

1. The search and study of detailed records of the geomagnetic reversal process in rocks.
2. The statistical analysis of the distribution of reversals in time.
3. The study of amplitude and frequency characteristics of paleosecular variations in times of geomagnetic instability.

Possibilities and some results of geomagnetic reversal study, which are available due to detailed examination of "transitional" beds, are discussed in this chapter.

A detailed record of the reversal process in transitional beds is likely to occur in different types of rocks, including continental and oceanic sediments and effusive rocks. However, the record and preservation of typical aspects of geomagnetic field behaviour at the moment of reversal in transitional beds is requires a combination of several independent factors. A regime of sufficiently rapid and uniform sedimentation (or lava outflow) and field reversals should be consistent in time; the existing conditions should establish the origin and preservation of primary magnetization of a rock; the effects of secondary remagnetization should be slight, etc. Discontinuous sedimentation and magmatic activity are the main causes explaining the failure in the search for transitional beds between zones of normal and reversed magnetization. Finally, since the time during which a reversal of the field occurs is relatively short (10^3–10^4 years), it is difficult to obtain a detailed record of a reversal.

Many paleomagnetologists have made attempts to discover these transitional beds in rocks of different age and to describe the behaviour of a field at the moment of reversal. At present, there are numerous data available for transitional beds in the entire scale of reversals for the Phanerozoic. The Cenozoic reversals studied are most abundant, for the last 5 m.y. in particular, whereas the Mesozoic reversals are less well known (Petrova et al. 1972; Burakov et al. 1976; Fuller et al. 1979; Gurary 1977a, 1981; Burov et al. 1977; Boronin 1979; Herrero-Bervera and Helsley 1983). Data on the early Paleozoic reversals are mainly based on the examination of transitional beds in the Cambrian and Ordovician sections to the south of the Siberian platform (Khramov et al. 1974; Rodionov 1969; Kravchinsky 1970; Gurary 1969; Metallova et al. 1980; Khramov and Rodionov 1980b). The transitional beds between the normally and reversely magnetized strata in lavas, intrusions and sedimentary rocks of different origin, which have been divided into transitional zones, have been examined in Europe, Asia, Australia, North America and in regions of the Atlantic, Pacific and Indian Oceans. The results of these examinations, describing changes in the parameters of the geomagnetic field at the moment of reversal, should be used to determine whether any common regularities exist in the geomagnetic field behaviour during its reversal. Several reviews of this problem have been published (Creer and Ispir 1970; Petrova et al. 1972; Gurary 1977b; Khramov 1978; Fuller et al. 1979; Vadkovsky et al. 1980; Bucha 1983; Hoffman 1983).

It is imperative to elaborate common methods in the study and presentation of data to overcome the difficulties in obtaining and interpreting the results on the transitional zones, thus permitting a comparison of the details of different reversals. A transitional zones (judging from the published data) should be continuously sampled when studied. To prove the reliability of the results, it is necessary to use a combination of laboratory procedures while defining the directions of primary magnetization, studying the minerals which are the carriers of magnetization J_n^o and recognizing the origin of magnetization. Graphs of angular parameters, declination D and inclination I, show a pattern of field transition. The parameters Φ_a (the ancient latitude) and Λ_a (the ancient longitude of the virtual geomagnetic pole), generalize the change in the angular elements of the field. These characteristics and the values J_n, $\varkappa$ and J_n/J_{rv} disclose the peculiarities of each reversal. The change in factor $Q_o = J_n^o(\varkappa H)$ along a section is often used as an estimation of the ancient field intensity change. Factor Q_o is known to indicate only relative changes in the field amplitude and for a comparative analysis, it is necessary to use one of the common methods, for instance, the H_e-method (stepwise remagnetization) to calculate the field value. At the moment the expression of the morphology of geomagnetic reversals is accepted in terms of the equivalent paths of VGP which correspond to changes in the direction of a geomagnetic field in the sites of reversal as this allows comparison of results from different parts of the world.

The reliability of the pattern of the field change would then be established by the similarity of results from transitional zones of the same reversal found in different sites of a region along the sections represented by rocks of different genesis, tectonic regime, etc.

3.3.3 Morphology of the Early Paleozoic Reversals

The first works on early Paleozoic reversals were conducted in the transitional beds of the Upper Cambrian and Ordovician sequences to the south of the Siberian platform and reviewed in the work of Khramov et al. (1974). As a result, three types of reversals were recognized: "sudden" reversals in the field polarity; reversals accompanied by a significant dispersion in the field directions; reversals with a gradual looplike rotation of the field.

The reversals in the epochs of unstable polarity of a geomagnetic field in the late Cambrian were examined by Rodionov (1969) in certain sections of red beds in the Verkholena Formation; by Kravchinsky (1970) in the Middle Lena River region and by Gurary (1969) in the same strata in the Angara River region. The late Cambrian reversals are reversals of a single type with a gradual polarity transition. These reversals are characterized by the following common features: a displacement of a virtual geomagnetic axis that antedates the reversal, a looplike movement of a pole, a residual misalignment of a virtual geomagnetic axis and a short-time restoration of the previous polarity prior to the final reversal of the field (Fig. 3.10). This nature of the field reversal gives much wider zones of paleomagnetic instability in a section than those of the polarity transition. In this case a long-time displacement of the virtual geomagnetic axis, observed in the process of reversal, might be the cause of systematic error in the estimation of paleomagnetic pole coordinates. A detailed bed-by-bed tracing of paleomagnetic directions in sections with frequent polarity change might favour identification of these intervals of paleomagnetic instability related to "anomalies" of paleomagnetic poles adequate to that established in an exposure of the Upper Cambrian strata near the village of Khamra (Khramov et al. 1974). The late Cambrian reversals are accompanied by a significant decrease in the intensity of a geomagnetic field; according to data of Kravchinsky and Davydov, the intensity of a field at the moment of reversal decreases by a factor of three to six. The evaluated changes of intensity of the geomagnetic field (by the stepwise remagnetization method) in two parallel sections in the Lena River, sampled by

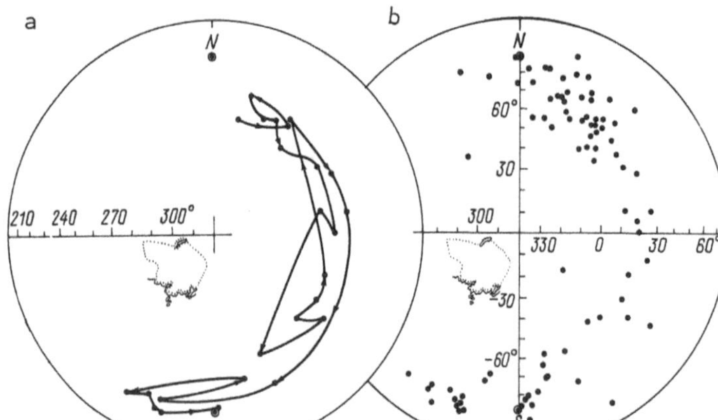

Fig. 3.10a,b. Virtual geomagnetic poles during the late Cambrian transition $n_5 \rightarrow r_1$. (Khramov and Rodionov 1980b). **a** The VGP trajectory, data from Khamra village, Lena River; **b** distribution of poles for the $n_5 \rightarrow r_1$ reversal from all sections studied in the Lena River region

Rodionov, confirmed the above statement (Metallova et al. 1980). In the zones of paleomagnetic instability in these sections the intensity of the geomagnetic field appears to be two times less than that of the zones of the bordering normal and reversed magnetization zones.

Geomagnetic reversals in the early and middle Ordovician, unlike those of the late Cambrian, are typified by a series of sharp reversals of the field without any intermediate NRM directions. They represent reversals with several rhythms (a complete rhythm includes two subsequent polarity transitions; complete or incomplete). The transition N → R in the early Ordovician, examined by Rodionov in the Lena River 2 km below the mouth of the Mandra River, exemplifies this type of reversal. From the red bed section, with a thickness of 9 m, a continuous column was examined that characterized 327 stratigraphic levels (the thickness of a stratigraphic level is 25 mm). The rock lithology in this section varies: the lowermost parts are built up largely of siltstones with subordinate beds of sandstone and limestone; the middle part is typified by an alternation of siltstone, sandstone and marl; in the upper part of the section sandstones and siltstones are dominant. The values of the diagnostic variables of a field (D, I, Φ_a, J_n^a/J_{rv}) were smoothed out over five levels by gliding averages.

The pattern of change in the angular elements of a field along the vertical stimulates the division of the section examined into three parts (Fig. 3.11). The lower part of the section (2 m) contains rocks of reversed magnetization; in the middle part (3 m) a frequent alternation of evident and nonevident zones of normal and reversed magnetization is registered; the uppermost part (4 m) exhibits normal magnetization.

A field of stable, reversed polarity in the lowermost part of the section passes upward into a field of normal polarity through a series of complete reversals with a significant drop in intensity in the middle part of the strata. This field of normal polarity is unstable to a certain extent in its initial stage as seen from the scattering of D and I parameters at a still low intensity of the field. An increase of the J_n values (and probably also of the field intensity) is registered in the intervals of reversed and normal polarity prior to and after the reversal. The polarity reversal in this section embraces an interval of 3-m thickness, extending over 3×10^5 m.y. (proceeding from the thickness and time of formation of the whole sequence).

Thus, the late Cambrian and Ordovician reversals are of a different type, and they occur during significant decrease in the field intensity. The estimated durations of the early Paleozoic reversals vary from 10^4 to 3×10^5 m.y. (a reversal denotes all changes in the field between its two stable states against a background of diminished intensity).

3.3.4 Reversals at the Paleozoic-Mesozoic Boundary

A frequent alternation of the zones of different polarity at the Paleozoic-Mesozoic boundary forms the so-called critical Illawarra interval that is traced conspicuously in the Upper Permian and Lower Triassic sections of all continents. Ten magnetozones are recognized within this interval. The lower five correspond to the Upper Permian Tatarian stage according to the classification of this zone (up-section i.e. from bottom to top): R_1P, N_1P, R_2P, N_2P and R_3P; the subsequent five zones: N_1T, R_1T, N_2T, R_2T, N_3T are compared with the Indusian, Olenekian, and lowermost Anisian stages of Triassic. These frequent reversals of a geomagnetic field and the little paleontologic control of the paleomagnetic zones divided in this very short time interval present difficult problems in their correlation even within a single region. A number of successive late Permian and early Triassic reversals of a geomagnetic field was examined in the Upper Permian and Lower Triassic in the northeastern part of the Russian platform and in the Mangyshlak Peninsula (Burov et al. 1977; Boronin 1979; Slautsitais 1973).

The types of reversals in the late Permian-early Triassic epoch are the same as those of the early Paleozoic. "Spontaneous" field reversals have been recorded in the transitions $N_1P \rightarrow R_2P$ in the

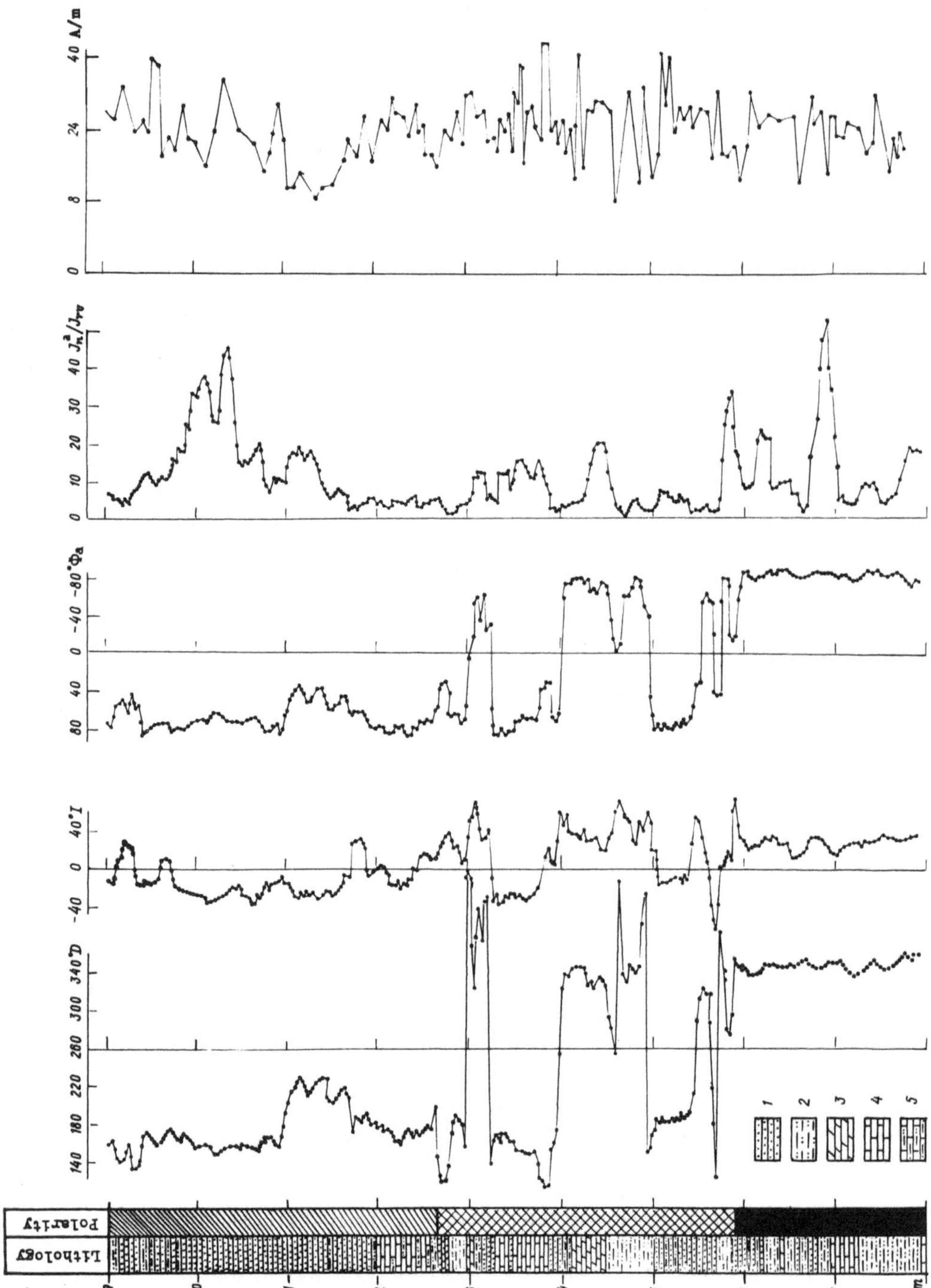
Polarity
Lithology
79
80
81
82
83
84
85
86
87
88 m
1
2
3
4
5

Permian section near the Goloshubikha village at the Volga riverside and $R_1T \rightarrow N_2T$ in the lower Triassic section in the Mangyshlak Peninsula (Burov et al. 1977; Boronin 1979; Slautsitais 1973).

Two reversals of this type that bound the N_1T zone at the base of the Triassic portion of the section were examined in steeply dipping (60°–80°) red sandstones and siltstones of the Dolnapa Formation in the Sarysu sequence in the northern slope of the Western Karatau Range (Mangyshlak). Detailed sampling was done in 30-m-thick strata with an average interval of 0.3 m (Fig. 3.12). This spacing between the adjacent samples in this particular section corresponds to a time interval of 3500 years. High paleomagnetic stability of the rocks examined (Burov et al. 1977) and a combination of methods applied in the separation of the direction of primary magnetization (compensation method, $\tau = 2.5$ months, then temperature cleaning, $t = 500°C$) gave reliable results.

A field of stable reversed polarity (the R_3P zone), recorded in the lowermost part of the strata, passes without any significant disturbance into a field of normal polarity which is reflected by the N_1T zone of 21-m thickness by a rotation of the J_n^o vector through 180°. Then the field again becomes reversed by the same mode, as evidenced by the R_1T zone in the upper part of the section. Both reversals occur at relatively low values of the Q_o factor most likely indicate a decrease in the field intensity which does not reach a normal value even in a stable period of the normal polarity field. Higher values of Q_o, and possibly of the field intensity, have been recorded for the pre- and postreversal intervals (with a constant magnetic susceptibility $\varkappa$ in the section).

A similar character of reversals was recorded in studies of the same transitions in a parallel section in the Tushchibek ravine in the southern slope of the Western Karatau Range. In both sections the reversals bounding the N_1T zone in the lowermost Triassic are represented by quick reversals to the opposite polarity against a background of low field intensity. A similar character of the field reversal was registered in the early Triassic transition $R_1T \rightarrow N_2T$ in the same Sarysu section (Slautsitais 1973).

Reversals with several rhythms and a distinct scattering of J_n directions have been registered in the Permian section in the northeastern part of the Russian platform (the Vyatka and Yug Rivers) as well as in the Mangyshlak Peninsula (Burov et al. 1977). Details of one of these reversals ($R_2P \rightarrow N_2P$), recorded in the Permian section near the village of Putyatino at the Vyatka riverside, were quite distinctly registered during an analogous field reversal in a section that is situated at a distance of more than 1000 km from the first one. The reversal $R_2P \rightarrow N_2P$ in the Tushchibek section in Mangyshlak, as well as in the Putyatino section, includes two complete rhythms of field reversals between the zones of stable polarity and is accompanied by a significant drop in the field intensity (Burov et al. 1977).

A gradual rotation of the J_n^{o1} vector (Fig. 3.13) was registered for the reversal $n_1R_3P \rightarrow r_2R_3P$ (in the Medvezhy Vzvoz exposure) that bounds from above the subzone of normal polarity n_1 inside the R_3P zone in the Upper Permian section in the northeastern part of the Russian platform. A band of movement of the virtual geomagnetic poles calculated from this reversal is situated near the paleomeridian of the actual pole (Fig. 3.14). A superposition of a "slow" component of the ancient magnetization and of a recent J_n component might distort the reversal pattern. Correspondence of this reversal record in two adjacent sections (Boronin 1979), however, shows the dominance of a "quick" component (synsedimentary) and thus, the validity of the reversal features is recognized.

A common peculiarity of all late Permian reversals studied is a bandlike distribution of the virtual geomagnetic poles along the ancient meridian. This band lies in the area of paleomeridians of 10°–40° W. In certain cases the rocks with values of magnetic susceptibility several factors higher have been recorded prior to and after the reversal.

Baag and Helsley (1974a) observed reversals of the same types in paleomagnetic studies of the late Triassic transitional beds of the Moenkopi Formation in North America. Each of the six reversal

◄

Fig. 3.11. Changes of paleomagnetic parameters in the early Ordovician transition, the Mandra section, Lena River. (After Khramov and Rodionov 1980b). From *left* to *right*: meters from the top of the section downward; lithology: *1* sandstone; *2* clay, claystone; *3* marl; *4* limestone; *5* clay limestone. Polarity information; *D* declination; *I* inclination of ChRM; Φ_a latitude of the virtual pole; J_n^a/J_{rv} stable to viscous magnetization ratio; H_a paleointensity of the geomagnetic field. Latitude of the virtual pole Φ_a is estimated relative to the mean pole $\Phi = -42°$, $\Lambda = 127°$. *D*, *I*, Φ_a were determined using the gliding average method with an averaging interval of five points, H_a (right) was determined using the method of stepwise remagnetization

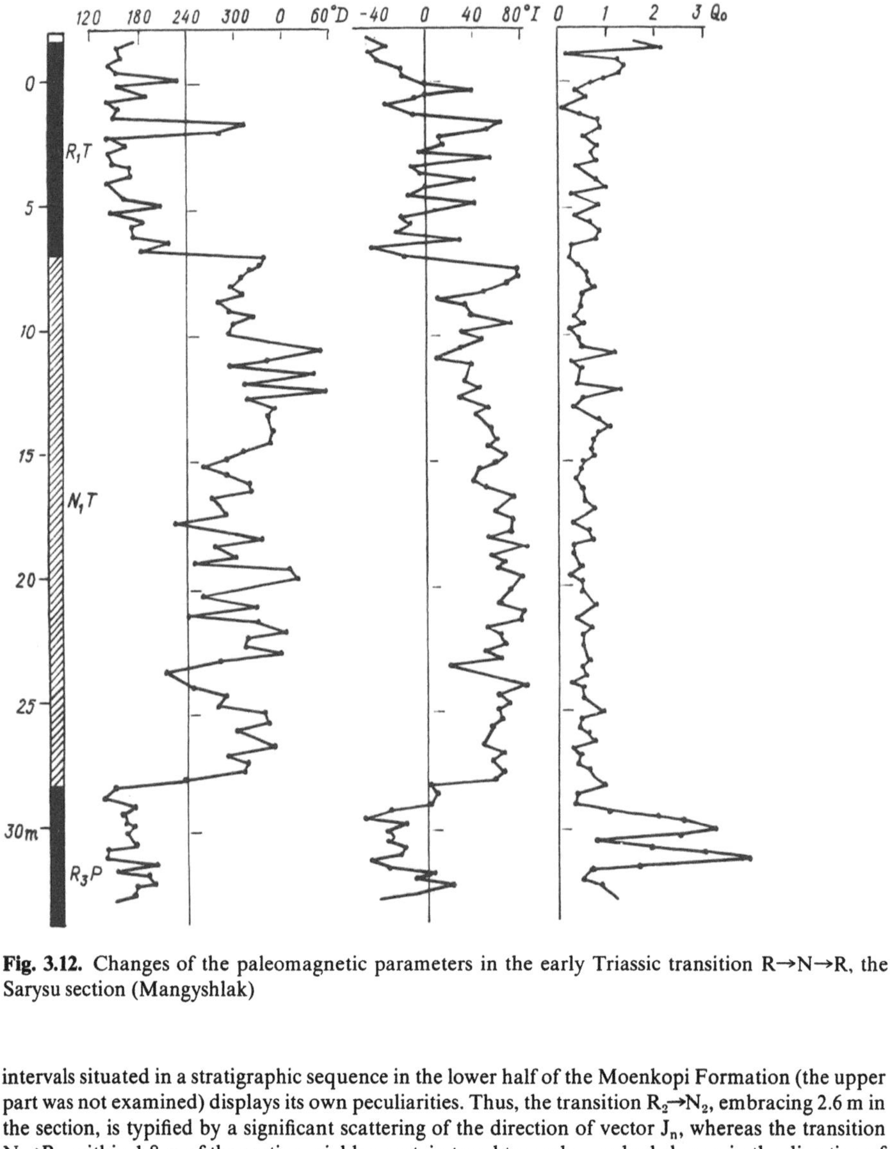

Fig. 3.12. Changes of the paleomagnetic parameters in the early Triassic transition R→N→R, the Sarysu section (Mangyshlak)

intervals situated in a stratigraphic sequence in the lower half of the Moenkopi Formation (the upper part was not examined) displays its own peculiarities. Thus, the transition $R_2 \rightarrow N_2$, embracing 2.6 m in the section, is typified by a significant scattering of the direction of vector J_n, whereas the transition $N_2 \rightarrow R_3$, within 1.8 m of the section, yields a certain trend toward a gradual change in the direction of J_n. In both cases this transition is accompanied by a significant decrease of J_n. The transitions $R_1 \rightarrow N_1$, $N_3 \rightarrow R_4$ and $R_4 \rightarrow N_4$ were registered due to a quick reversal of vector J_n through 180°.

The total number of paleomagnetic zones in the lower half of the Moenkopi Formation (the results of examinations in natural outcrops) according to data of Helsley (1969) is nine, that is as many as in the Upper Permian and Lower Triassic sediments of the Russian platform (Khramov et al. 1974). As the authors do not deny a probable late Permian age for the lowermost part of the Moenkopi Formation (Helsley 1969; Baag and Helsley 1974a), this fact shows not only the planetary scale of these zones and the validity of reversals of a geomagnetic field, but it might also be useful in the correlation of stratigraphic scales of these two distant regions.

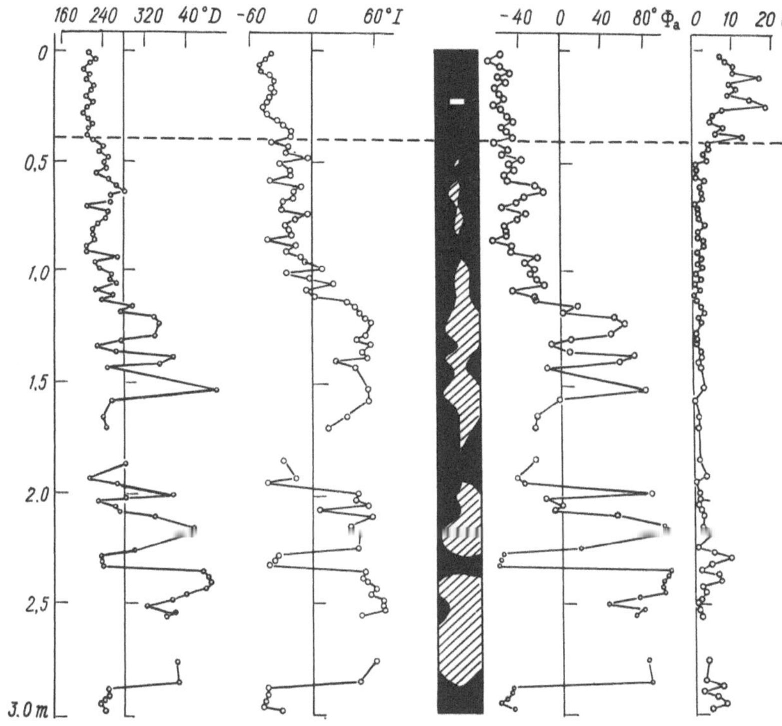

Fig. 3.13. Changes of the paleomagnetic parameters in the late Tatarian transition of the Medvezhy Vzvoz section, Yug River. (After Burov et al. 1977). *D* Declination; *I* Inclination of ChRM after cleaning at 250° ÷ 330°C; Φ_a latitude of virtual geomagnetic pole estimated relative to the mean pole $\Phi = 48°$, $\Lambda = 165°$; Q_o Koenigsberger ratio for ChRM

Behaviour of vector J_n^o in the transitional intervals of a portion of the stratigraphic scale examined changes from zone to zone in the sections on the Russian platform as well as in North America. This might be used as an additional criterion in the correlation of unipolar zones. In fact, the transition $R_2 \rightarrow N_2$ in the Moenkopi Formation, with a distinct scattering of vectors J_n, displays a similarity with the transition R_2P-N_2P in the Putyatino and Tushchibek sections. Moreover, the "trend" of vector J_n recognized in the transition $N_2 \rightarrow R_3$ in the Moenkopi Formation is similar to that of the vector behaviour in the Medvezhy Vzvoz section. An abrupt field reversal toward an opposite polarity in the transitions $N_3 \rightarrow R_4$ and $R_4 \rightarrow N_4$ in the Moenkopi Formation is also typical of the transitions $N_1T \rightarrow R_1T$ and $R_1T \rightarrow N_2T$ in many exposures in the Triassic portion of the Mangyshlak Peninsula.

The validity of this correlation is supported by estimations of the reversal duration at the Paleozoic-Mesozoic boundary, which appear to be very similar for such remote regions. According to determinations in different transitional intervals at the end of the Paleozoic, the time of the field reversal is estimated to be from 2×10^3 to 2.2×10^4 years (in the sections of the European part of the USSR). These values are in good agreement with data of Baag and Helsley who also gave time estimations of each transition in the Moenkopi Formation and who suggested that the field reversal might occur for a period of 1×10^4 to 5×10^4 years.

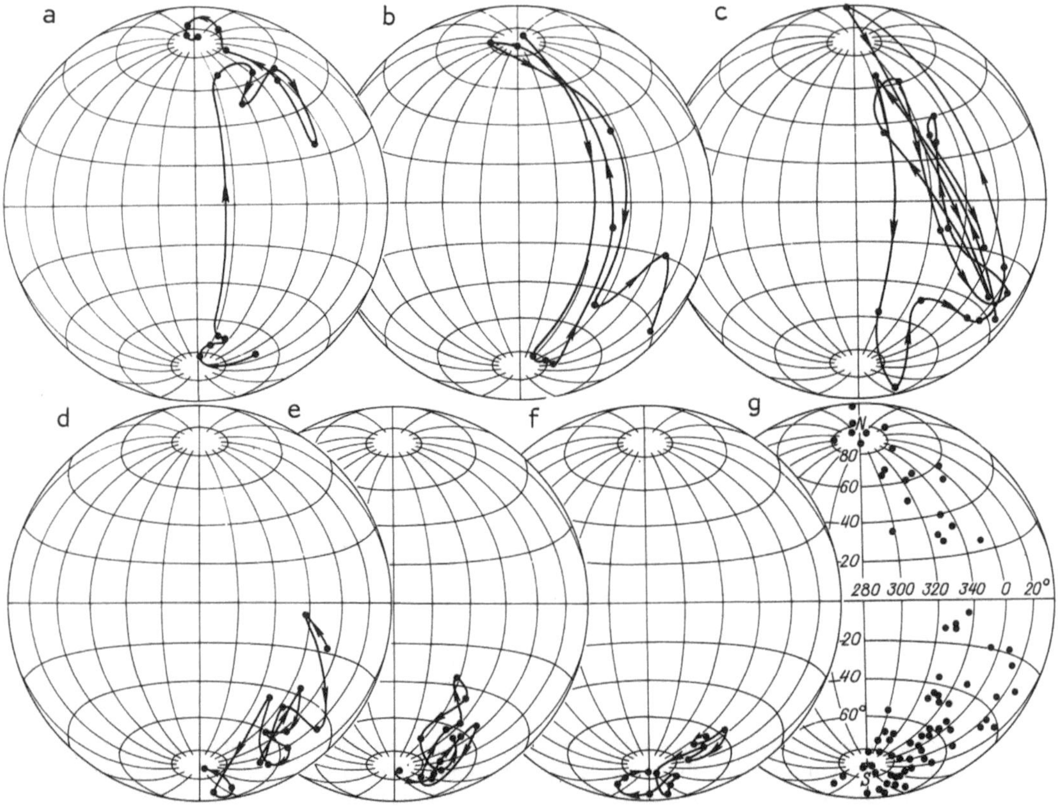

Fig. 3.14a-g. Virtual geomagnetic poles during the late Permian reversal studied at Medvezhy Vzvoz (Yug River). Poles are calculated relative to the paleogeographic net; longitudes are arbitrary. **a-f** Successive parts at the VGP trajectory during the reversal n_1R_3P–r_2R_3P; **g** distribution of the VGP's for this reversal

Analysis of the field changes from data on transitional regimes at the end of the Paleozoic and during the Mesozoic (Burov et al. 1977; Boronin 1979; Slautsitais 1973) suggests that at the end of the Paleozoic the reversals with distinct rhythmicity of destruction and resetting of a field, and the reversals with a gradual rotation of vector J_n^o toward an opposite polarity, were the most typical. In the early Mesozoic the reversals typified by a quick reversal of a field were dominant.

3.3.5 Reversals in the Late Mesozoic-Cenozoic

Data on the transitional beds of the Mesozoic-Quaternary period are quite sufficient to be analyzed. Reviews of this type have already been published (Creer and Ispir 1970; Lavley 1970; Petrova et al. 1972; Burakov et al. 1976; Gurary 1977b; Khramov 1978; Fuller et al. 1979; Bucha 1983). Similar types of reversals of the geomagnetic field characterize young reversals and the reversals of the Paleozoic and Mesozoic. Thus, Nazarov and Pogarskaya made a detailed study of the late Cretaceous geomagnetic reversal in the Upper Cretaceous (Santonian) rocks in Turkmenia. This examination revealed that transitions R→N and N→R, which bound a subzone of normal magnetization, exhibit abrupt field reversal toward an opposite polarity (Pogarskaya et al. 1976).

Interesting data were obtained on the behaviour of the geomagnetic field during a reversal in the transitional beds of Cenozoic strata of Western Turkmenia (Gurary 1977a). In a section of Miocene strata a series of successive reversals were examined: N→R in the middle Sarmatian, R→N in the late Sarmatian and N→R at the boundary of the Gauss-Matuyama paleomagnetic zones. Distinct features of each reversal are the path and pattern of movement of the virtual geomagnetic pole. Thus, during a reversal in the middle Sarmatian, the pole was moving along a complex, looplike trajectory in the area of Africa-Atlantic Ocean. The R→N reversal of the late Sarmatian shows a rather quick movement of the pole from the Southern to the Northern Hemisphere. The Gauss-Matuyama reversal, examined in four parallel sections at a distance of 180 km from each other, is typified by a virtual geomagnetic pole situation in a band of 120°-160°E [according to data of Koči, the transitional poles in the Gauss-Matuyama reversal in sections of the Federal Republic of Germany and Czechoslovakia (Bucha 1973; Koči 1973, 1974) and in two late Pliocene reversals in Iceland (Einarsson 1957a; Sigurgeirsson 1959) are approximately located in the same longitudes]. These data show that the pole movement during the reversals, occurring different times, occurs in different areas of the world. This movement is peculiar and permits one to distinguish noncoeval reversals.

A comparison of the paleomagnetic characteristics of reversals reveal common features that are typical of the major Cenozoic reversals.

1. The virtual geomagnetic poles move along a limited band of longitudes.
2. A reversal occurs at a lower intensity compared to that of the stable periods.
3. A drop in the field intensity precedes a significant disturbance in its inclination and declination.
4. The time interval of a reversal varies between (0.5 and 4) $\times 10^4$ years.

3.3.6 Morphology of Geomagnetic Reversals: Conclusions

The field "geometry" at the moment of reversal suggests that the rhythmicity in the destructive process and resetting of a field is responsible for the change in the field direction; the most distinct rhythm yields a period of $n \times 10^3$ years. The type of reversal and its peculiarity is determined by the number of rhythms; one to several complete reversals have been registered in the sections examined. The most typical are the following reversals:

1. A reversal follows a single rhythm and represents a "sudden" field reversal to an opposite polarity, accompanied by a short interval of decreased field intensity. It is necessary, however, to note the possible incompleteness of the field change pattern due to the discontinuity of rock formation, to the presence of masked hiatus, erosion, etc.
2. A reversal consisting of several rhythms: after a decrease in the field intensity and certain lowering of its stability, a "sudden" field reversal is observed. The

subsequent one-two rhythms represent abrupt and complete changes in the field polarity. After a last resetting (sometimes incomplete) of the field of the previous polarity, the change in the field direction results in a smoother pattern, with a gradual decrease in the amplitude of angular element oscillations and with an approximation of their values to those typical of the new polarity.

3. A reversal with a distinct "trend" in the directions of vector J_n is accompanied by a significant decrease in the field intensity. The J_n vectors occupy an intermediate position between the stable directions of N and R polarity, gradually moving from one group of directions to the other.

The principal characteristics of the geomagnetic field reversals are the following:

1. A polarity reversal in the geomagnetic field is accompanied by a decrease in the field intensity by a factor of two to ten and more.
2. After the beginning of decrease in the intensity and before a change in its direction, an increase in the scattering of local directions of a field is observed.
3. The virtual geomagnetic pole movement, representing a change in the field direction during a reversal, occurs within a limited band of longitudes; a looplike pole oscillation observed in its displacement has an irregular nature.
4. A reversal covers a time interval of 1×10^3 to 2×10^5 years.

These peculiarities in geomagnetic field behaviour are the most common in most cases of reversal records, however, they are not always observed. Thus, opposed to the data for a significant decrease in the field intensity during the reversal, are the measurements revealing abnormally high intensity fields within a field reversal interval, prior to and after the reversal. Together with the data indicating a longer duration of the decrease in intensity than a reversal of field angular components, there are examples of a synchronous changes in intensity and direction or of a decrease in intensity that occurs during a shorter time interval than the change in direction. Certain transitions record a movement of the virtual geomagnetic pole not in a meridional, but in a latitudinal direction.

The regularities in the geomagnetic field change during a reversal are better known in their qualitative aspect, its quantitative characteristics mentioned above are only rough approximations. The methods of determination of the ancient field intensity that are available permit the evaluation of only its relative changes. Any estimation of the reversal duration might include significant errors, for they are based on sedimentation rate values or on the intensity of volcanic activity. The paths of the virtual geomagnetic poles during a reversal are drawn proceeding from a suggested dipole field during that period, although it is appropriate. The differences between the reversal types might be a result of incomplete record of the process of field reversal in rocks.

Thus, the analysis of data on the change of the geomagnetic field during reversals, resulting in certain conclusions on the characteristics of this field, shows a significant uncertainty in these conclusions. This is most likely due to the lack of data, the amount of which is insufficient for solving this complex problem, and furthermore, to the lack of criteria and the control of their reliability. This almost 20-year period of reversal studies should be perhaps regarded as a preliminary stage, which has disclosed the difficulties arising in these studies. In particular, it became evident that it is very important to examine coeval transitional zones in different sections and regions. This might provide an answer to the problem of the existence of reversals of different types and allow the detection of the field intensity behaviour during the reversals.

3.4 Time Structure of the Geomagnetic Field Between the Reversals

3.4.1 Events and Excursions of the Geomagnetic Field

For a long time it was generally accepted that the last reversal of a geomagnetic field occurred 0.7 m.y. ago at the boundary of the Brunhes and Matuyama magnetic epochs. In 1967, however, a short-time event of reversed polarity in the 9,000–20,000 year interval of the Brunhes epoch, called Laschamp, was recognized while examining young lavas in France.

It was concluded that within a range of 9,000–20,000 years ago there was a short interval in which a geomagnetic field was of a reversed polarity (Bonhommet and Zähringer 1969). Thus, the terms "event" and "excursion" appeared in the works on paleomagnetology, although they had different meanings in the works of different authors. Here, we denote an event as a short (less that 10^5 years) interval of a geomagnetic polarity that is reversed in relation to the bordering intervals. An excursion of a geomagnetic field is called here a large-scale fluctuation [when the latitude of a virtual geomagnetic pole (VPG) lies below 45°] when it did not transform into a complete reversal. It is to be noted that it is sometimes difficult to distinguish an event and an excursion due to the high noise level.

Upon paleomagnetic examination of young (13,000-14,000 years old) strata in Sweden, Mörner (1974) established in this interval a polar excursion, Gothenburg (12,000 years ago). Later, anomalous fluctuations of the paleomagnetic field were registered in rocks of equivalent age in other regions (Verosub and Banerjee 1977). Another event of reversed polarity with a duration of 6,000 years (from 108,000 down to 114,000 years), called Blake, was identified by Smith and Foster (1969) from study of marine and oceanic columns.

It is difficult to determine at the present, how many events and polar excursions of normal polarity have been recorded in the Brunhes epoch. Thus, Japanese workers, on the basis of remanent magnetization of the lacustrine sediments of the Biwa Lake, revealed the occurrence of three or even more events of reversed polarity and of two polar excursions. The age of these events (104,000-117,000; 176,000-186,000; 292,000-298,000 years) was estimated by extrapolation according to the rate of sedimentation and C^{14} dating (Yaskawa et al. 1973).

Pospelova (1971) defined six horizons of negative magnetic inclination while examining a core of the Pleistocene strata in the Ob' River Plateau region. As only the magnetic inclination was measured, it was impossible to classify them into events or excursions. Zudin (1973) traced six narrow intervals of magnetization of anomalous direction in loess loam in the Ob' River Plateau region that were formed in the Brunhes epoch. Gamov and Pen'kov (1976) recognized six events of reversed polarity in a single section of the loess piles in the Tajik depression. Kulikova registered a polar excursion that occurred approximately 40,000 years ago in the Kargapolovo section in Western Siberia, which was also later recorded while studying strata of the same time interval in Moldavia (Kulikova and Pospelova 1976; Kulikova 1980).

Thus, it may be concluded that the geomagnetic field of the Brunhes epoch is of a complex structure, including several events of reversed polarity and some polar excursions. The most detailed studies were conducted in the Okhotsk Sea (Pospelova 1976b) and in the western part of the Atlantic region (near Haiti) (Denham et al. 1977), in the Pacific Ocean (Ninkovich et al. 1966), in Italy (Creer and Readman 1980) and the south of the European part of the USSR (Zubakov et al. 1982). These studies verified the complex structure of the Brunhes epoch, and established the existence of the Blake event and of several excursions in the second half of this epoch, which were established irrespective of the examination of the young strata sections in the continents.

Which of the events and polar excursions of the Brunhes epoch are valid? To answer this question, the following requirements are necessary: (a) the precise age; (b) tracing along several distant sections; (c) the synchroneity of magnetization of the rocks must be absolutely certain. Despite the fact that the material accumulated upon the examination of the magnetic field of the Brunhes epoch is sufficient, not a single polar excursion (or event), which was observed, can be considered reliably established. An event of geomagnetic field in the interval of 10,000–14,000 years ago (Gothenburg?) and in the interval of 100,000–120,000 years ago (Blake?) are the most reliable, because the reversed and anomalous directions of magnetization were recognized in more than 50 sections in the corresponding beds and in more than 30 sections throughout the world, respectively.

Unfortunately, the events and polar excursions recognized in the USSR (up to six in one section) do not have any distinct age reference; besides, we have no data to determine whether these were events or polar excursions.

The Pleistocene strata, made up of many lithologic varieties, certainly contain rocks possessing magnetization of reversed and anomalous directions. In some cases this magnetization might reflect a short-time change of the geomagnetic field in the Brunhes epoch, i.e. events of reversed polarity or excursions. Further examinations of the geomagnetic field in the Brunhes epoch are needed to clarify the problem of the validity of these events and their stratigraphic importance.

Excursions differ from events, first, by their short duration (less than 10^4 years), and second, by common preservation of the polarity of a geomagnetic field (see Sect. 3.1.3). This short duration is suggestive of the fact that any event or excursion of a geomagnetic field should be recorded in a minor portion of sections of a certain age range due to the discontinuity in sedimentation and in accordance with the theory of probability. Unfortunately, it reduces the stratigraphic importance of events and excursions, but to a certain extent it mitigates the requirements of their validity. The intensity of a paleomagnetic field during an excursion probably does not decrease as much as it does at the event boundaries (Verosub and Banerjee 1977). It is not certain whether the excursions can be considered as incomplete reversals; moreover, there is no available data to establish whether the excursions are fluctuations of a dipole field or whether they are to be entirely ascribable to a nondipole field. These problems might be clarified by the search and examination of excursions of more ancient epochs, down to the early Paleozoic. In this respect only preliminary results are available.

3.4.2 Paleosecular Variations of the Geomagnetic Field

Secular variations of the geomagnetic field were discovered by direct observations. These observations, however, started just recently (on a geologic time scale) and provide information only on the variations within periods of not more than several hundreds of years. The archeomagnetic studies widen to a great extent the range of harmonics of secular variations that are accessible for examination. However, only these paleomagnetic studies allow the recognition of variations with periods of 10^4–10^5 years.

Paleosecular variations in paleomagnetology are important in two aspects. On the one hand, their study brings about a better understanding of the mechanism of the geomagnetic field generation. According to Braginsky (1978), secular variations can be divided into three groups: (1) torsional (with periods of 10–60 years) (2) MAC waves (200 to several thousands of years); (3) the dynamo oscillations proper (about 8,000 years). Thus, the identification of secular variations with similar periods is an experimental corroboration of this theory. Besides, the recognition of periodic changes in the paleomagnetic series might be useful as a basis for detailed paleomagnetic stratigraphy. This type of work has already been conducted (Kochegura and Balesta 1979; Creer 1977). On the other hand, when related with tectonic problems, the paleosecular variations are the background "noise" from which the "signal", i.e. a situation of the paleomagnetic pole, is to be separated.

The studies of paleosecular variations can be divided into two groups. The first group consists of works related to the examination of magnetization of the dyke series or lava flows with an unknown time succession (Irving 1964). In this case the scatter of mean directions J_n on dykes (or flows) is separated against a background of general scattering. In general, this is done by the two-tier analysis of Watson (1956). This scattering might be due to different reasons, but the principal one is paleosecular variation. An angular standard deviation is taken as a measure of the paleosecular variation amplitude:

$$\theta_{63} \approx 81\ K^{-1/2}, \tag{3.1}$$

where K is the precision parameter associated with the mean directions J_n on dykes or lavas, (Watson 1956).

The second group includes studies of time (or frequency) characteristics of paleosecular variations. Igneous as well as sedimentary rocks are useful in this case. Besides Alekseeva et al. (1979) directed attention to a soil-pyroclastic cover as a useful material due to its precise age reference by the radiocarbon method.

Paleosecular variations are well known in Holocene strata. Some works (Creer et al. 1972; Creer and Kopper 1976; Thompson 1978) have made a great contribution to this knowledge; sediments of some European and North American lakes dated by the radiocarbon method have been examined in these publications, resulting in certain interesting conclusions. First, records of paleomagnetic field declination yield much better agreement with each other than those of inclination. This is probably explained by the fact that a parameter such as the inclination of remanent magnetization suffers much more from "noise" than declination (errors of inclination, sedimentation, etc., were mentioned earlier). Thus, the value of the signal/noise ratio for declination curves should be higher. Second, as a result of this work a period of 2700 years in paleosecular variations, recorded in the European lacustrine deposits, was established with sufficient reliability. Third, a westward drift was confirmed. And fourth, their results are in accordance with the model of Alldredge and Hurwitz (see Sect. 1.2.4).

Petrova and Burlatskaya (1979) analyzed data obtained from direct observations and upon archeo- and paleomagnetic examinations. The conclusion on the discreteness of a spectrum of secular variations is the most important result in this analysis. It provides a fundamental basis for the separation of paleosecular variations against a background of various disturbances forming a continuous spectrum.

Even employing a collection of all available data on paleosecular variations of ancient epochs (these are very few), only two conclusions have been reached. First, the amplitude of paleosecular variations in ancient epochs did not differ greatly from that of recent secular variations. Thus, in the Upper Carboniferous and Lower Permian rocks of the Donbas (Khramov et al. 1974), the amplitude of paleosecular variations calculated by two-tier analysis equals 15°, thus agreeing with the angular standard deviation of a recent geomagnetic field for a latitude $\varphi = 10°$ (the paleolatitude of the Donbas in the late Carboniferous-early Permian). The consistency in the materials of Baag and Helsley (1974b) is also satisfactory. Second, in different epochs the geomagnetic field drifted westward as well as eastward. While analyzing paleosecular variations, it is convenient to examine the behaviour of three groups of parameters: (1) the intensity and direction of a nondipole field; (2) the geomagnetic dipole moment (dipole oscillations); (3) the change in orientation of a geomagnetic dipole relative to the central axial dipole. Several models of paleosecular variations were suggested with different relations of the contributions of these three groups of parameters. Thus, in model A of Irving and Ward (1964), the dipole field is considered fixed and secular variations are related to a nondipole field. The amplitude H_N of a nondipole field is assumed constant and its direction from point to point changes randomly. Then, an angular standard deviation (in degrees) is obtained (Irving 1964):

$$\theta_{63} = 46.8(1 + 3\sin^2 \varphi)^{-1/2}(H_N/H_e), \tag{3.2}$$

where φ is the geographic latitude of the site; H_e is the value of a dipole field at the equator.

In model B of Creer et al. (1959) the secular variations, on the contrary, are represented by the wobbles of the axis of the central dipole. If the directions of this axis are distributed according to Fisher near the mean direction with the precision parameter K_1, then the precision parameter of field directions in a given point is

$$K = K_1(1 + 3\sin^2\varphi)\,(5 - 3\sin^2\varphi)^{-1} \tag{3.3}$$

Model C of Cox (1962) represents the superposition of models A and B: a nondipole field is added to a "dipole wobble". A general precision parameter of virtual poles is calculated by the formula:

$$1/K = 1/K_N + 1/K_D. \tag{3.4}$$

The most universal is model D of Cox (1970) that includes all three effects. The dipole and nondipole fields change in amplitude and direction. The complete angular dispersion of directions in this case equals

$$\theta_{63}^2 = (5/2)W_D^2 S_D^2 + W_N^2 [H_N(\varphi)/H_e]^2 (1+r^2)^{-3/2}, \tag{3.5}$$

where

$$W_D = [(1/5(5+3\sin^2\varphi)\ (1+3\sin^2\varphi)^{-2}]^{1/2};$$
$$W_N = (1+3\sin^2\varphi)^{-1/2};$$

S_D^2 is the angular dispersion of the dipole wobble; r is the ratio of the dipole oscillation amplitude to the mean dipole moment.

Baag and Helsley (1974b) proposed model E in which dipole and nondipole fields are related to the correlation coefficient C_{DN}. Then, the complete angular dispersion of the virtual poles would be

$$S^2 = S_D^2 + S_N^2 + 2C_{DN} S_D S_N. \tag{3.6}$$

Yukutake and Tachinaka (1968) examined the nondipole part of a geomagnetic field and divided it into stable and drifting components. In accord with the above authors, McElhinny and Merrill (1975) proposed another model of paleosecular variations: model M, in which a nondipole field is also subdivided into stable and drifting parts. Model M exhibits a better consistency with the recent latitudinal distribution of the angular dispersion in field directions if, as suggested, the amplitude of the "dipole wobble" is 9°.

3.5 Regularities in the Spatial Distribution of the Paleomagnetic Field

3.5.1 Paleomagnetic Field; Procedures of Isolation and Analysis

Two mutually related tasks should be clarified when examining a model of the central axial dipole. The first task consists in the analysis of the structure of time changes of the geomagnetic field and of the mode of formation of the mean (paleomagnetic) field, i.e. the analysis of secular variation morphology and geomagnetic field reversals in the geologic past. The second task consists in the analysis of the spatial structure of the paleomagnetic field in certain geologic epochs. The paleomagnetic data on the Quaternary period were repeatedly examined when a model of a central axial dipole was employed. It has been established that the paleomagnetic field in this period is very similar to the field of a central axial dipole (Opdyke and Henry 1969; Khramov and Sholpo 1967). This conclusion was not essentially affected when certain peculiarities in the temporal behaviour of geomagnetic field excursions and anomalous deviations were detected due to the quite short time span of these events.

Equations (1.3)–(1.5) give a one-to-one correspondence between the direction of a paleomagnetic field at a given point and the corresponding paleomagnetic pole. Experimental errors in determinations of the paleomagnetic pole and the possible presence of nondipole field components result in the dispersion of paleomagnetic poles even if these are calculated for a small area and for the same rock series. This dispersion might be estimated by the following values:

θ is the angular standard deviation from a mean pole; K is the pole precision parameter and A_{95} is the radius of the confidence circle for a mean pole:

$$K=(N-1)/(N-R);\ \theta=\arccos(R/N)\approx 81K^{-1/2};$$
$$A_{95}=140N^{-1/2}K^{-1/2}, \qquad (3.7)$$

where

$$R=[(\sum_1^N \sin\Phi_i)^2+(\sum_1^N \cos\Phi_i\cos\Lambda_i)^2+(\sum_1^N \cos\Phi_i\sin\Lambda_i)^2]^{1/2},$$

thus, the geometrical sum N of the radii vectors are the projections, their projections on the Earth's surface are the paleomagnetic poles with a geographic latitude Φ_i and longitude Λ_i.

Coordinates Φ and Λ of the mean paleomagnetic pole are estimated according to the rule of vector addition:

$$\sin\Phi=(\sum_1^N \sin\Phi_i)/R;\ \mathrm{tg}\Lambda=(\sum_1^N\cos\Phi_i\sin\Lambda_i)/(\sum_1^N\cos\Phi_i\cos\Lambda_i). \qquad (3.8)$$

With the few errors in individual paleomagnetic determinations, values θ, calculated for a set of paleomagnetic poles of a certain epoch, might be regarded as a measure of the nondipole part of a geomagnetic field of this epoch. Apparently, $\sin\theta$ is an estimation of the relative contribution of nondipole components of a paleomagnetic field.

Since a model of central axial dipole approximates a paleomagnetic field only at averaging intervals of the order of 10^5 years and more, it is necessary that every paleomagnetic determination is suitable for this averaging. Thus, to obtain data free of secular variations and the nondipole part of the ancient geomagnetic field, the paleomagnetic collection should represent a significant stratigraphic range, i.e. the samples should be taken from different geologic bodies (sheets, flows, intrusions) formed at different time intervals in a range of not less than 10^5 years.

The effects of ancient magnetic anomalies related to the Earth's crust might easily be excluded by observations covering sufficiently extensive areas. The small scale of these anomalies, the relatively rare occurrence of great amplitude anomalies among them and a zero mean value of the anomalous field in the anomaly area, all these facts, contribute to an easier solution of the problem. Practically, the possible effect of the ancient magnetic anomalies might be evaluated and eliminated by determinations in three to four sites (exposures, sections) distributed in an area of the order 100×100 km.

This procedure of averaging implies, as the main objective, the exclusion of temporal deviations and, thus, should be carried out at the level of beds (flows, intrusions). The mean direction of an ancient component of magnetization J_n^a of a formation (group) is calculated by the addition of direction cosines of the mean vectors J_n^a for every bed. The question of the minimal and optimal number of temporal levels (beds) then arises. Statistics should not be considered as legitimate, if the number of vectors and, consequently, the beds, are less than ten; everything else depends on the requirements of the determination precision, vector dispersion and the nature of changes of J_n^a from bed to bed.

Assuming a tolerable error in the determination of the mean direction of vector J_n^a, then $\alpha_{95} = 5°$ (for paleotectonic tasks, for instance, this precision is quite sufficient) and two limiting between the bed precision parameters $K = 10$ and 300 (proceeding from the actual performance of conducted work). Taking into account that $\alpha_{95} = 140N^{-1/2} K^{-1/2}$, we obtain the necessary number of beds; in the first case, $N = 78$ and in the second, $N = 3$. At the same time, it is evident that $N = 30$ is optimal at any precision parameter K, because, if it is increased further, the error α_{95} decrease very slowly; if the desired precision is not achieved, it is necessary to find a means of reducing the dispersion of vectors.

This calculation, as any other calculation, is based on the laws of distribution of random deviations, and is valid only for a random sampling of N beds from their whole set in the formation. Thus, sampling is inadmissible, if the samples taken from these beds represent a minor portion of a formation: then long-time variations would not be excluded. Besides, there is a risk in sampling a series of beds that were deposited at a much faster rate compared to the mean rate of sedimentation of this formation. Special studies have shown that this difference might amount to two to three orders.

3.5.2 Paleomagnetic Field During the Carboniferous and Permian and the Model of the Central Axial Dipole

The applicability of the model of a central axial dipole in the examination of the paleomagnetic field for ancient epochs was specially tested many times. The most impressive results were obtained while analyzing the spatial distribution of a paleomagnetic field in the late Paleozoic (Khramov et al. 1974). The Carboniferous and Permian systems include the most carefully inspected paleomagnetic determinations, thus providing details of the sections studied. Of 320 determinations 70 were of the lower and middle Carboniferous, 118 were of the upper Carboniferous and lower Permian and 132 were of upper Permian and Permo-Triassic. These data were obtained from all continents, except Antarctica, but their distribution is not equally dense. The most abundant determinations have been made from Europe and North America. Generally, paleomagnetic data from the Carboniferous and Permian are quite voluminous and, as a rule, have satisfactory reliability and good precision.

Analysis of the global-scale regularities in the distribution of the Earth's magnetic field during the Carboniferous and Permian periods is difficult because of the necessity to choose one of two rival concepts: fixism or mobilism. Additionally, even accepting the mobilistic idea, it should be noted that mobilistic constructions do not yet possess the desired high precision: reconstructions of the relative situation of continental blocks in the past have varied considerably in the works of different authors (see Chap. 4). Thus, territories appear to be constrained to the epi-Hercynian platforms, for they represent the rigid portions of the Earth's crust since the late Paleozoic and did not undergo deformation, as recognized in both concepts.

The North-Eurasian epi-Hercynian platform is the most extensive and best known in detail from the point of view of paleomagnetic studies. The paleomagnetic directions and paleomagnetic latitudes demonstrate good agreement within this platform (Fig. 3.15), and change regularly from west to east. Paleomeridians, exhibiting an almost recent orientation in western Europe, i.e. northerly but becoming more easterly in direction, the more easterly the site is located. At the same time, paleolatitudes change from the equatorial to temperate and boreal, extending to the extreme east of the platform 50°N in the early-middle Carboniferous and 70°N at the Permian-Triassic boundary.

This pattern shows a qualitative correspondence the field of a central dipole with a geomagnetic pole in the northwestern part of the Pacific Ocean. Quantitative characteristics of this correspondence can be estimated by the precision parameter K = 12–104 (for different epochs), which is seen in the distribution of the paleomagnetic poles of northern Eurasia around the mean pole (see Appendix). Probable deviations of the ancient field from the field of a central dipole might be evaluated by a plot (Fig. 3.16) demonstrating the dependence of the paleomagnetic inclinations I from the paleomagnetic latitude of the site, calculated from the mean paleomagnetic pole of the epoch.

The same regularities are observed in the distribution of paleomagnetic directions for other platforms, but these are not as distinct because the intervals of paleomagnetic latitude in these platforms are much narrower than those in the North Eurasian platform and because the number of sites of the paleomagnetic determinations is less. Nevertheless, the paleopole precision parameter K = 42–52 for the North American platform shows that in this case the extent of correspondence of the ancient field to the field of a central dipole is approximately the same as for Eurasia.

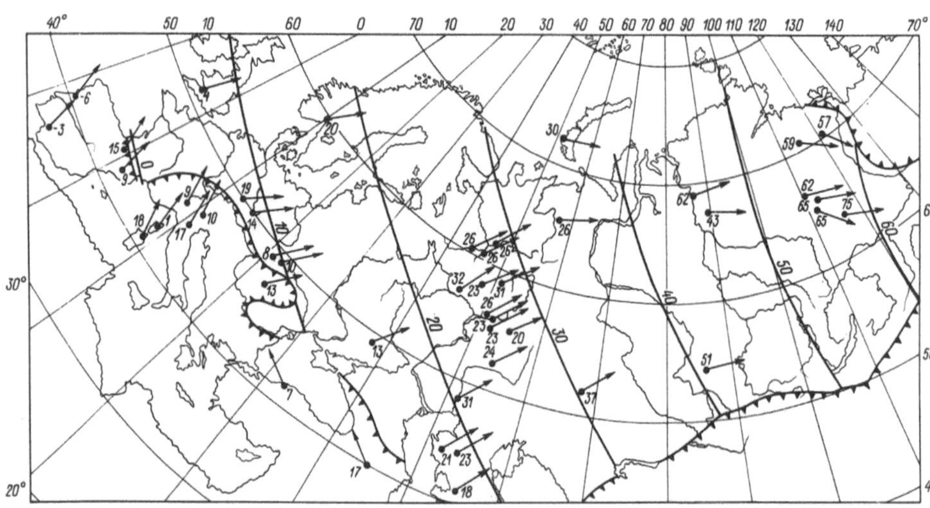

Fig. 3.15. Paleomagnetic directions and paleolatitudes in northern Eurasia for the late Permian

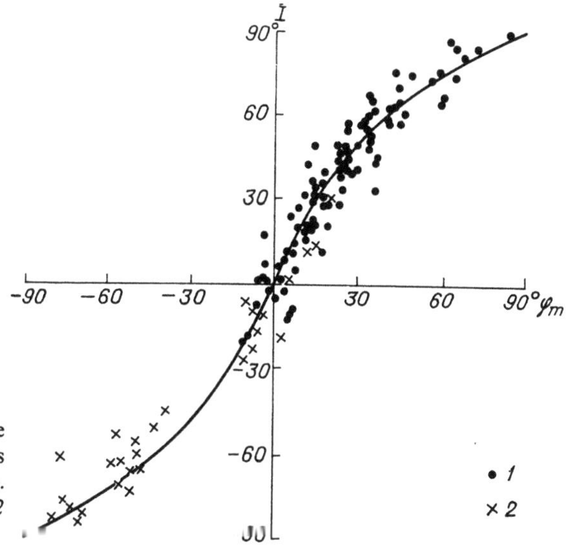

Fig. 3.16. Divergences of the late Paleozoic geomagnetic inclinations from the axial dipole field inclination. (Khramov et al. 1974). *1* Eurasia; *2* other continents

Within the southern parts of Europe and Asia, forming a zone of the Tethys seas and Mongolian-Okhotsk fold belt, the mosaic-like Chinese platform and the arcs of the Japanese Islands, the consistency of paleomagnetic directions and poles is much poorer than in the northern part of Eurasia. This scatter of paleopoles might be caused by horizontal tectonic deformations (rotations of rock mass) that affected the scatter of paleodeclinations much more than that of paleoinclinations (see Fig. 3.15).

Thus, the following conclusions have been made:

1. The paleomagnetic field within each epi-Hercynian platform approximates closely to the field of a central dipole.
2. Dipole axes for each platform are various; these discrepancies exceed the errors in the determination of their situation.
3. In the tectonically active zones where post-Permian movements are registered, the consistency of paleomagnetic declinations is worse than it is on the platforms; the agreement of paleomagnetic inclinations is approximately the same as in the platforms.

For the well-studied North Eurasian and North American platforms, the paleopole precision parameter K = 38–90 (for the late Carboniferous-early Permian interval), $\theta = 9$–13 and a nondipole component, 16–22% of the total field, correspond to these observations. As the errors in a single determination of this evaluation are ignored, the 16% might be considered as a possible upper limit of the nondipole component of a field. These values appear to be underestimated only when the ancient anomalies were much more extensive than the continents.

The question arises whether the axis of a dipole, approximating to a constant part of the geomagnetic field in time, coincided with the Earth's rotational axis. Naturally, this answer cannot be found from data on the Earth's magnetism, since information from other branches of geophysical and historical geology are involved. These may be grouped as follows:

1. Distribution of sediments and other indicators of paleoclimate (laterite, arid red beds, salt, glacial formations).
2. Distribution of past cold- and heat-loving fauna and flora.
3. Directions and peculiarities of paleowinds and paleocurrents that depend on the position of the Earth's rotational axis.
4. Distribution of active zones in the crust that could have been associated with the "critical parallels" during deformation of the Earth's ellipsoid under effects of changes in the rotation velocity.
5. Orientation of asymmetry of annual rings of tree stems in situ.

The data directly related to the late Paleozoic (except for the scant data on the fourth and fifth groups) will be briefly discussed here.

Irving (1964) analyzed the distribution of the most typical paleoclimate indicators (halogenes, coals and glacial formations) along the paleomagnetic latitudes and concluded that this distribution is easily explained proceeding from the supposition that the paleomagnetic latitudes coincided with the paleogeographic ones. This means that the axis of a geomagnetic dipole was directed along the Earth's rotational axis. It should be noted that the paleoclimatic pole determined by Strakhov (1963) for the late Paleozoic lies just in the middle of the paleomagnetic poles of Europe, Africa and North America for that time (Khramov et al. 1974). The distribution in paleomagnetic latitudes of the paleoclimate indicators, such as coal-bearing strata, has been studied in detail for all periods (see Chap. 6). Analysis of these distributions also supports the suggestion of a coincidence of the paleomagnetic and paleogeographic axes during a major portion of geologic history, including the Carboniferous and Permian periods.

The directions of trade winds in North America and Europe determined from dune studies, and also data on the peculiarities in cross-bedding structures that are governed by the Coriolis forces affecting paleocurrents, yield good agreement with the situation of the paleomagnetic equator (Khramov et al. 1974). All the examined data permit the following conclusion: the Carboniferous and Permian results fully support the hypothesis of a central axial dipole.

3.5.3 Asymmetric (Anomalous) and Axially Symmetric Nondipole Fields

All data on the Carboniferous and Permian agree with the hypothesis of a central axial dipole, which does not conflict with the physical theories of the Earth's magnetism. From the viewpoint of the central axial dipole hypothesis, the divergences of paleomagnetic poles determined for different continents might be explained by the continental drift. Other models of the paleomagnetic field, however, should be examined; they can also explain these divergences.

Possible causes of intercontinental discrepancies in paleomagnetic data might be: continental magnetic anomalies of long duration, an axially symmetric nondipole field or a field of a noncentral axial dipole. The assumption of long-duration, magnetic anomalies makes more assumptions necessary and, thus, is hardly acceptable. Rusakov (1969) was the first to examine this hypothesis. He concluded that, in general, this hypothesis does not contradict modern, physical theory of the Earth's magnetism or paleomagnetic data. However, it was shown later that this hypothesis challenged the known reliability of certain statements (Khramov 1970). The hypothesis of continental anomalies denies, in particular, a latitudinal dependence on the amplitude of paleosecular variations, typical of a dipole field. This relationship corresponds exactly with the supposition of secular variations in the Carboniferous and Permian caused by a rotation of an equatorial dipole, the magnetic moment of which was made one-fifth of the moment of the axial dipole. The observations for Gondwanan continents indicate that the amplitude of variations are typical of the latitude φ_m calculated from the inclination I at a site (Fig. 3.17). Thus, Australia in the late Paleozoic actually was at high geomagnetic latitudes and inclinations of the order 70°–80° cannot be explained by a continental anomaly (Khramov et al. 1974; Khramov 1970).

The assumption of an axially symmetric nondipole geomagnetic field appears more appropriate. Earlier, Cox and Doell (1960) took into consideration that the Permian paleomagnetic meridians of North America and Europe in the Northern Hemisphere cross at a point situated between the region of observation and a corresponding northern paleomagnetic pole. Later, Irving (1964) and then Rusakov (1969) established that the paleomagnetic meridian of Australia intersects with the paleomeridians of North America and Europe in the same regions. Furthermore, Van Hilten (1963) discovered that the paleomeridians drawn from different points of the same continent reveal similar features, though to a less extent. New data verify this regularity (Fig. 3.18). It is notable that the Carboniferous, Permian and Triassic paleomagnetic meridians of all southern continents also cross in the same manner.

The intersection of paleomeridians in the same area evidences an axial symmetry of a geomagnetic field in the late Paleozoic. The inconsistency of the observed paleoinclinations and those calculated from the dipole formula depends on two alternative suppositions: (1) a symmetric field relative to the axis, diverging significantly from a dipole field, and thus asymmetric relative to the equatorial plane; (2) the Earth's radius was shorter than that of Recent time (paleoinclinations in the Northern Hemisphere change more rapidly than expected according to the dipole law).

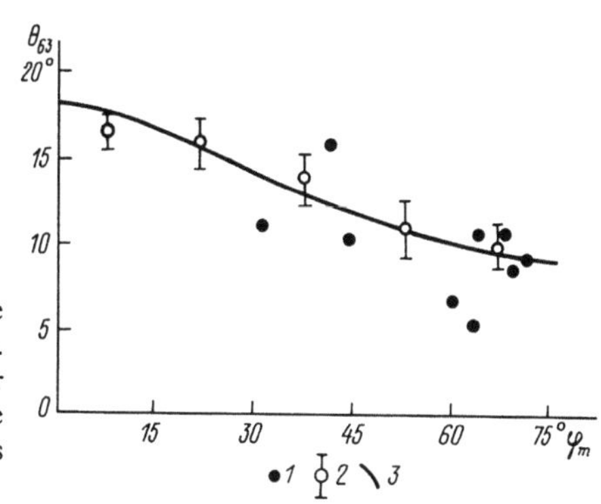

Fig. 3.17. Amplitudes of the secular variations in the Carboniferous and Permian for Gondwana. (Khramov et al. 1974). *1* Observed values for Gondwana; *2* mean values in 15° intervals for the northern continents; *3* variation according to Cox's model A

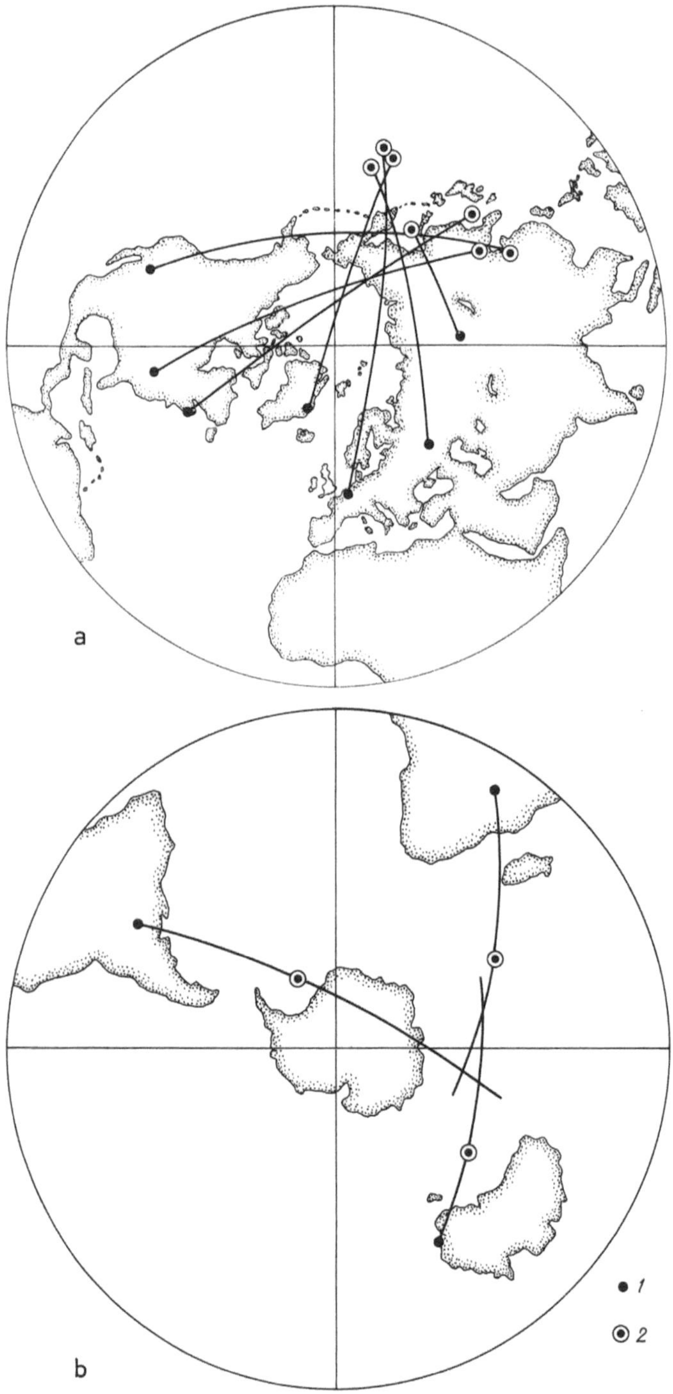

Fig. 3.18a,b. The Permian paleomeridians. (Khramov et al. 1974). **a)** Gondwanan continents; **b)** northern hemisphere. *1* Centre of the site; *2* mean paleomagnetic pole

3.5.4 Model of Noncentral Axial Dipole and Its Testing

Let us examine the simpliest axially symmetric field that would appear suitable for an approximation of a geomagnetic field in the Carboniferous and Permian without the assumption of differentiated movements of continental blocks relative to the Earth's rotational axis and without an hypothesis on global expansion (Khramov et al. 1974). On a sphere with a radius R the inclination I and field intensity H of an axial dipole with a magnetic moment M placed at a distance a from the sphere centre are distributed in the following way with respect to the geographic latitude φ :

$$H = MR^{-3}(1 + 3\sin^2\varphi - 8b \sin\varphi + 4b^2)^{1/2}(1 - 2b \sin\varphi + b^2)^{-2}$$
$$\mathrm{tg}\, I = [2\mathrm{tg}\varphi - 3b\sec\varphi + b\,(2b - \sin\varphi)\mathrm{tg}\varphi]/[1 - b(2b - \sin\varphi)]; \qquad (3.9)$$
$$D = 0,$$

where $b = a/R$ is a northward relative displacement of the dipole.

A model of the noncentral axial dipole was often used for the interpretation of paleomagnetic data (Briden et al. 1970; Khramov 1970; Wilson and McElhinny 1974). Two features of the field of this model need to be considered (Fig. 3.19).

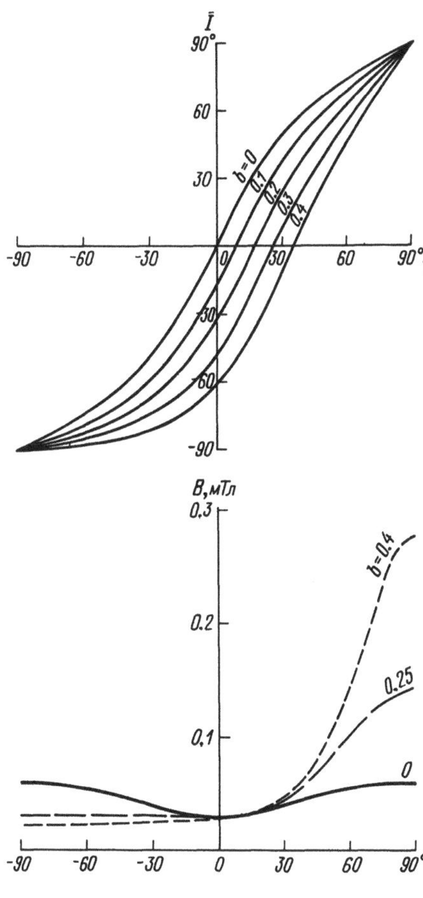

Fig. 3.19. Inclination I and intensity B of the offset axial geomagnetic dipole versus the latitude for different northward displacements b of the dipole. Values b are expressed as a fraction of the Earth radius. (Khramov et al. 1974)

1. In the latitudinal interval O°–30°N at any displacement b (from 0 to 0.4), the values of H are almost the same.
2. Angular slopes of curves I (φ) in an inclination interval from –40° to +40° are fairly similar. Therefore, the distribution of paleoinclinations in over a wide interval (of the order of 80°, i.e. about 5000 km along a meridian) obeys the law of a central axial dipole. Consequently, if the observations were conducted in the territory where the inclinations change from –40° to +40°, the fields of noncentral and central axial dipoles are not distinguishable when the dipole displacement amounts to even 0.4R, i.e. 2500 km from the centre of the Earth.

Just this distribution of a field at b = 0.25–0.4 closely resembles the pattern of a geomagnetic field in the Carboniferous and Permian. Actually, in the continents of the Northern Hemisphere (North America, Europe) paleoinclinations from –30° to +50° have been recorded and their distribution corresponds fairly well to the law of a central dipole; in the Southern Hemisphere (South America, southern part of Africa, Australia), on the contrary, large negative inclinations have been registered that do not agree with those extrapolated from the Northern Hemisphere according to the law of a central dipole. Unfortunately, only initial studies are in progress on the history of intensity changes of the Earth's magnetic field in the distant geologic past; the obtained determinations are few and their reliability is low (see Sect. 3.6). Thus, just the following facts should be mentioned in favour of an assumption of a field of a noncentral axial dipole displaced northward in the Carboniferous and Permian:

1. Small paleoinclinations in the Northern Hemisphere and large, negative inclinations in the Southern Hemisphere.
2. Intersection of paleomeridians of almost all continents in a single area indicating an axial symmetry of a field.
3. A closer situation of the northern polar area, defined by the method of paleomeridian intersection, to any northern continent and a more distant situation of the corresponding southern polar area to the southern continents compared to that expected from the values of paleoinclinations in the continents.

It can be concluded that the hypothesis of a noncentral axial dipole yields a satisfactory consistency for the paleomagnetic data of the late Paleozoic, but a requirement for this consistency is the maximal theoretical displacement of a dipole from the centre of the Earth to the ancient north. Models of quadrupole fields examined by Briden et al. (1970) and models of several dipoles show poor correspondence to a geomagnetic field in the late Paleozoic.

All the above stated factors favour the probability of an assumption that a stable part of the major geomagnetic field in the Carboniferous and Permian represents a field of central axial dipole. Paleoclimatic data and the striking agreement of paleomagnetic data for all southern continents in absolutely independent reconstructions of Gondwanaland are also arguments for this assumption.

Thus, all paleomagnetic data on the Carboniferous and Permian can be regarded as supporting the hypothesis of central axial dipole. At the same time, the probability of a dipole displacement cannot be excluded at the moment. The two solutions discussed are not alternatives: several solutions with various combinations of scales of dipole displacement and continental drift exist between them. It seems probable that the scales of the events might be selected in such a way that they would correspond to the observations most fittingly. It is obvious, however, than an exact solution can be found only upon analyzing these events from a historical viewpoint, i.e. upon interpretation of paleomagnetic data for all periods.

3.5.5 Spherical Harmonic Analysis of Paleomagnetic Data

Spherical harmonic analysis is the most objective method for studying of the spatial structure of a geomagnetic field. However, this analysis, elaborated by Gauss in its classical form (see Sect. 1.2), is not directly applicable to paleomagnetic data, because determinations of intensity of an ancient geomagnetic field are as yet too few and episodic to characterize a paleomagnetic field, excluding the secular variations. A modification of spherical, harmonic analysis involving angular elements of geomagnetism, i.e. declination D and inclination I, has been specially developed (Adam et al. 1975; Beňkova et al. 1973).

An assumption is required for the first coefficient of expansion of a magnetic potential [see Eq. (1.2)] which defines the field of a central axial dipole, not equal to zero, $g_1^o \neq 0$, in the spherical, harmonic analysis (SHA) using angular elements D and I. The results of the analysis of D and I are represented in the form of relative Gauss coefficients:

$$G_n^m = g_n^m / g_1^o \text{ and } H_n^m = h_n^m / g_1^o .$$

The first question that appears concerns the unambiguity of SHA using angular elements. It was shown, if for the entire surface of the Earth, the direction of the field, determined by two angles, D and I, is known, then the obtained values of G_n^m and H_n^m are unambiguous (Kono 1973). However, if only one of the angles, D or I, is known, then a reliable, unique solution is not possible. And moreover, the ambiguity in the determination of G_n^m and H_n^m by using only declination D is tested (Kono 1973). Therefore, the SHA of paleomagnetic data is accomplished by using values of inclination I or both angles, D and I.

An analysis by means of inclination I includes the following: in an expression that relates I to the field components in the Cartesian coordinates X (northern), Y (eastern) and Z (vertical):

$$tg^2 I = (X^2 + Y^2)/Z^2 ;$$

X, Y, Z are substituted into spherical functions and then all terms are divided by g_1^o. If value I in N points is known and $N \gg K$ (where K is a number of unknown coefficients G_n^m and H_n^m), we have a system of N inhomogeneous equations of the second order. Its solution is too awkward and thus a linearization is applied. The terms of the second order and higher are neglected according to the assumption that the geomagnetic field is essentially of dipole nature and therefore $|G_n^m|$, $|H_n^m| < 1$. The system of linear equations obtained is solved by the least square method and then, for making the values of G_n^m and H_n^m more precise, a method of iteration is applied (Adam et al. 1975; Ben'kova et al. 1973).

This technique of SHA of the paleomagnetic field of the last two geomagnetic epochs, Brunhes (normal polarity) and Matuyama (reversed polarity), revealed the general dipolar nature of a paleomagnetic field. This field, with an error of 5°, is approximated by a dipole field oriented along the Earth's rotational axis and insignificantly (200–400 km) displaced from its centre (Adam et al. 1975).

Thus, the initial data for the hypothesis of a central axial dipole were apparently valid. Similar results were obtained for the Neogene (Table 3.4).

Table 3.4 Results of spherical harmonic analysis of the paleomagnetic field. (After Adam et al. 1975; Ben'kova et al. 1973)

Period, epoch	Paleomagnetic pole Latitude	Longitude	Offset of geomagnetic centre, km	Harmonic amplitude ratio C_2/C_1	Mean inclination difference between observed and synthetized fields
Q (Brunhes)	89°	135°	491	0.14	5°
N_2+Q_1 (Matuyama)	87°	203°	343	0.11	4°
Q (Total)	87°	254°	454	0.16	5°
N	83°	254°	659	0.19	10°
T	65°	158°	1055	0.33	17°
T(r)[a]	89°	196°	824	0.22	10°
P	77°	122°	1032	0.31	28°
P(r)	87°	337°	863	0.24	7°
C_2+P_1	57°	172°	1250	0.38	26°
C_3+P_1(r)	79°	234°	867	0.24	8°

[a]Index r denotes analyses using mobilistic reconstructions.

The closeness of the paleomagnetic and geographic poles ($\Phi = 83°$–$89°$), the small displacement of the magnetic centre from the Earth's centre (400–700 km), the dominance of the first harmonics ($C_2/C_1 \ll 1$) and the slight difference between the observed field and the field synthesized by only two harmonics, all indicate the validity of a model of a central axial dipole for the geomagnetic field during the last 25 m.y.

The applicability of a central axial dipole model for the paleomagnetic field of more ancient epochs was also specially tested by the SHA technique (Adam et al. 1975; Ben'kova et al. 1973). Paleomagnetic fields of different geologic epochs were analyzed for the situation of the present continents as well as for mobilistic reconstructions independent of paleomagnetic data. Evidently for the Permian-Carboniferous, Permian and Triassic periods, the analysis using reconstructions approximates the geomagnetic field to the field of a central axial dipole much more than it does using the present geographic grid. The pole latitude increases from

57°–65° to 79°–89°, the displacement of the geomagnetic centre decreases from 1000–1200 km to 800 km and the ratio of the second to the first harmonics and differences between the observed and synthesized fields decrease 1.5–4 times.

Thus, the use of mobilistic reconstructions simplifies, to a great extent, the global structure of the paleomagnetic field which approximates well (with an error of 7°–10°) with a field of a dipole oriented along the Earth's rotational axis and somewhat displaced from its centre. An unexpected feature in the results was a small, but constant, asymmetry for almost every epoch of the paleomagnetic field: the magnetic centre of the Earth is always displaced toward the western part of the Pacific Ocean (Ben'kova et al. 1973). This asymmetry does not greatly affect the geotectonic aspects of paleomagnetic data interpretation, but it might be of great help in the study of the Earth's magnetism. More complicated models than that of a central axial dipoles were recurrently applied in the interpretation of paleomagnetic data (Briden et al. 1970; Coupland and Van der Voo 1980). However, it is not as yet certain whether this asymmetry is caused by the geomagnetism proper or whether it is caused by the effects of inaccuracy in reconstructions, peculiarities in the distribution of the initial data, etc. Further studies are imperative in this field. It should also be mentioned that a model of a central axial dipole fulfills two basic requirements of the modern dynamo theory of the Earth's magnetism: the general subaxial symmetry of a geomagnetic field and the absence of its precise symmetry for each separate time moment.

3.6 The Intensity of the Geomagnetic Field in the Phanerozoic

Preliminary studies on the history of changes of the geomagnetic field intensity H_a in the geologic past are still in progress. Determinations of H_a display an irregular coverage in the geologic time scale: the majority falls in the late Cenozoic, Triassic and late Paleozoic, whereas very few determinations for the early Paleozoic and Precambrian are available. Results of the first determinations of H_a are included in the work of McElhinny (1973). They revealed (despite great scatter) an obvious trend of decrease in the field H_a back to the early Paleozoic from the present and an increase in the Precambrian back to the present level. These determinations were carried out on dated magmatic rocks in the Canadian Shield and cover an interval of 2700 m.y. (Smith 1970). Thus, the assumption that the Earth's core was formed largely in the early stages of the Earth's evolution appears to be substantiated (Schwarz and Symons 1969). The reliability of this conclusion, however, is not satisfactory, because the determinations on which it is based are very few and they were carried out, as a rule, on very old and altered rocks. The Phanerozoic, specially its Meso-Cenozoic part, is much better known from the viewpoint of H_a changes, which is mainly due to new determinations done by Bol'shakov and colleagues, who used modified and more refined methods (Bol'shakov and Solodovnikov 1980a; Bol'shakov et al. 1978; Bol'shakov and Solodovnikov 1980b). The determinations of geomagnetic field intensity available on the Phanerozoic rocks in the USSR are given in Table 3.5

The redeposition method, which has been tested and reliably established for slightly lithified, terrigenous sedimentary rocks, and for thermomagnetized

Table 3.5 Geomagnetic field intensity in Phanerozoic time (USSR data)

Sampling area	Age	Method of determination	B_a, μT Range	Mean	Reference
Armenia	Q	Thellier, He	38–48	45	Petrova et al. 1979
	N_2	Thellier, H_e	30–34	32	Petrova et al. 1979
	P-Q	Thellier	21–52	40	Bol'shakov and Solodovnikov 1969
	N-Q	Thellier	28–29	29	Minasyan 1973
	N	Thellier	16–52	38	Bagina et al. 1973
	N	Thellier	–	36	Bol'shakov and Solodovnikov 1980a
Trans-Carpatian	N	Thellier	20–60	35	Tsykora et al. 1973
Armenia	K_2	Thellier	28–32	31	Bol'shakov et al. 1978
	K_2	Thellier	26–31	28	Bol'shakov et al. 1981
	K_2	Thellier	–	34	Bol'shakov and Solodovnikov 1980a
	K_2	Thellier	–	19	Bol'shakov and Solodovnikov 1981b
	J_3	Thellier	11–21	19	Bol'shakov and Solodovnikov 1980b
	J_3	H_e, Shaw	15–26	20	Petrova and Bagina 1984
Cis-Urals	T_{2-3}	Redeposition		28	Slautsitais 1969
Donets Basin	T_1	Redeposition	36–37	37	Slautsitais 1969
Tunguska	T_1	Thellier	35–45	40	Bol'shakov and Solodovnikov 1975
Taimyr	T	Thellier	–	34	Bol'shakov and Solodovnikov 1975
Norilsk region	T_1 }	Stepwise re-	{ –	20	Metallova and Shashkanov 1978
Norilsk region	P_2 }	magnetization	{ –	30	Metallova and Shashkanov 1978
Volga region	P_2	Q_o	25–32	28	Burov and Boronin 1973
Cis-Urals	P_2	Redeposition	25–31	28	Shmeleva 1963
Mangyshlak	P_2	Redeposition	41–47	44	Slautsitais 1969
Donets Basin	P_1-C_3	Redeposition	47–49	48	Khramov pers. comm.
Moscow region	C_2-C_3	Redeposition	42–44	43	Khramov pers. comm.
Moscow region	C_3	Anisotropy	39–43	41	Ivkin and Aparin 1980
Tikhvin region	C_1	Redeposition	20–24	22	Shmeleva 1963
Main Devonian field	D_3	Redeposition	0.6–3.7	1.9	Pogarskaya and Komissarova 1977
Minusinsk region	D_3	H_e	1.4–5.0	2	Pogarskaya and Komissarova 1977
Minusinsk region	D_1	H_e	2–9	6	Komissarova pers. comm.
Mugodzhary	D_2	H_e, Van Zijl	1.7–11.8	7	Burakov et al. 1984
Dniester River	D_1	H_e	1.0–3.4	2.2	Pogarskaya pers. commun.
Dniester River	S_1	H_e	5.3–6.7	6	Pogarskaya pers. commun.
Lena River	O_{2-3}	Redeposition	–	18	Komissarova and Rodionov pers. commun.
Lena River	O_1 }	Stepwise re-	{ –	40	Metallova et al. 1984
Lena River	$Є_3$ }	magnetization	{ 24–71	38	Metallova et al. 1980
Volyn' region	V	Redeposition	43–70	67	Komissarova and Chumakov pers. commun.

magmatic rocks, is called the Thellier method. The results obtained by other methods can also be controlled by these data. The precision of every method (see Sects. 2.7.2–2.7.7) is determined by the deviation of some H_a values from the mean value, but the major factor is the recognition of the nature of magnetization J_n and of the preservation grade of the primary magnetization.

Data on the field paleointensity of the upper Cenozoic rocks (from 15 m.y. to the late Quaternary) yield H_a values similar to the recent field intensity (Bol'shakov and Solodovnikov 1980a). All determinations of H_a were done separately on rocks of normal and reversed magnetization; they verified the equivalence (according to intensity and variations) of the normal and reversed geomagnetic field. These data are however, insufficient for the construction of a good model of changes of geomagnetic field intensity in the Phanerozoic, but the principal trends are obvious (Fig. 3.20). In the Vendian and late Paleozoic-Triassic the maxima of the geomagnetic field have been recorded. At that time the field was higher in comparison to the recent. The highest H_a values were obtained for the Vendian (the intensity was almost twice the present) and the late Carboniferous-early Permian interval (1.5 times higher). In the post-Vendian period the geomagnetic field intensity started to decrease and in the late Cambrian it equalled the recent one. In the middle Ordovician the geomagnetic field decreased to one-half of the recent field. In the middle Paleozoic the lowest values were recorded, the intensity in the Devonian, estimated by different methods and on various rocks was, on the average, 12–15 times lower than the recent one. Since the end of the middle Paleozoic, the geomagnetic field intensity started to increase and reached its maximum in the late Paleozoic. In the late Jurassic the lowest values for the Mesozoic-Cenozoic were recorded, they are two times lower than those of the recent field. Since the end of the Cretaceous the geomagnetic field intensity has become similar to the recent intensity. These results are indicative of the cyclic nature of changes of geomagnetic field intensity: the minima and maxima of H_a are repeated with an interval of about 220 m.y.

In general, the changes in the geomagnetic field intensity, registered for the magmatic rocks of the Canadian Shield by Thellier's method (Schwarz and Symons 1969), are roughly similar to those obtained for the USSR. The H_a values in the Canadian Shield are higher compared to other determinations, however, the

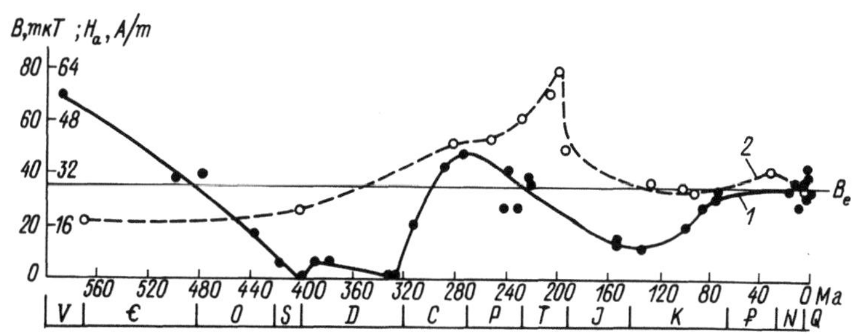

Fig. 3.20. Changes of geomagnetic field intensity during the Phanerozoic. *1* Extrusive and sedimentary rocks from the USSR (data from Table 3.5); *2* Canadian Shield intrusive rocks. (After Schwarz and Symons 1969)

minimum is recorded in the early-middle Paleozoic and an increase in the field intensity, in the late Paleozoic-early Mesozoic.

The evaluations of the geomagnetic field intensity are few and irregular for different time intervals and, thus, only certain preliminary conclusions can be made:

1. The geomagnetic field intensity changes with geological time. This change is of a cyclic nature with a period of about 220 m.y.
2. The maxima of H_a are outlined at the boundaries of the Vendian-Paleozoic, late Paleozoic-early Mesozoic, Neogene-Quaternary and the minima in the middle Paleozoic and late Jurassic.
3. The highest values of H_a were obtained in the USSR for the Vendian-Paleozoic boundary (twice as high as the present); the lowest values were obtained for the late Devonian (6–10% from the Recent). Possibly all of the H_a values are somewhat underestimated, because of the "ageing" of magnetization J_n (its partial destruction during the evolution of a rock). The increase of H_a in the early Paleozoic-Vendian indicates that a uniform lowering of H_a values does not occur due to ageing of the rocks; this could be a direct indication that the ageing of magnetization J_n increases for ancient rocks.
4. Many workers, who have made evaluations of H_a at the boundaries of the magnetic zones (in the transitional zones) have noted a decrease of the geomagnetic field intensity in comparison with the value at the steady field (see Sect. 3.3).

The regularities in the temporal changes of the geomagnetic field intensity H_a in the geologic past are very likely to be related to the processes in the Earth's interior and in particular, at the core-mantle boundary. Therefore, the knowledge of these regularities is of great implication for the physical theory of the Earth's magnetism and of the general endogenetic process. Besides, the geomagnetic field intensity (its mean value and amplitude of oscillations at certain time intervals) seems to be an important parameter in the paleomagnetic geochronologic scale. Further improvement of the methods in the determination of H_a and the accumulation of data on the regularities of this parameter change, for the ancient epochs in particular, are imperative in future.

CHAPTER 4

Paleomagnetism and Plate Tectonics

4.1 Paleomagnetic Studies of Plate Movements and Fold Belt Tectonics

4.1.1 Application of Paleomagnetic Methods in Tectonics

Tectonic movements are examined according to their effects, i.e. according to relative and absolute mass transport. To interpret this transport, the original position of these masses must be known. For this purpose the sea level and horizontal surfaces of the bedding are the most reliable reference data. Therefore, vertical movements and folding are the best known in tectonics. The interpretation of relative, horizontal displacements is more difficult because a detailed differentiation of a section is also necessary. Common geologic methods are almost useless in recognizing relative and absolute horizontal rotations and absolute horizontal displacements, i.e. movements relative to the geographic grid. Probably, this difficulty in defining the horizontal movements explains the fact that many tectonic experts deny the fundamental importance of those movements in the formation of the actual image of the Earth.

Paleomagnetic observations are quite important in tectonics just because of their usefulness in defining such horizontal movements and, thus, this deficiency in tectonic reconstructions is avoided by these means. Paleomagnetic methods involve two factors in the studies of horizontal tectonic movements. The first one is the regularity of a paleomagnetic field (on a global scale it is expressed in the dipole character of the field) and the presence of records of a geomagnetic field in rocks. These are the components of natural remanent magnetization that can be recognized and compared with certain stages in rock formation. The second factor is the existence of geomagnetic reversals with a certain regularity in the past, which have been recorded in rock sequences as zones of normal and reversed magnetization.

Thus, any movements associated with the rotation of rock masses would necessarily bring about corresponding rotations of paleomagnetic vectors. Latitudinal displacements of massives would also result in the inconsistency of paleolatitudes calculated from paleomagnetic inclinations, when compared with the present latitudes of the massives. Shear deformation and the lithospheric growing process would result in a pattern change of isochrons, which are marked by the boundaries between the zones of normally and reversely magnetized bodies. Just this provides the possibility of using paleomagnetic methods in the study of horizontal tectonic movements of a different scale: folding, horizontal deformations in fold zones and island arcs, continental drift and seafloor spreading.

Thus, the tectonic movements, which have been studied using paleomagnetic methods, are generally divided into three classes: local, regional and global. The following aspects determine the designation of these movements to one of these classes (Bazhenov 1980).

Local Movements. (1) Block size is no more than several tens of kilometers. (2) A block could be move as an entity or could be deformed. (3) Block movements are not necessarily parallel to the Earth's surface; masses are rotated along an axis of any direction. (4) Paleomagnetic data are usually available only for a certain short interval of geologic time. (5) It is quite sufficient to confirm the predeformational age of an isolated NRM component. (6) The paleomagnetic field within the block is homogeneous. (7) The interpretation of results is based on the paleomagnetic directions (the directions of isolated NRM components, i.e. ChRM). (8) The following movements are studied: body shape changes, any rotations, tilting and rotations around nonvertical axes. (9) Only relative movements are determined.

Regional Movements. (1) Block size amounts to hundreds of kilometers. (2) Blocks are not necessarily rigid; deformation could be possible in any part of an area. (3) Displacements occur only parallel to the Earth's surface, which means the rotational axes are vertical. (4) Paleomagnetic data are usually available for two to three (rarely more) periods of time. (5) In certain cases the exact age of the isolated NRM component must be known, in others, its prefolding age is sufficient. (6) The paleomagnetic field within the block is practically homogeneous. (7) The interpretation of results is possible according to paleomagnetic poles and directions. (8) The following movements are studied: body shape changes, rotations, latitudinal displacements; relative as well as absolute (relative to the geographic grid) movements are examined. (9) The amplitudes of the movements under study vary over a wide range.

Global Movements. (1) Block size amounts to thousands of kilometers. (2) Blocks are rigid; their shape and size change as a result of fracturing and collision of blocks. (3) Displacements of blocks occur only parallel to the Earth's surface, whereby the rotational axis extends through the centre of the Earth. (4) Paleomagnetic data are available for large intervals of geologic time with more or less equal coverage. (5) The exact time of formation of an isolated component of rock magnetization must be known. (6) The paleomagnetic field within the block varies greatly. (7) The interpretation of results is done by analyzing the positions of the paleomagnetic poles, therefore, the dipole character of the paleomagnetic field is a necessary requirement. (8) The following movements are studied: latitudinal displacements, the determination of the relative position of blocks. (9) Amplitudes of movements under study amount to thousands of kilometers; absolute displacements are usually determined.

The same classification is applicable to the tasks which arise in the investigation of tectonic movements, i.e. local, regional and global tectonic tasks can be distinguished. However, this classification does not include, e.g. the dating of the tectonic event according to the secondary NRM components. It deals rather with problems, which can possibly be solved by the examination of geomagnetic reversals. The global scale tasks, including the examination of seafloor spreading processes, and the local tasks, related to the formation of folds, shears and faults are examples.

4.1.2 Principles of Paleomagnetic Methods in Solving Tectonic Tasks

The new global tectonics, or the conception of mobile lithospheric plates, are known to be based on three phenomena: seafloor spreading, continental drift and plate collision accompanied by subduction, formation of fold belts, island arcs and their deformation. The quantitative studies of these phenomena are now possible due to the progress in paleomagnetic methods.

Two separate paleomagnetic methods for solving global tectonic tasks are available. They allow the determination of the rate and geological time span in seafloor spreading and the reconstruction of the position of the continental lithospheric blocks through geological time.

The first paleomagnetic method is the second from the viewpoint of priority and is based on the geomagnetic reversal phenomenon recorded in the magnetically active layer of the seafloor during spreading (see Sect. 3.15). The geochronologic scale of geomagnetic reversals, compiled from studies of dated rocks, permits the establishment of the time reference of each linear marine anomaly. Thus, the isochrons of the oceanic floor can be drawn the history of seafloor spreading, associated with the drift of continental plates, can be reconstructed. When reconstructing the relative position of blocks for any moment of geologic time, the corresponding isochrons of seafloor need to be determined (Le Pichon 1968; Le Pichon et al. 1973; Karasik 1978).

The second method is based on the apparent wander of paleomagnetic poles that, according to the principle of a central axial dipole (see Sects. 1.2.3 and 3.5), simply reflects a drift process, i.e. a change in the lithospheric block position relative to the rotational axis of the Earth. When paleomagnetic poles of different lithospheric blocks and the rotational axis of the Earth correspond, it is possible to determine the position of these blocks relative to the pole at any time. Thus, the knowledge of the changes in these positions from one epoch to another allows the reconstruction of the relative position of continental blocks (Irving 1964).

The basis of paleomagnetic methods used in solving regional and local tectonic tasks, e.g. deformations associated with the development of fold belts, island arcs and local structures, is quite simple and evident: any deformations of geologic bodies that occurred after their remanent magnetization in a geomagnetic field would produce the same deformation in the pattern of paleomagnetic directions "imprinted" in these bodies. To solve these tasks, the deformed and displaced bodies must be returned to the positions in which the paleomagnetic directions become parallel.

The model of a central axial dipole, as indicated above, is the basis for solving global tectonic tasks. The geomagnetic field is regarded as the sum of two fields: secular variations and the paleomagnetic field. The field of secular variations is changing quickly (in terms of geologic time) and its mean integral value for several ten or hundred thousands of years is zero (it is not zero at any moment of time). The paleomagnetic field is the field of a dipole (of normal or reversed polarity) located in the centre of the Earth and oriented along its rotational axis.

When tested, for different geologic epochs, this model appeared to be a good first approximation (see Sect. 3.6).

4.1.3 Principles of Analysis and Selection of Paleomagnetic Data

Detailed paleomagnetic studies provide data on the geomagnetic field history and are used successfully for analyzing the regularities in the distribution of the ancient geomagnetic field and its evolution. It seems convenient to tabulate the results of paleomagnetic studies for the purpose of this analysis.

Since 1960 Irving, and later McElhinny, regularly published tables of paleomagnetic data compiled according to the information obtained from scientific works appearing in the time period between the publications. The data tabulated in the early issues were later included in monographs (Irving 1964; McElhinny 1973) and the tables that appeared still later were published separately after the publications mentioned above (McElhinny 1972; McElhinny and Cowley 1977, 1978, 1980). These new tables, however, do not contain data from the USSR. All results available through 1975 are also known from general catalogues (Irving and Hastie 1975; Irving et al. 1976a, 1976b, 1976c). These catalogues are the most comprehensive, however they do not include all of the data available from the USSR for that period.

The first table of data from the USSR (111 determinations) was compiled by Kalashnikov (1961); the next, more comprehensive one (303 determinations) was published in the work of Khramov and Sholpo (1967). Then, it was deemed necessary to regularly publish tabulated, paleomagnetic data from the USSR with more detailed comments compared to those accepted in other countries. We have already published five issues of these tables (Khramov 1971, 1973, 1975, 1979, 1982).

Paleomagnetic determinations represents a combination of vectors of ancient magnetization obtained within an area of about 100 × 100 km on coeval formations of different origin, irrespective of the quantity of independent data (single determinations) and irrespective of how many workers obtained this result in a given region.

Single determinations should fulfill the following requirements (Khramov 1973, 1982).

1. The geologic age of the body under study should be known up to an epoch range.
2. The time interval under determination should be sufficiently long for averaging secular variations.
3. The number of time levels represented (usually the number of samples, dykes or lava flows) should not be less than ten.
4. The radius of the circle of confidence α_{95} for the mean direction of primary magnetization should not exceed 20°.
5. The isolation of an ancient (characteristic) component of magnetization should be carried out by at least two separate methods (unless statistical and magnetic data on the high paleomagnetic stability of rocks are available).
6. There should be no evidence of an essential age difference between the time of magnetization and that of rock formation.
7. All statistical values for every determination and all coordinates should represent a concordant system.
8. Local tectonics should be considered.

Determinations that do not fulfill these requirements should be rejected; an exception is allowed only for the data from poorly studied regions or for an epoch with few available determinations, but those results are to be specified.

Paleomagnetic determinations embracing more than three, single determinations may be obtained by one of two methods. The first one is the vectorial addition of directions of mean magnetizations J_n^a for every determination and calculation of α_{95}, Φ, Λ values from the obtained values D, I and R. The second method is the averaging of poles, i.e. the vectorial addition of unit radii vectors having coordinates Φ_i and Λ_i. In these cases the value A_{95} is indicated, which is the radius of the circle of confidence of the mean pole. If we have three, or less, single determinations, then the mean pole must be recalculated by giving unit weight to the observations from each sample.

As a result of the classification of data tabulated by the above-mentioned workers and of recent published data in accordance with the principles of rejecting and averaging, we have compiled a generalized table of the mean paleomagnetic poles for the continents (see Appendix), which has become the basis for paleomagnetic reconstructions of several epochs. The tables of the new and most reliable, previous data, which appeared in the book edited by McElhinny and Valencio (1981) made its compilation much easier.

In our table (see Appendix) the mean paleomagnetic poles calculated for several geotectonic units (plate, platform, block, massif) are given in the order of their geologic age from Paleogene to Cambrian (the Neogene paleopoles are not included for their coordinates are very close to the Recent ones). Statistical data are also given: the precision parameter K, the radius of the circle of confidence A_{95} and the number of paleomagnetic determinations N used in the calculations. Coordinates of the mean paleopoles Φ and Λ were estimated by averaging the paleopoles of every determination according to formulae (3.7) and (3.8).

The number of mean paleopoles for a certain epoch depends on the number of reliable and ascertained coeval paleomagnetic determinations within the limits of this epoch. In some cases we have several groups of poles for the same time interval due to several reasons. Thus, the unsolved problem of the middle and early Paleozoic for the North European plate is due to the existence of two groups of poles corresponding to field I (similar to the late Paleozoic field) and to field II (different from it). Divergences of the poles I and II are clearly seen from the examples of groups (S–D) II and DI, SI.

Poles for the southern part of the British Isles, of Brittany and Normandy form a special group because in the pre-Carboniferous epochs they diverge from the poles of the remaining part of the North-European plate. It has been suggested that this divergence is indicative of the existence of a separate Armorican plate at that time (Van der Voo et al. 1984; Van der Voo 1983). In addition to the groups of poles I and II for the Carboniferous, Devonian and late Silurian, a group of the early Silurian poles S_1III became known recently from four intrusive bodies. This group of poles is greatly displaced eastward relative to the early-middle Paleozoic poles of the British Isles (Piper 1978a).

The early Paleozoic poles of the Bohemian Massif also constitute a separate group that differs sharply from coeval poles of the North-European plate. The pre-Middle Carboniferous poles of the Urals are also divided into two groups: A and B. Group A includes data mainly obtained in the areas west of the Greenstone Belt and the Major Fault; these data are close to the direction obtained for the North-European plate. Group B represents determinations in the areas adjoining the fault.

Paleomagnetic determinations for the Scythian-Turan plate and Kazakhstan block are examined separately.

The early Paleozoic poles of the Anabar block and Aldan Shield, among the Paleozoic poles of Siberia, are also grouped separately in order to establish the relative position of these blocks in the early Paleozoic. The poles of the Altai-Sayan block, represented by six groups from the early Cambrian to Permian, are also examined separately.

Paleomagnetic determinations for the Silurian and Devonian of North America are also divided into groups. Poles from the northeastern part of New England and the Canadian Maritime area also make up a special group, for there is evidence that in the Silurian-Devonian time this region did not adjoin the North American craton; it was situated at more southern latitudes (Van der Voo et al. 1984).

Poles of the Silurian and Devonian for this region are combined in group A because of the few available determinations. Besides, there is another group B that includes three determinations from the plutons of New Brunswick and Newfoundland with extremely different polar coordinates. When establishing the ancient magnetization several components, differing in age, are sometimes obtained; – these are also regarded as separate determinations. Such determinations are present in group B and partially in group A.

The poor internal consistency of certain Paleozoic poles from South America is probably explained by the ambiguity in age determinations. Some paleomagnetic results on the Devonian, Permian-Carboniferous and Carboniferous are not used in calculating the mean paleopole because the age of the strata that yield this information is debatable.
The poles situated at very high latitudes (75°–85°) and at longitudes 155°–159° belong mainly to this group.

Numerous paleomagnetic determinations for Southern Europe, for some parts of Asia during the Cretaceous and Paleogene, within separate blocks and microplates, and a good inner consistency of data permitted the calculation of the mean paleomagnetic poles of these epochs for each geotectonic unit.

In epochs with few paleomagnetic determinations, data on boundary time intervals are sometimes included when calculating mean paleopoles on plates (thus, certain determinations on the late Triassic and early Cretaceous are included in the mean paleopoles of Jurassic).

Thus, tabulated data are the factual material for plotting a geographic grid of paleolatitudes for past epochs, which is the basis for the entire paleogeography, pole wander, continental drift and horizontal deformations of the Earth's crust in fold belts. Analysis of tabulated paleomagnetic data is of importance in establishing a paleomagnetic chronostratigraphic scale. These data are also useful in determining the geologic age of certain formations by paleomagnetic methods.

4.2 Apparent Polar Wander Path for Major Plates; Formation and Break-Up of Supercontinents

4.2.1 Paleomagnetic Poles and Their Wandering

According to the principle of a central axial dipole, a change in paleomagnetic directions in any part of the Earth's surface is just a result of (reflection of) local, regional and global tectonic movements. Thus, the paleomagnetic polar wander path, in general, is not the actual path of migration of geographic poles on the Earth's surface, therefore, they are called the apparent polar wander paths. Representation of the horizontal tectonic movements, on a global scale in particular, in the form of an apparent polar wander path, however, is very attractive, for it is simple and obvious. By comparing these paths for crustal blocks, we can estimate the scale, nature and time of relative horizontal movements of these blocks. Thus, the first necessary approach in paleotectonic (global and regional) reconstructions have been made.

Paleomagnetic poles, obtained as a result of different determinations for one and the same block and for the same geologic epoch, show a dispersion due to many causes: e.g. random errors in paleomagnetic determinations, ignorance of local tectonics, errors in rock and rock magnetization dating and also in the actual displacements of poles for the time interval under examination. Thus, the averaging of poles over very long time intervals (about an epoch range) becomes unavoidable, when drawing the paths through the mean poles. Just these mean poles are given in the Appendix; they were used for plotting the polar wander paths, and, thus, these paths are to be regarded as smoothed and generalized. The maps of the northern, or Pacific, (λ = 90–270E) hemisphere, where the northern pole was displaced in the late Precambrian and Phanerozoic time, are very suitable for plotting the apparent polar wander paths for the northern continents. As to the Gondwana continents and subcontinents (South America, Africa, India, Australia and Antarctica), it is preferable to compare the apparent polar wander paths of the southern pole.

4.2.2 Apparent Polar Wander Paths for Major Plates

Here, the apparent polar wander paths (APWP) for the continental parts of the major lithospheric plates of the Northern Hemisphere are examined: North Europe, East Siberia and North America (Fig. 4.1). These paths, forming a typical "spiderlike" pattern, appear rather simple, thus indicating that the recent, relative position of the plates is a result of past horizontal movements. Several loops can be recognized against the background of a general northward displacement of paleomagnetic poles from the equatorial areas of the Pacific Ocean. These are registered in the intervals of late Triassic-Paleogene and Cambrian-Silurian for North America, for the same periods, and during the Devonian-Middle Car-

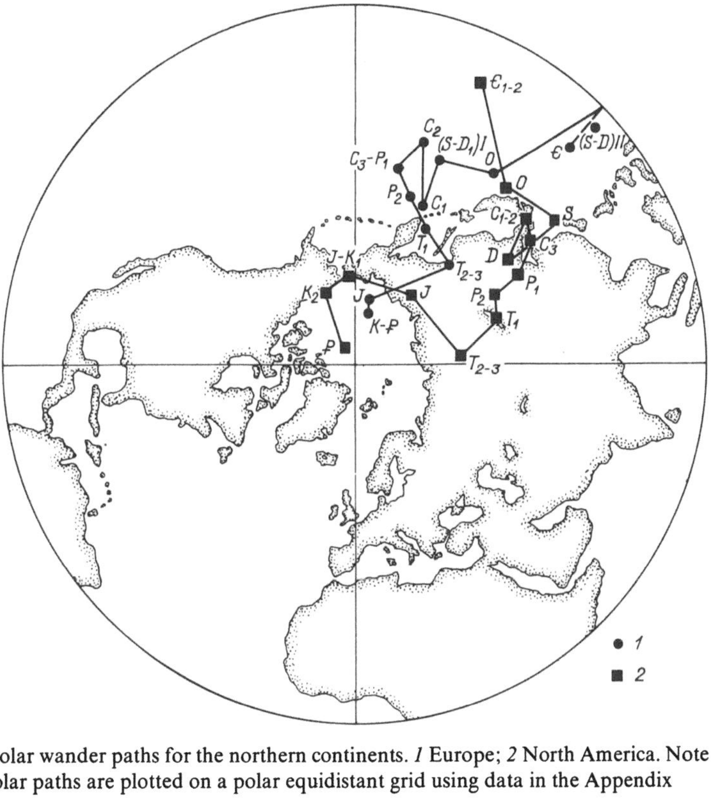

Fig. 4.1. Apparent polar wander paths for the northern continents. *1* Europe; *2* North America. Note to Figs. 4.1 to 4.3: polar paths are plotted on a polar equidistant grid using data in the Appendix

boniferous for Northern Europe and for Eastern Siberia (see Sect. 4.3), in the Ordovician-Devonian interval. The position of these loops in the geochronologic scale can probably be explained by a changes in the character and direction of the lithospheric plate motion.

The apparent polar wander paths for the southern continents (Fig. 4.2) differ essentially from the paths for the northern continents in particular by a much greater amplitude in their early Paleozoic parts. This reflects a larger scale of drift of the southern continents, especially India and Australia.

The polar wander paths for smaller lithospheric blocks and deformed fold belts are much more complicated (Fig. 4.3), and related to regional and local horizontal movements (see Sect. 5.1-5.3).

4.2.3 Reconstruction of the Relative Positions of Two Continents

In essence, paleomagnetic determinations give the relative position of sites according to the ancient latitude and relative to the poles, and give the orientation of the areas relative to the meridian. If only paleomagnetic data for one and the same epoch are available, it is not possible to obtain a difference in the longitudinal distance, if these sites are located in different continents (plates, blocks) that later underwent relative displacements. However, this problem can be solved if it is known that both blocks were once parts of a single plate, i.e. their relative position did not change for a certain period of time in the past. There are two methods of solving this task: the methods of Irving (1964) and Graham (Graham et al. 1964).

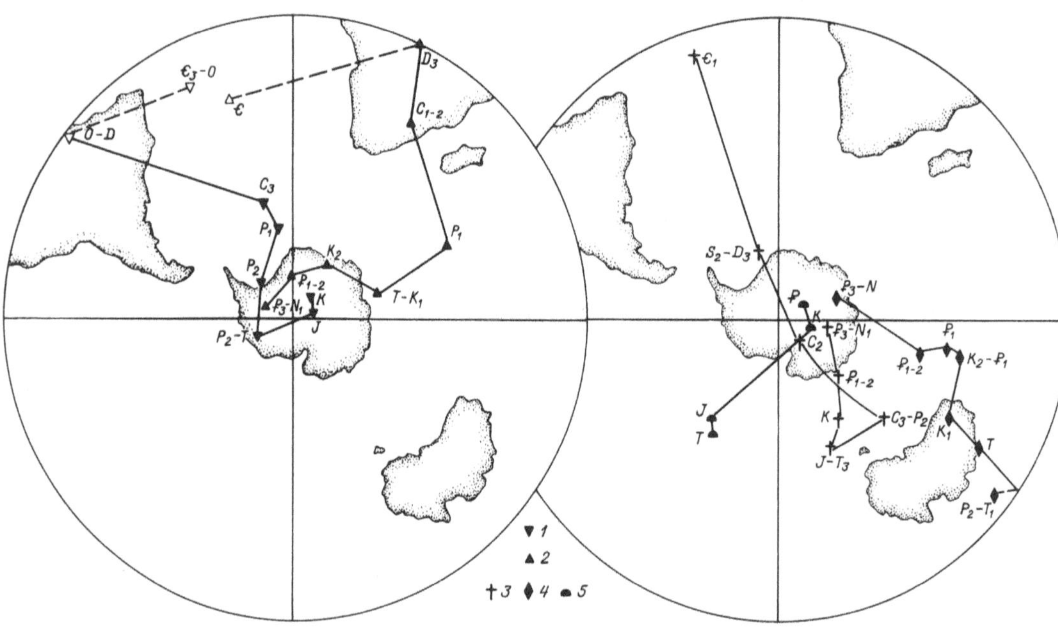

Fig. 4.2. Apparent polar wander paths for the southern continents. *1* South America; *2* Africa; *3* Australia; *4* India; *5* Antarctica

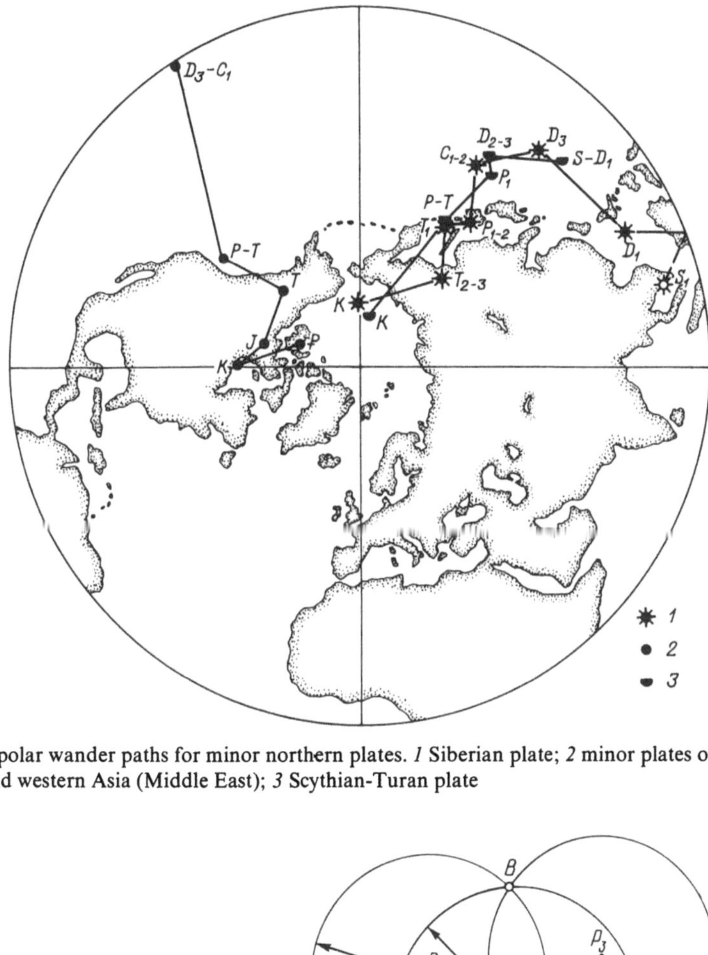

Fig. 4.3. Apparent polar wander paths for minor northern plates. *1* Siberian plate; *2* minor plates of southern Europe and western Asia (Middle East); *3* Scythian-Turan plate

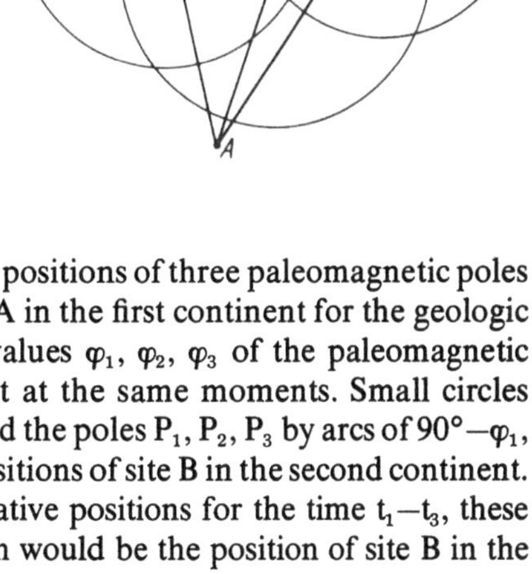

Fig. 4.4. Determination of the relative ancient position of two sites *A* and *B* of paleomagnetic observations. (After Irving 1964)

Irving's method is as follows (Fig. 4.4). The positions of three paleomagnetic poles P_1, P_2 and P_3 are estimated relative to site A in the first continent for the geologic time moments t_1, t_2 and t_3 and also the values φ_1, φ_2, φ_3 of the paleomagnetic latitudes of site B in the second continent at the same moments. Small circles circumscribed on the Earth's surface around the poles P_1, P_2, P_3 by arcs of $90°-\varphi_1$, $90°-\varphi_2$ and $90°-\varphi_3$ specify the possible positions of site B in the second continent. If both continents did not change their relative positions for the time t_1-t_3, these circles would cross at the same point, which would be the position of site B in the second continent relative to the first one. As the ancient meridian orientation in the second continent is also known from paleomagnetic observations for the moments t_1, t_2, t_3, the rotation of the second continent relative to the first one can be obtained.

Graham's method consists in matching parts of the polar paths and in matching the displacement of the continents closely related to these paths. For example, there are two series of determinations of poles for two continents, A and B, in the epochs 1,2 . . . 10 (Fig. 4.5). If the continents drifted separately during the epochs 1, 2, 3, and then collided in epoch 4 and later moved together relative to the pole during epochs 4, 5, 6, 7, then, the apparent polar wander paths would coincide in the portion 4–7 (Fig. 4.5a). Assuming in epoch 7, a break-up of continents occurred, which later drifted relative each other (epochs 8–10); then, as a result the recent position of continents and polar wander paths related to them, would appeared similar to that shown in Fig. 4.5b. Starting from point 10, portions of the polar wander paths are being sequentially superimposed and in portion 4–7 these paths coincide (Fig. 4.5c). Thus, we have the relative position of continents A and B, the time of formation 4 and the break-up 7 of the supercontinent AB.

4.2.4 Formation and Break-Up of Supercontinents

For the first time this method of overlapping of polar wander paths was applied in the study of the formation and break-up processes of Pangaea. The correlation of polar paths for all continents in the paleogeographic reconstructions of Wegener-Du Toit showed that a concordance of portions of these curves occurred in the Silurian-late Triassic interval (Fig. 4.6). In pre-Silurian time at least two groups of continents existed: Gondwana and Euramerica (McElhinny 1973).

The break-up of Pangaea appears to have started with the separation of the Indo-Australian block and of the northern continents (Laurasia) from West

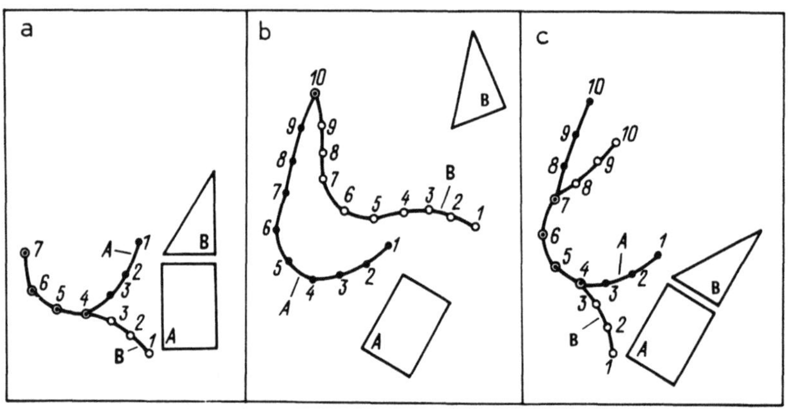

Fig. 4.5.a-c. Reconstruction of ancient positions of *A* and *B* continents which were parts of a single plate. 1–10 Successive paleomagnetic pole positions for continents A and B. **a** Positions of the continents and poles at the end of the interval 4–7 of their common movement; **b** Recent positions of the continents and apparent polar wander paths; **c** Reconstruction of the continent position by fitting the APWP's in the interval 4–7

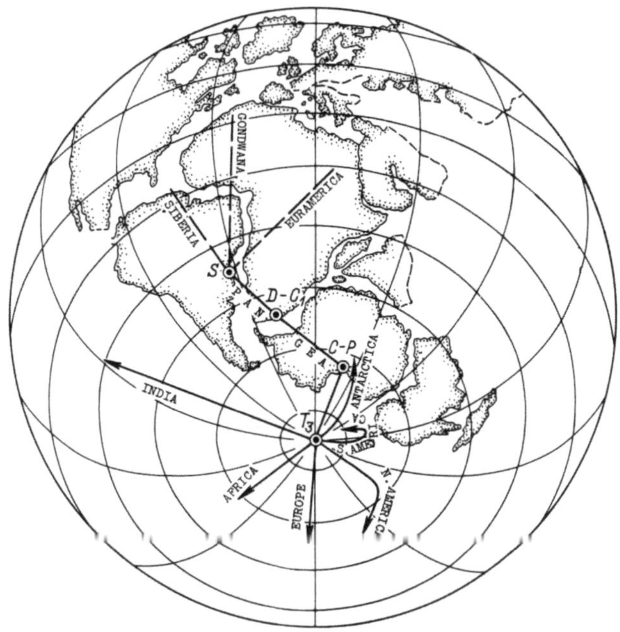

Fig. 4.6. Schematic apparent polar wander paths on the reconstructions of Pangaea. (After McElhinny 1973). APWP's for Siberia, Gondwana and Euramerica prior to the formation of Gondwana and Mesozoic APWP's for India, Africa, Europe, North America, South America and Antarctica are shown

Gondwana at the beginning of Mesozoic; the separation of North America from Eurasia started in the Cretaceous period.

From the comparison of the apparent polar wander paths, it follows that the formation of the Eurasian supercontinent is likely to have been a complex, multiphase process that terminated at the end of the Mesozoic-Oligocene with the linking of lithospheric blocks across the closed Tethys Ocean, India and certain structures of the northern part of the USSR.

These stages of Alpine consolidation of Eurasia are discussed in Chap. 5, here paleomagnetic data on the earlier formation stages of Eurasia are given.

4.3 Paleomagnetic Poles and Movements of the Russian and Siberian Platforms in the Paleozoic

4.3.1 Apparent Polar Wander Paths of the Russian and Siberian Platforms

The history of the late Proterozoic and Paleozoic horizontal movements, resulting in the formation of the recent structure of a major portion of Northern Eurasia, is a very important problem in global tectonics and is far from being solved. This explains to a great extent the ambiguity and uncertainty of global reconstructions, for the Devonian epochs in particular. Several greatly varying reconstructions are available (Briden et al. 1974; Bucha 1975; Morel and Irving 1978; Gorodnitsky and

Zonenshain 1979). When comparing them, it appears evident that the key problem lies in the relative position of the Russian and Siberian platforms which were the pre-Paleozoic "cores" of the North European and East-Siberian platforms.

Paleomagnetic methods were applied for the first time in studies of the Paleozoic history of horizontal movements of the Russian and Siberian platforms at the beginning of the 1960's, when the first results of paleomagnetic determinations for the Paleozoic of these platforms were obtained and the time series analyzed (Khramov and Sholpo 1967). It was demonstrated that these series might be consistent if both platforms are assumed to move not only relative to a geographic pole, but also assumed to change their relative position significantly, at least during the early Paleozoic. Later, an alternative hypothesis was proposed, thus explaining the divergence of the early Paleozoic paleomagnetic poles by the wide occurrence of late Paleozoic remagnetization in the Russian platform and adjoining regions (Creer 1970a). As to the problem of the relative movement of the Russian and Siberian platforms in the early Paleozoic, this remains open to debate (Khramov et al. 1974).

The actual knowledge of the paleomagnetism of Paleozoic formations of the Russian and Siberian platforms permits the examination of their horizontal movements in more detail. In the paleomagnetic data catalogue (see Appendix) 165 paleomagnetic determinations are given for the Paleozoic and early Triassic within the epi-Baikalian platforms and 105 determinations for the adjoining regions: the Urals belts, Kazakhstan block, Scythian-Turan young plate, Altai-Sayan fold area. These data were analyzed and selected in accordance with the accepted, minimal criteria of reliability (see Sect. 4.1). Determinations for each tectonic unit are combined into groups of time intervals of the order of one or two geologic epochs and the mean paleomagnetic pole is estimated for each tectonic unit (Fig. 4.7).

Paleomagnetic poles for the early Paleozoic of the Siberian platform show a good internal consistency ($A_{95} = 3°–8°$). Their gradual westward migration is seen within the temperate southern latitudes of the western part of the Pacific Ocean and then, after the Middle Cambrian, northward across Australia to the equator. The Middle Paleozoic poles are few and show poor agreement ($A_{95} = 12°–24°$). This is probably due to erroneous age data, since nonfossiliferous red beds and intrusions were mainly studied. Permian and Triassic poles are more numerous ($N = 27$) and have good internal consistency ($A_{95} = 4°–8°$); the mean pole is situated in the sea of Okhotsk, to the east of Sakhalin Island.

Paleomagnetic poles for the Altai-Sayan area are few, but still show an obvious southward position in the Middle Paleozoic and a westward one in the early Paleozoic relative to the coeval poles of the Siberian platform.

Paleomagnetic poles of the late Paleozoic and the beginning of Triassic for the Russian plate are fairly numerous ($N = 52$) and show internal consistency. The middle and early Paleozoic poles also show a good internal convergence; however, as determinations are few and late Paleozoic remagnetization is possible, it is necessary to specially examine the problem of the pre-middle Carboniferous portion of the polar path.

The group of the early Carboniferous poles included in the Appendix belong to the end of Visean and Namurian age. One determination (not given in the Table) for the Tournaisian red beds in the Moscovian basic gave a pole $\Phi = 7°$, $\Lambda = 133°$ (Khramov et al. 1974, p. 137). A similar pole was estimated for an "abnormal" group of vectors that were isolated by thermal cleaning of some Upper Devonian red beds in the Valdai region: $\Phi = -8°$, $\Lambda = 139°$ (Khramov et al. 1974, p. 136). These poles are very similar to the coeval poles of the Urals group A. Detailed paleomagnetic studies of the Ordovician rocks of the Baltic region also gave the same "abnormal" group of vectors, corresponding to the pole $\Phi = -13°$, $\Lambda = 133°$ (Khramov and Rodionov 1977). Just one early Ordovician pole for the Urals lies still farther westward. The post-Ordovician poles of the Urals group A practically coincide with the coeval poles of

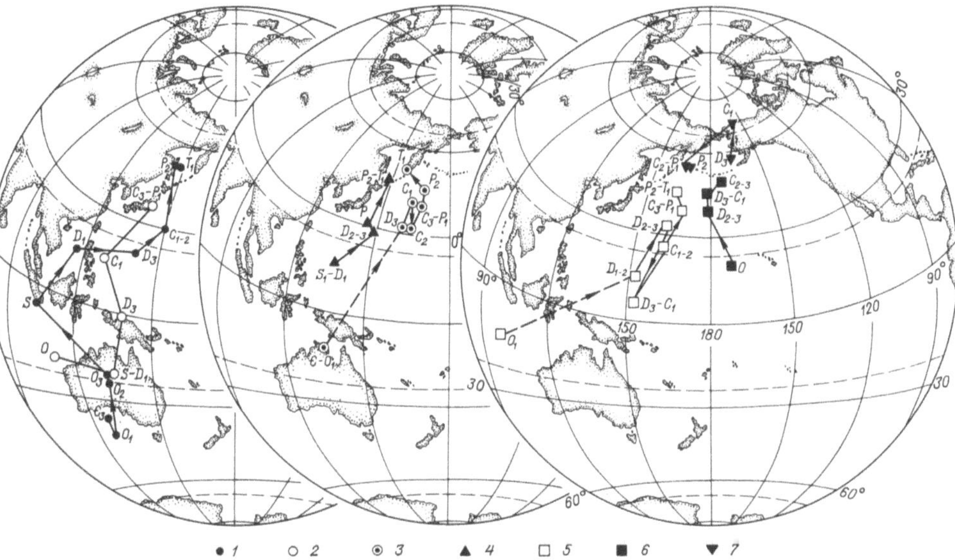

Fig. 4.7. Apparent polar wander paths for the tectonic units of the USSR. (After Khramov and Rodionov 1980a). *1* The Siberian platform; *2* the Altai-Sayan region; *3* the Russian platform; *4* the Turan and Scythian plates; *5* the Urals, group A; *6* the Urals, group B; *7* the Kazahstan block

the Russian platform. Most determinations included in group A were obtained for rocks situated westward from the Urals Major Fault and its Greenstone Belt. Thus, in a tectonic sense, they belong to the pole group of the Russian platform. In contrast, all sampling sites of group B belong to the Greenstone Belt and are situated close to the Major Fault. Geologic and paleomagnetic data show a wide distribution of sinistral displacements, thrusts and counterclockwise rotations of rock masses (Ivanov 1968). This is probably the explanation of the eastward divergence of the apparent polar wander path for group B.

Paleomagnetic determinations for the Scythian-Turan plate were done on dislocated formations. The southwestward divergence of poles for these plates, relative to the coeval poles of the Russian platform (see Appendix), might be explained by a clockwise rotation of blocks with dextral displacements.

The paleomagnetic poles of the Kazakhstan block, on the contrary, are significantly offset northeastward. The polar path curve from the Devonian to the end of Paleozoic lies to the north of the polar path for the Urals group B. This review of the series of paleomagnetic determinations that are included in the Appendix makes it possible to draw standard, apparent polar wander paths in the Paleozoic for the Siberian and Russian platforms.

The Siberian Platform. A curve drawn through the poles of the Siberian plate can be regarded as an apparent polar wander path (see Appendix). This series is quite reliable in the Cambro-Ordovician and Permo-Triassic portions. The Middle Paleozoic portion of the polar path is of a rough character though it receives good confirmation in the paleomagnetic poles for the Altai-Sayan area, which form a similar curve, though displaced southward (see Fig. 4.7).

The Russian Platform. Only one point $\Phi = -13°$, $\Lambda = 135°$ in the Cambrian-early Ordovician is available for drawing the early Paleozoic portion of the polar path. Certain confirmation of this location of the pole comes from the Ordovician data in both Urals groups as they correspond to the same paleolatitudes of the Urals, but with a rotation of paleomagnetic directions by 35° clockwise for group A, and for by 55° counterclockwise for group B. In the early Silurian epoch, taking different signs and approximately equal rotation angles of the Eastern Urals blocks and Scythian-Turan plate into consideration, the pole $\Phi = 8°$, $\Lambda = 147°$ might be accepted for the Russian platform, which is the mean pole between the poles $\Phi = 13°$, $\Lambda = 169°$ and $\Phi = 6°$, $\Lambda = 128°$ (see Appendix). For the late Silurian-early Devonian, we accept pole $\Phi = 15°$, $\Lambda = 151°$ and then a sequence of poles Φ, Λ: 31°, 161°; 19°, 155°; 46°, 158°. Thus, the poles of the North-European plate for the Ordovician and Silurian of group I are neglected because they correspond to the late Paleozoic remagnetization field. In addition to the Russian platform, the remagnetization processes of late Paleozoic were widely distributed in the Urals area and within Hercynides of Central Europe (Khramov et al. 1974; Bucha et al. 1976).

4.3.2 Paleozoic Drift of the Siberian and Russian Platforms

The problems that arise in studying the horizontal movements of the Russian and Siberian platforms are, as always, related to the ambiguity of information on the middle Paleozoic and insufficient paleomagnetic data on the early Paleozoic of the Russian platform proper. Nevertheless, taking the data on the regions surrounding the platforms into consideration, it is possible to construct a conformable model of their horizontal movements (Khramov and Rodionov 1980a).

At the beginning of the Cambrian, the Siberian platform was situated in the Southern Hemisphere and oriented with its present southern margin to the north. Since that time the plate has been moving to the north and at the end of Ordovician (Fig. 4.8), it shifted from the latitudinal belt 30°–10°S into the belt 5°S–15°N, that is, it drifted 25° (about 3000 km). Later, the platform continued moving towards the northern pole with its western margin (now the eastern margin), during Permian-Triassic it passed by the pole, thus, acquiring a reversed orientation relative to the pole. Furthermore, its western (Verkhoyansk) margin becomes its eastern margin and its eastern (Yenisei) margin, its western.

During the early Paleozoic, the Russian platform was situated in the Southern Hemisphere in low latitudes. At the end of the Silurian, the equator ran along the Urals. Later, the Russian platform generally moved to the north with a clockwise rotation. During the Tournaisian and in the middle Carboniferous, certain recurrent movements with southward displacement to the equator have been registered.

In the late Silurian-early Carboniferous interval, the Altai-Sayan fold area rotated clockwise relative to the Siberian platform. In the Silurian the movements within the Urals fold belt also started and they probably terminated only in the early Permian. These movements were accompanied with a counterclockwise rotation of some blocks.

The Scythian-Turan plate in the middle and late Paleozoic appears to have moved with the Russian platform. The divergence of polar wander paths is explained by the clockwise rotation of the masses which was concurrent with the

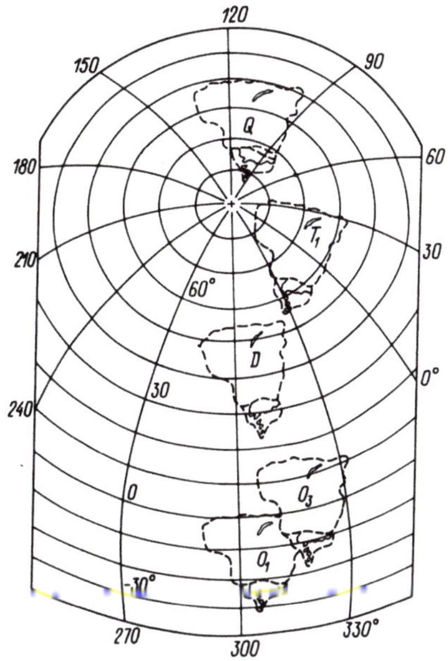

Fig. 4.8. Siberian platform drift since the Paleozoic. (Khramov and Rodionov 1980a)

dextral displacements (Shevlyagin 1982). A slow northward drift of the Kazakhstan block has been recorded for the middle and late Paleozoic; its movement in the early Paleozoic is not yet certain.

The data in the Appendix show that all of the tectonic units examined merged together only at the end of Paleozoic, thus forming the North-Eurasian lithospheric plate.

Notable is the fact that the movement of the Siberian platform since the late Cambrian up to the present might be represented as a drifting along a great circle almost without any noticeable rotation (see Fig. 4.8). This movement is so simple that it prompts the hypothesis that the Siberian platform was part of a great plate during this entire time interval. It is not yet known whether this plate embraced some parts of eastern Gondwana and blocks of southeastern Asia, but the proximity of the early Paleozoic poles of Siberia and Gondwana is commonly noted, (Khramov et al. 1974). From the viewpoint of the relative movements of plates, the late Proterozoic is typified by a moving away of the Siberian, and probably the Kazakhstan blocks, from the Russian platform. In the late Precambrian, the Russo-Baltic and Siberian plates were situated more closely. Their breaking up probably started at the middle Riphean-late Riphean age boundary.

In the early Paleozoic, the Siberian and Russian platforms were more widely separated than they are now. Later, their oblique convergence started concurrent with a northward movement of these plates, which was accompanied by a transverse compression of the Urals geosyncline. Just at that time the Altai-Sayan area moved clockwise and joined on to the Siberian platform in the early Carboniferous. The Russian and Siberian platforms appear to have reached their maximum, proximity in Permian time. About the same time an oblique convergence of the Kazakhstan block and the Urals is recorded, together with the counterclockwise rotation of certain blocks within the Ural limits. In the Triassic the Siberian platform, with its present northern margin, moved slightly away from the Russian platform. Probably, it was also the time when Taimyr moved away and the Yenisei-Khatanga rift trough was formed (Gusev 1974). The Scythian-Turan plate might have undergone internal shifts in the Paleozoic as well as in the Mesozoic, but it is also evident that this plate formed a single unit with the Russian platform at the beginning of Mesozoic.

It was noted above that the interpretation of paleomagnetic data proposed here is far from being exact in the estimation of character and scale of the relative drift of the Russian and Siberian platforms. It is not possible to refute completely the alternative approaches proposed earlier, though they seem to be hardly probable. Further studies are needed to solve the problem as a whole.

4.4 Paleomagnetic Poles and Plate Kinematics

4.4.1 Reconstruction of Plate Positions

The main goal of the interpretation of paleomagnetic pole data in terms of plate tectonics is the conversion of the apparent polar wander paths into the movement of the plates proper, calculation of kinematic parameters of this movement and finally, production of paleotectonic and paleogeographic reconstructions.

To find an appropriate solution for this task we have examined at the beginning, the movement of a rigid body on the surface of a sphere (Le Pichon et al. 1973). According to Euler's theorem, any movements of a rigid body on the surface of a sphere are just rotations. The movement on a plane, however, is different, for any transportation of a body in this case might be resolved into two components: parallel transfer and rotation proper, whereas on a sphere, parallel transfer is impossible. When a combination of rotation and transfer is referred to (usually this occurs when regional and local tectonic tasks are examined), a combination of rotations around a nearby pole and a distant rotational pole is most often meant, which is equivalent to a specific rotation.

Assume that the paleomagnetic pole, with Φ and Λ coordinates, for a certain period of geological time and the present position of a corresponding plate are known. It is possible that this paleomagnetic pole can be resolved into a geographic one by an infinite set of rotations. Rotational poles should lie on a great circle that is equidistant from the same geographical and paleomagnetic poles (northern and southern) (Le Pichon et al. 1973). The plate should be rotated in the same manner, for the pole and the plate are closely related. Thus, when the rotation has been carried out and paleomagnetic information has been incorporated, the location and orientation of the plate relative to the paleogeographic grid can be determined. The paleolongitude, however, remains uncertain, for any rotation of plate around the Earth's rotational axis does not change either the paleolatitude or plate orientation, therefore, as a result, it satisfies the paleomagnetic data.

In practice, when undertaking paleoreconstructions on plots, we use a rotation around the axis running through the point where the paleoequator crosses the present equator, i.e. through the point with the latitude $\varphi_p = 0°$ and longitude

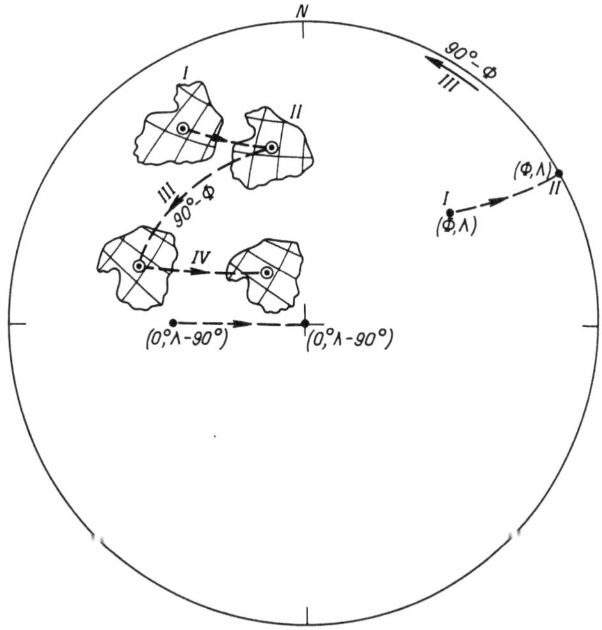

Fig. 4.9. Graphic method of paleomagnetic reconstructions. *I* Plate and pole positions relative to the actual geographic grid; *II* shifting of the pole to the edge of projection and corresponding shifting of the plate along the present latitudes (rotation around the pole N); *III* shifting of the paleomagnetic pole to the N pole, i.e. rotation of the plate around the centre of projection through an angle 90°−Φ; *IV* shifting of the plate along the paleolatitudes

$\lambda_p = (\Lambda - 90°)$ (Fig. 4.9). After conducting similar operations for all lithospheric plates, an intermediate scheme is obtained. Then, this scheme is easily resolved into a paleogeographic reconstruction, when the necessary rotations of the plates around the paleogeographic pole are carried out, i.e. the plates are then moved along the paleoparallels until a consistent pattern is obtained. Errors are inevitable in the coordinates of the mean paleomagnetic poles for the plates and in some cases, it becomes necessary to make minor corrections within the limits of the confidence circle radii A_{95} for the mean paleopoles.

The most convenient for reconstructions is an equal-area equatorial projection of a sphere, i.e. Lambert's projection (Schmidt's grid) with a diameter of 240° (Fig. 4.10). Outlines of plates and continents and also the 10° lines of the present geographical are converted and transferred onto the paleogeographic grid by the same two rotations that were used for the paleomagnetic poles. To make this operation easy and rapid, a series of oblique, equal-area projections with an axial inclination of 10° to 90° are drawn. These are used as projections of the present geographic grid in the paleogeographic system of coordinates. The above-mentioned drawing should be produced only for two points, i.e. for the present pole and for one point on a plate. Then, choosing a suitable oblique grid, both points of this oblique grid are matched with the corresponding points on the paleogeographic (equatorial) projections. This picture and the position of the oblique grid is then used as the present geographic grid and thus it is easy to plot contours and any points of a plate in the paleogeographic system of coordinates, i.e. in the reconstruction.

When the analytical methods for similar reconstructions are known, computers can be used (Minster et al. 1974; Galushkin and Ushakov 1979).

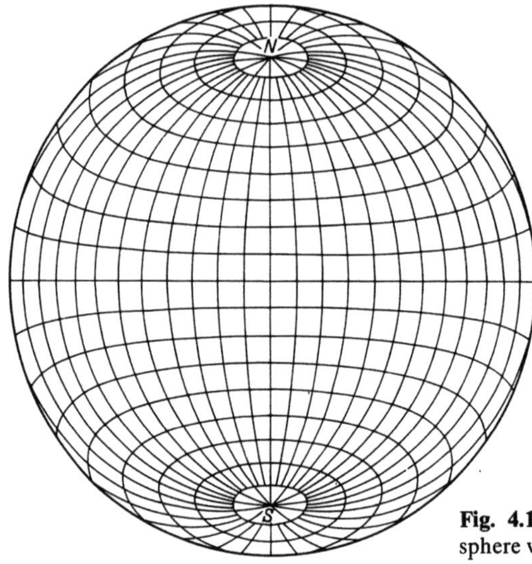

Fig. 4.10. Equal-area equatorial projection of a sphere with a projection diameter of 240°

4.4.2 Determination of Kinematic Parameters of Plate Movement

The reconstruction described above by two sequential rotations, first, around the point (0°, Λ–90°) and then, around the pole (φ=90°) for each plate does not, however, provide the kinematic parameters of plate movement. Three parameters are commonly used in plate tectonics: the latitude and the longitude λ_p of the rotational pole and the angle ϑ, through which one of the plates is to be rotated in order to appear in the former position relative to the other plate. The methods for determining these parameters are fundamentally developed by using the orientation of transform faults and matching the linear magnetic anomalies in oceans (Le Pichon et al. 1973; Minster et al. 1974). However, published information on the paleomagnetic pole data used for this purpose is not available, and no attempts are known for elaborating the corresponding method. Meanwhile, the estimation of rotational parameters using paleomagnetic pole data should be duly appreciated, for it permits the comparison of these parameters with those determined by linear anomalies and, thus, the quantification of the consistency of separate reconstructions.

First, a common rotational pole of two plates should be determined in order to solve this task. Assume it is represented by two plates with paleomagnetic poles P_1 and P_2 with coordinates $\Phi_1\Lambda_1$ and $\Phi_2\Lambda_2$ (Fig. 4.11). It has been pointed out that the rotational pole, transferring the paleomagnetic pole to a geographic one and the corresponding plate to its paleogeographic position, should lie on an arc of a great circle equidistant from both poles, i.e. on an arc joining point $M_1[(90° + \Phi_1)/2, \Lambda_1]$ and point N_1 (0°, 90°–Λ_1). For the second plate, the other arc can be plotted on a great circle which will extend through point $M_2[(90° + \Phi_2)/2, \Lambda_2]$ and point N_2 (0°, 90°–Λ_2). This is also the locus of the rotational poles that brings the paleomagnetic pole into correspondence with the geographic poles. The point of intersection P_o (φ_p, λ_p) of these arcs is a rotational pole that gives both the relative position of the

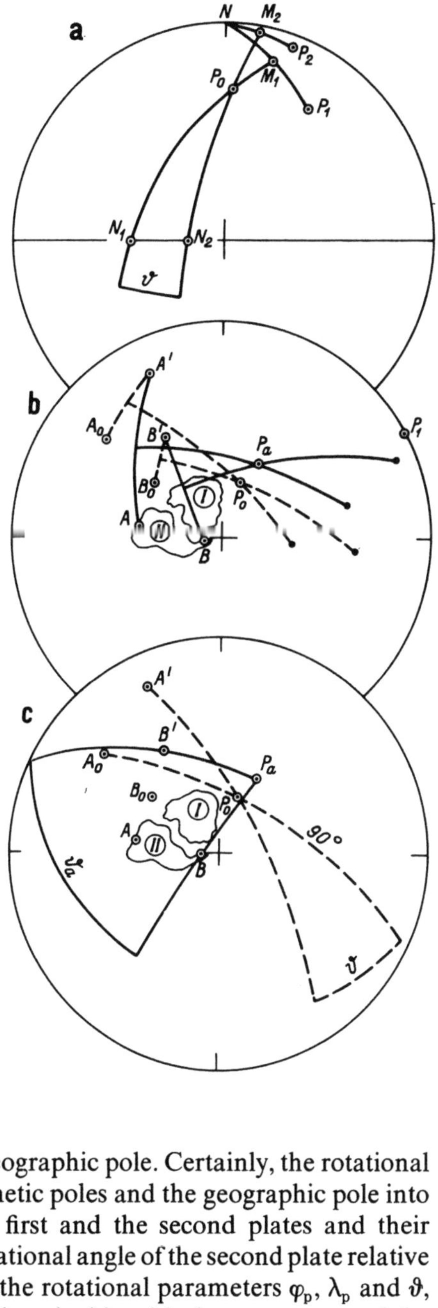

Fig. 4.11a-c. Determination of the kinematic characteristics of the relative movement of two plates. **a** Determination of a common apparent pole P_o of the rotation of two plates which have paleomagnetic poles P_1 and P_2; **b** determination of the pole P_a of the absolute rotation of plate *I* and of the pole P_o of the rotation of plate *II* relative to plate *I*; **c** determination of the angles of absolute ϑ_a and relative ϑ plate rotation

plates and their situation relative to the geographic pole. Certainly, the rotational angles ϑ_1 and ϑ_2 that bring the paleomagnetic poles and the geographic pole into coincidence, would be different for the first and the second plates and their difference would yield an angle ϑ, i.e. a rotational angle of the second plate relative to first one (Fig. 4.11a). It is evident that the rotational parameters φ_p, λ_p and ϑ, measured in this way, would not generally coincide with the parameters of the relative movement of plates calculated from linear anomalies.

For the complete solution of the task, the previously conducted paleomagnetic reconstructions should be employed. The procedure of determining the parameters of relative and absolute rotations is as follows (Fig. 4.11b,c). After reconstruction within the limits of one of the plates, for example II, we take two distant points A and B. By operations in a reverse succession to the one shown in Fig. 4.9, we determine the position of points A′ and B′, when plate I is transferred into the recent position. Using pairs AA′ and BB′, by a method similar to the one applied for the determination of a common rotational pole of two plates, we determine pole P_a and angle ϑ_a of rotation, i.e. the pole and angle of the absolute rotation of plate I. It is easy to determine, by the same method, pole P_0 and angle ϑ for the rotation of plate II relative to plate I, if we plot the present position of points A_0 and B_0 of plate II and find pairs A_0A' and B_0B' (Fig. 4.11b).

4.4.3 Paleomagnetism of Seamounts and Movement of Oceanic Plates

The main deficiency of paleomagnetic observations lies in the fact that they cover, as a rule, only the continental parts of the major lithospheric plates, yet the most extensive of these is the Pacific plate which is entirely of oceanic type. Sampling for paleomagnetic studies in oceans is constrained: sediments obtained by corers and lava flows of the islands characterize only the Neogene-Quaternary time span. This is absolutely insufficient for obtaining reliable data on the movement of oceanic parts of the plates. Drilling is quite expensive and for paleomagnetic studies, the core needs to be oriented in a horizontal plane; then, drilling is still impossible on a shelf and the upper parts of continental slopes.

Thus, indirect methods were developed for obtaining paleomagnetic data by studying the anomalous magnetic field. Three methods are known for the determination of the magnetization vector of the rocks making up the oceanic floor (Le Pichon et al. 1973; McElhinny 1973): (1) according to the relief and magnetic field of seamounts; (2) according to curves of an anomalous field over the sources of linear magnetic anomalies; (3) according to anisotropy of an anomalous field determining by mapping.

All of these methods are useful when three conditions are observed:

1. The magnetization of an anomaly-forming body is entirely remanent or the Koenigsberger's ratio Q_n is known and constant. The NRM direction and, thus, the position of the paleomagnetic pole can then be determined with satisfactory accuracy (with Q_n value unknown) only if $Q_n > 10$.
2. The natural remanent magnetization was acquired during 10^4–10^5 years, so that paleosecular variations were averaged and excluded, and geomagnetic reversals did not disturb the homogeneity of body magnetization.
3. The secondary magnetization (including the viscous type) does not contribute more than several percent to the NRM value.

All these conditions are observed for oceanic basalts that cooled quite rapidly in water, thus acquiring high and extremely stable thermoremanent magnetizations. Values of $Q_n \approx 100$ occur close to the axial zone and they decrease to $Q_n \approx 20$ in rocks with an age of 10 m.y. and older; viscous magnetization can be neglected as it is quite insignificant (Le Pichon et al. 1973); Demenitskaya et al. 1978).

Magnetization during 10^4–10^5 years and its homogeneity are best exemplified in small seamounts that are composed of basalts, forming conelike and flat-peak volcanic structures (Le Pichon et al. 1973). Large seamounts could have been active during intervals up to 10^7 years and the basalts forming them could have originated in epochs of different polarity of the geomagnetic field, which makes the interpretation of an anomalous magnetic field very difficult. Thus, the magnetic field of small seamounts provides the principal paleomagnetic data on the movement of oceanic parts of the lithospheric plates.

The direction of homogeneous magnetization J of a seamount is calculated in accordance with the results of magnetic and bathymetric surveys. The shape of the seamount is commonly approximated by prisms, the sizes of which are determined from the isobath pattern. The direction of vector J is found by the selection of the best agreement of anomalous fields of calculated H_b and of observed H_a. The technique applied is as follows (Le Pichon et al. 1973). When the direction of a normal geomagnetic field H_T in the site i and sizes and position of all prisms are known "the matrix of shape coefficients" S is calculated in such a way that the projection H_{bi} of the intensity of the anomalous field on the direction of vector H_T in the site i equals element "i" of vector $H_b = SJ$. Then, the differences are calculated $\Delta H_i = H_{ai} - H_{bi}$, where H_{ai} is the observed anomalous field in the site i. If measurements are taken at n sites, then a combination of ΔH_1, H_{ai}, H_{bi} values is obtained. Vector J_i is chosen in such a way that the residual field is the least (method of least squares). Some results of paleomagnetic studies of seamounts are given in Table 4.1.

This method permits the establishment of a great (30°–40°) northward displacement of the Pacific plate in post-Cretaceous time. This movement was approximately in the same direction during the whole Cenozoic as seen from a comparison of the coordinates of the paleomagnetic poles 1–13 in Table 4.1. An

Table 4.1 Paleomagnetic poles deduced from the study of seamounts

No.	Seamounts groups	Age, m.y.	Number of seamounts	Pole position Φ	Λ	A_{95}	Reference[a]
1	Tripod	10	7	87°	90°	14°	1
2	Midway	30	1	75°	12°	–	1
3	Nanpo	25-38	3	74°	321°	10°	2
4	Kuriley Island	38-55	2	59°	300°	–	2
5	Californian	40	6	71°	354°	13°	1
6	Line Island	41	4	78°	17°	4°	4
7	Abbot	40-45	1	76°	5°	–	4
8	Line Island	72-85	4	63°	355°	13°	5
9	Hawaiian	84-89	9	60°	357°	8°	3
10	Musicians	80–100	4	60°	339°	10°	3
11	Equatorial	110–130	4	67°	355°	15°	3
12	N.Japanese	100–140	5	51°	352°	14°	3
13	S.Japanese	150	4	59°	302°	14°	3
14	Wake Island	150	4	53°	341°	18°	3
15	Guinean	100	5	66°	252°	19°	1
16	New England	160	7	76°	108°	21°	1

[a] *1* Le Pichon et al. 1973; *2* Demenitskaya et al. 1978; *3* Harrison et al. 1975; *4* Sager 1983; *5* Sager and Keating 1984.

important contribution is also the common movement of oceanic and continental parts of the North American and African plates, which is confirmed by the closeness of corresponding paleomagnetic poles (see poles 14 and 15 in Table 4.1 and poles in Appendix). Unfortunately, this method of seamounts has constraints, because the oceanic floor is young and thus, only refers to post-Triassic time.

4.5 Global Reconstructions for the Paleozoic and Mesozoic Epochs

4.5.1 Global Reconstructions for the Mesozoic and Paleogene

The horizontal movements of lithospheric plates for the last 200 m.y. were principally characterized by the dispersal of the late Paleozoic-early Mesozoic Pangaea and by the separation of continents resulting in their present arrangement (see Sect. 4.2.4). This fact, generally confirmed by different methods, provides us with a useful tool for making paleomagnetic reconstructions for all periods, and even epochs, within the Mesozoic and Cenozoic (Khramov et al. 1974; Briden et al. 1974). We have the initial and the final states, i.e. two pictures of the continental positions. (1) In the early Mesozoic, the so-called Pangaea A or Wegener's Pangaea (Wegener 1929), (2) and the present arrangement and hence, one can easily imagine the direction and scale of movement of some continental blocks. Therefore, using mean paleomagnetic poles for each block and each epoch of Mesozoic and Cenozoic, it is possible to make global reconstructions for these epochs. This method, however, allows certain arbitrariness in the longitudinal situation of continents, and thus, to diminish this uncertainty, we apply an assumption of minimal plate movement.

The reliability of these reconstructions depends greatly on the correctness of one of the reference configurations, i.e. the early Mesozoic Pangaea. According to the Pangaea hypothesis, the relative position of its parts, which later formed actual continents, did not change significantly during the late Paleozoic and Triassic. It is only the situation of a coherent continental block changing relative to the pole of Earth's rotation, which provided a means to check the whole conception (McElhinny 1973). Thus, if the relative movements of continental blocks in the late Paleozoic and Triassic were few or even lacking, the corresponding portions of APWP's, plotted according to data from different continents, do not converge on a present map, but they do on a paleogeographic map. When the reconstruction is incorrect, portions of these paths would not match. Besides, we have the possibility of a separate paleomagnetic reconstruction of Pangaea by converging the late Paleozoic-Triassic portions of APWP's, which are firmly related with the corresponding continents, i.e. by applying Graham's method (see Sect. 4.2.3). At the same time, we can obtain the interval over which the Pangaea of this configuration existed.

The testing of the Pangaea hypothesis and the reconstruction of its outlines was done for the Permian period (Khramov et al. 1974). The model of the best consistency of shelf margins for the continents around the Atlantic Ocean and the Du Toit's reconstruction for the southern continents was thus accepted. Voluminous data on the coherency of the North African structures and of areas occupied in the past by the western part of Tethys Ocean were also considered; this coherency is supported by the tilt grouping of the paleomagnetic poles. It appeared that the Permian poles for different continents in the reconstruction, done in this way, are rather closely packed. Spherical harmonic analysis of the Permian and Triassic paleomagnetic field for this reconstruction gave a dipole structure for the field (Ben'kova et al. 1973), also indicating the acceptability of this model.

Thus, Pangaea A, a slightly modified Wegener's Pangaea, may be taken as a reference configuration for the Mesozoic and Paleogene paleomagnetic reconstructions (Figs. 4.12–4.17). The well-known reconstructions for the Mesozoic and Paleogene (Briden et al. 1974; Gorodnitsky and Zonenshain 1979) were done by a different method, i.e. by matching the isochrons of the oceanic floor, obtained from an analysis of linear magnetic anomalies. This method gives only the relative positions of the whole continental configuration, which can then be oriented by matching the paleomagnetic poles of this epoch with the geographic pole. Comparison of these reconstructions with the paleomagnetic ones reveals their general similarity and provides good confirmation of the objectivity of both methods and of the initial data. There are, however, certain differences.

Firstly, the reconstructions given here were done mainly for those intervals that were separated by stages of accelerated APWP's for almost every continent (see Sect. 4.2) and not done formally for equal intervals of geologic time. Second, the reconstructions involve the separation of paleomagnetic provinces of Eastern

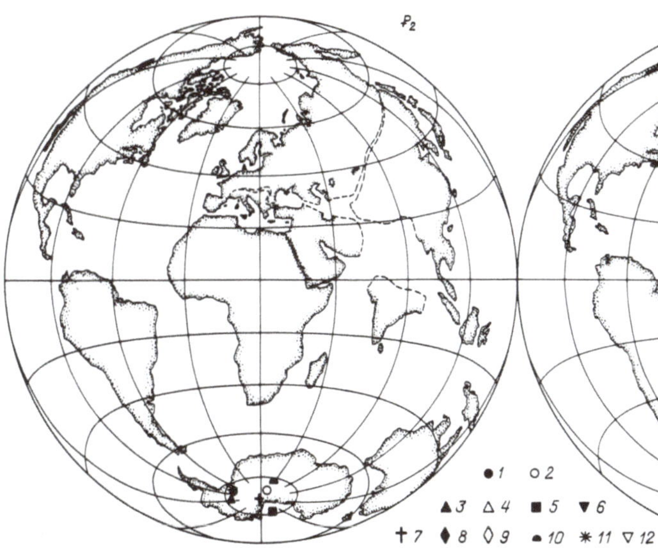

Fig. 4.12

Fig. 4.13

Fig. 4.12. Paleomagnetic global base map for the Eocene. Paleomagnetic poles (in Figs. 4.12–4.26) for: *1* northern Europe; *2* southern Europe and western Asia; *3* Africa; *4* Madagascar; *5* North America; *6* South America; *7* Australia; *8* India; *9* southern Asia; *10* Antarctica; *11* Siberia; *12* Kazakhstan

Fig. 4.13. Paleomagnetic global base map for the late Cretaceous (symbols, see Fig. 4.12)

Fig. 4.14

Fig. 4.15

Fig. 4.14. Paleomagnetic global base map for early Cretaceous (symbols, see Fig. 4.12)

Fig. 4.15. Paleomagnetic global base map for the Jurassic (symbols, see Fig. 4.12)

Fig. 4.16

Fig. 4.17

Fig. 4.16. Paleomagnetic global base map for the late Triassic (symbols, see Fig. 4.12)

Fig. 4.17. Paleomagnetic global base map for the early Triassic (symbols, see Fig. 4.12)

Asia, which permits one to follow the Mesozoic drift of corresponding plates and to provide a basis for the refinement of the Tethys, Mongolian-Okhotsk and Himalayan fold belt histories. Third, the reconstructions vary in detail with respect to the different evaluations of scales of horizontal deformation within the Mediterranean and the evaluations of the rate of the Jurassic opening in the northern and southern Atlantic regions.

4.5.2 Global Reconstructions for the Middle and Late Paleozoic

Paleomagnetic poles of the late and middle Paleozoic on the Pangaean reconstruction extend as a rather narrow band, permitting one to draw a common APWP (Fig. 4.18). Such a trajectory might be continued up to Triassic times (see Fig. 4.6). It was concluded that Pangaea existed from the Silurian through Triassic and that, during that entire interval, the continental block movements were not of a significant scale (Khramov et al. 1974; McElhinny 1973). More important in this connection seems to be the displacement of the whole of Pangaea relative to the pole that crossed almost the entire Gondwana from northwestern Africa to Australia. This path agrees almost exactly with the trajectory of displacement of the Paleozoic glaciation centres of Gondwana after Maack (1957).

It was shown in the previous section that although the continents were more or less closely packed, the differentiated movements of large continental parts of lithospheric plates, for instance the Russian and Siberian platforms, occurred during the middle and late Paleozoic. Analysis of paleomagnetic series for all continents (see Appendix) shows a similar behaviour for other pairs of plates in the Northern Hemisphere and also of these plates and Gondwana. Reconstructions, in accordance with the principle of movement minimization and consistency of paleomagnetic poles, are shown in Figs. 4.18–4.24.

Fig. 4.18. Paleozoic drift of Pangaea. *1* APWP's; *2* centres of Gondwana glaciation. (After Maack 1957); *3* approximate megaplate boundaries (for more symbols see Fig. 4.12)

Fig. 4.19

Fig. 4.20

Fig. 4.19. Paleomagnetic global base map for the late Permian (symbols, see Fig. 4.12)

Fig. 4.20. Paleomagnetic global base map for the early Permian (symbols, see Fig. 4.12)

Fig. 4.21

Fig. 4.22

Fig. 4.21. Paleomagnetic global base map for the late Carboniferous (symbols, see Fig. 4.12)

Fig. 4.22. Paleomagnetic global base map for the early and middle Carboniferous (symbols, see Fig. 4.12)

Fig. 4.23

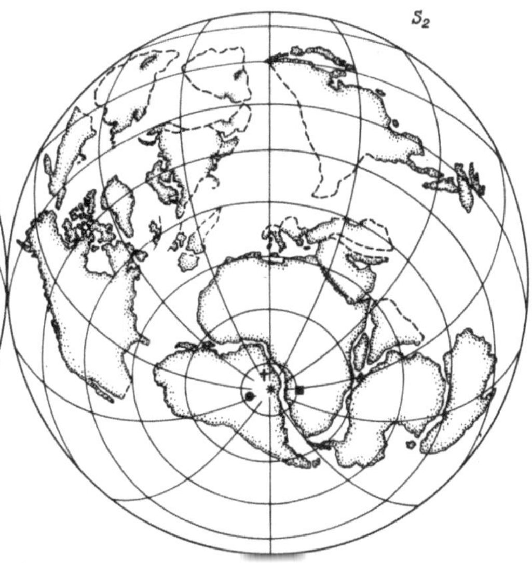

Fig. 4.24

Fig. 4.23. Paleomagnetic global base map for the late Devonian (symbols, see Fig. 4.12)

Fig. 4.24. Paleomagnetic global base map for the late Silurian (symbols, see Fig. 4.12)

In the late and middle Paleozoic reconstructions the key problem is the relationship of the northern continents and Gondwana. The evolution of these relationships proceeding from paleomagnetic data is needs to be confirmed by geologic evidence for the generation of the Caledonian and Hercynian fold belts and on the appearance and closure of small oceans, i.e. paleo-Atlantic, paleo-Asiatic and paleo-Tethys. The first paleomagnetic determinations for the Permian of southern Europe and southern Africa clearly showed the necessity of modifying the earlier accepted position of these regions relative to Northern Eurasia and North America (Briden et al. 1970; Morel and Irving 1978). This requirement was also confirmed by some results from an analysis of the Permian and Carboniferous paleomagnetic field based on Wegener's reconstructions (see Sect. 3.5). These results indicated that the best consistency of paleomagnetic data for the Permian and Carboniferous, in terms of a model of the central axial dipole, might be achieved proceeding from a supposition of the largest dextral shift occurring in the western part of the Tethys and in the Caribbean region at the Paleozoic-Mesozoic age boundary.

The reconstructions, including this supposed dextral shift, were done (Morel and Irving 1978) and this configuration was given the name Pangaea B. This configuration, on the whole, is better than the Pangaea A model. It agrees with the field of the central axial dipole and explains certain peculiarities in the distribution of paleoclimates and fauna in the southern continents; the history of the formation of Caledonides and Hercynides of Euramerica also acquires a logical basis and becomes more certain. However, it is still possible that this new model is just provoked by a mere error in the age correlation of the paleomagnetic series for the southern and northern continents. Taking into consideration the problems of global correlation of stratigraphic scales (e.g. the Permian of the Tethys and the Russian platform, not to mention southern Africa) and the relatively high rate of APWP in the middle and late Paleozoic, this seems to be a possible explanation. Thus, the problem of the late Paleozoic Pangaea B remains open to debate to a great extent. However, in our reconstructions we accepted only this model, not only due to its greater paleomagnetic reliability, but in order to attract attention to this problem.

4.5.3 Examples of Global Reconstructions for the Late Proterozoic and Early Paleozoic

Several detailed paleomagnetic determinations for the Siberian platform are surely the basis for solving the problem of global reconstructions for the early Paleozoic (Rodionov 1966; Khramov et al. 1974). The results obtained in cooperative paleomagnetic works in the Socialist countries also contribute to this problem (Bucha et al. 1976). The first global reconstructions for the early Paleozoic and the end of Precambrian were based mainly on results of these determinations (Bucha 1975; Briden et al. 1974). However, this problem is undoubtedly in the process of solution and further accumulation of paleomagnetic data are necessary, not only for the entire Siberian platform, but also for the Russian platform, in particular, and for the regions surrounding them. The status of this problem is discussed here.

Due to the divergence of APWP's on the Pangaea reconstructions for the northern and southern continents, it was concluded that two separate supercontinents – Gondwana and Laurasia – existed in the early Paleozoic (Creer 1970a). However, it appeared that quite numerous and internally consistent data for drawing the APWP for the early Paleozoic are available only for a part of Laurasia, i.e. for the Siberian platform (see Sect. 1.3). The scarcity of data for the remaining part of the territory is aggravated by specific difficulties: inconsistency of results for North America and Europe that seems to be explained by late Paleozoic remagnetization phenomena (Creer 1970a; Khramov et al. 1974), complexity of tectonic structures and of the evolution of Caledonian and Hercynian fold systems from which areas most paleomagnetic data were obtained. Additionally, the only method for an unambiguous reconstruction of the relative plate position consists in matching the APWP's and this is applicable only if the parts of converged curves correspond to an interval of common plate movement (see Sect. 4.2.3). It is evident that this method is unsuitable for plates with composite and rapid relative movements. Thus, paleomagnetic reconstructions for the early Paleozoic are inevitably ambiguous. Therefore, only a probable solution might be expected that could perhaps be obtained in accordance with the minimization of all movements and with reference to tectonic and paleogeographic relationships.

It is possible to overcome the longitudinal uncertainty in the early Paleozoic reconstructions if some additional constraints are placed on the nature of plate movements and their continental parts. Two such requirements in this respect seem most interesting: the immobility of the centre of continental masses and their movement in accordance with the field of the Coriolis force (Bucha 1975). Forsh, while developing his hypothesis of "planetary equilibrium" (Khramov et al. 1974), underlined the importance of the Coriolis force in the process of continental drift during the Phanerozoic (see Sect. 5.3).

Attempts to reconstruct Laurasia in the Ordovician exemplifies this complex problem (Khramov and Rodionov 1977). By matching the APWP's for the North American and East Siberian plates in their early Ordovician-early Silurian portions, two reconstructions were obtained (because of age uncertainties of the North American data). Longitudinal uncertainty in the early Paleozoic position of the Russian platform relative to the Siberian platform (see Sect. 4.3.2) supplied two further reconstruction variants. All of these reconstructions could be tested, but only if the following questions are answered: (1) Was extra-Alpine Europe a single plate since the early Paleozoic? (2) Which of the intervals within the early Paleozoic was a period of relative stability for all or some of the northern plates? (3) What is the effect of local tectonic rotations?

Alternative solutions might be obtained for a configuration of the assembly of three major northern plates at the change of polarity of the early Paleozoic poles for one or two plates. This change requires a different solution for APWP's, suggesting a quick displacement and a rotation of the North American and North European plates, or a sharp change in the direction of the East Siberian plate drift in the middle Cambrian and in the Silurian-Devonian. In the corresponding global reconstructions it becomes necessary to change the polarity of the paleomagnetic poles in the early Paleozoic for all plates, except East Siberia (Bucha 1975) or for Siberia (Gorodnitsky and Zonenshain 1979).

In the reconstructions presented here, an accepted variant corresponds to the principle of minimization of movements inside Laurasia and to a coherent Gondwana in the early Paleozoic (Figs. 4.25 and 4.26).

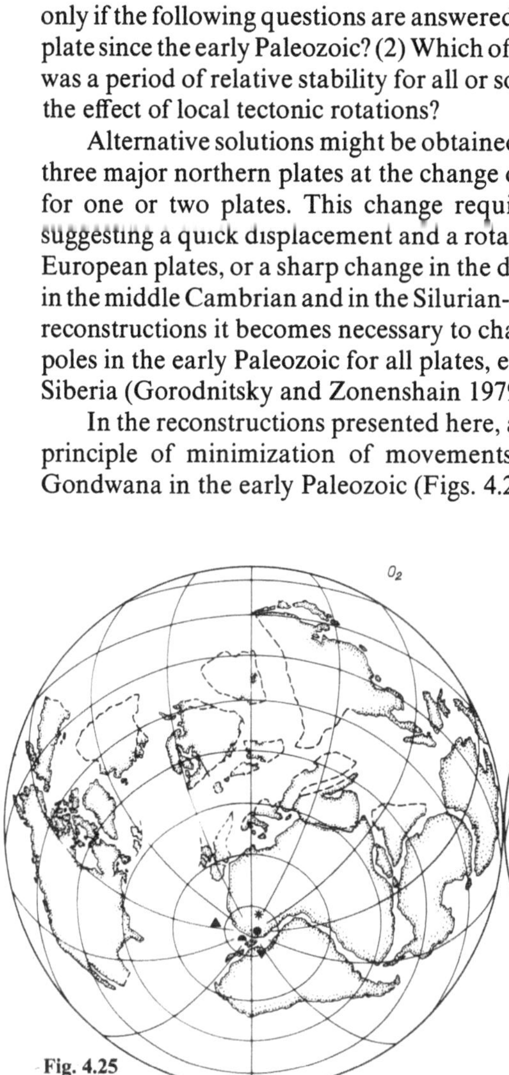

Fig. 4.25

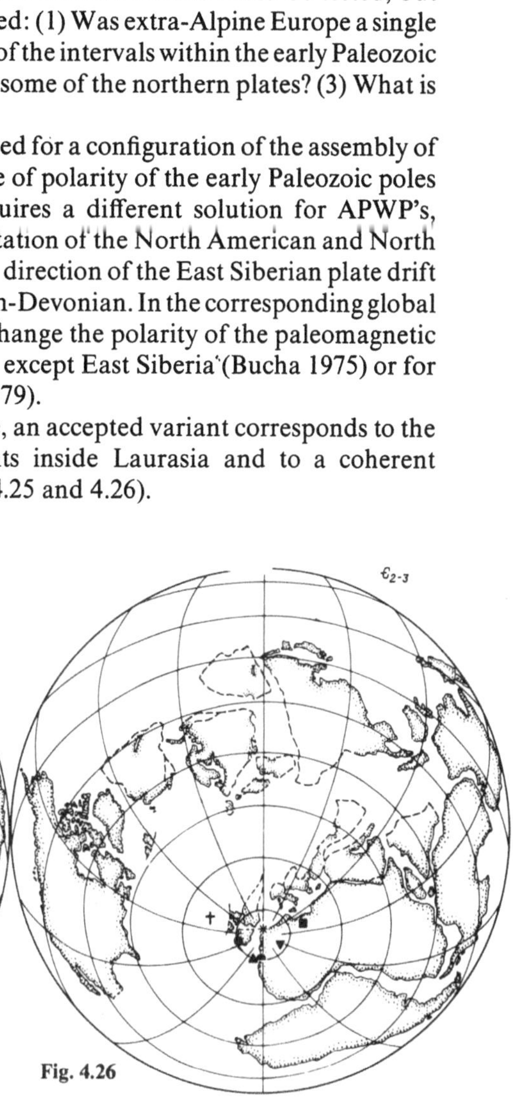

Fig. 4.26

Fig. 4.25. Paleomagnetic global base map for the middle Ordovician (symbols, see Fig. 4.12)

Fig. 4.26. Paleomagnetic global base map for the middle and late Cambrian (symbols, see Fig. 4.12)

It seems that a preference for one or the other solution can only be given after sufficiently complete paleomagnetic columns for the lower Paleozoic of all of the plates of Laurasia and Gondwana become available, as is the case for Siberia. Uncertainties in reconstructions will decrease with further accumulation and increased quality of paleomagnetic data. In this aspect, it is very important to improve methods of recognizing physical and chronologic components of the NRM in older sedimentary rocks. At the same time, this would help solve the problem of middle Paleozoic paleomagnetism and would aid in following the history of lithospheric plates at the beginning of the middle Paleozoic.

Plate tectonics during the Proterozoic encounter the same difficulties, are sometimes even greater. According to geohistorical and geostructural data, it is fairly obvious that Gondwana was a single entity in the late Proterozoic. The problem of the northern continents is open to debate. On the one hand, coherent Precambrian structures appear to exist on reconstructions of Pangaea. This has been shown, in particular, for the Russian and Siberian platforms when these were matched (Gusev 1973a/b). On the other hand, the dispersal of northern continents in the early Paleozoic seems to be inherited to some extent from the late Proterozoic. At present, paleomagnetic data for the Proterozoic are quite abundant. Their interpretation, however, is difficult due to many factors. First, these data are distributed over an enormously long time interval. Second, their time reference is complicated by extensive, superimposed processes. In order to get some idea of the Precambrian paleogeography and paleogeodynamics, paleomagnetic data were used (Morel and Irving 1978; Khramov and Rodionov 1980a; Piper 1982). The most important results gained in these attempts are as follows. First, it became evident that it is impossible to fit the APWP's in the Precambrian either in the classical Pangaea of Wegener or in any other similar continental crust configuration that formed a single unit for the whole of the Precambrian (Morel and Irving 1978; Piper 1982a). Second, it was demonstrated that apparently major changes of the relative positions of continents occurred repeatedly during the Precambrian and the processes of collision and dispersal of the continental masses seem to obey the 600 m.y. cycle of Wilson (Morel and Irving 1978). Finally, certain evidence was obtained for the formation of the Precambrian orogenic belts that are situated inside the platforms at the present time: sometimes these belts originated during plate collision (microplates); the Australian, Siberian and probably the Russian plates appear to follow this pattern (McElhinny 1981; Forsh 1981; Khramov and Rodionov 1980a). In other cases the origin of similar belts was not accompanied by any significant intracratonic displacements and cratons, such as North America and Africa, might actually be regarded as single units since the earliest Precambrian (McElhinny 1981; Brock 1981; Piper 1982a).

The available paleomagnetic reconstructions for the Precambrian epochs are constrained, conflicting and hypothetic in many aspects because of the incompleteness of the paleomagnetic series with resulting ambiguity of the polarity problem. Thus, Piper (1982) established a configuration in which all major shields formed a single continental body and the paleomagnetic poles of these shields run as a narrow band for the interval from the beginning of Proterozoic until the middle of Riphean. Piper concluded that a single supercontinent existed during this entire range, which dispersed only at the beginning of the Cambrian. He acknowledged, however, that the data for the 800–570 m.y. interval were discrepant and a different assumption is possible. In contrast, Schmidt and Embleton (1981) constructed a common APWP for Africa, Australia and North America with Greenland, proceeding from an isolated situation similar to the present one. These authors suggested that this paradox could be overcome by assuming that the Earth's radius 1600–1000 m.y. ago, earlier, was only one-half the present radius and then in the 1600–1000 m.y. span it increased to its present size (see Sect. 7.2). Actually, an interval of 1250–600 m.y. provides the possibility of constructing a model of tectonic plate motion with rates similar to those in the Phanerozoic without using the Earth's expansion hypothesis (Khramov 1983).

4.5.4 History of Horizontal Motions of the Major Lithospheric Plates in the Proterozoic and Phanerozoic: Conclusions

The global paleomagnetic reconstructions for the Proterozoic and Phanerozoic epochs examined here, and by some other workers (Scotese et al. 1979; Zonenshain 1976; Poorter 1981), are definitely open to other alternatives and versions. These reconstructions, however, constitute a good basis for a general model of the history of global-scale motions in the post-Archean time which can be divided into several stages (Khramov 1983).

The first stage was at the end of Archean, the beginning of Proterozoic (2700–2200 m.y.). The continental crust stabilized as a single supercontinent "Pangaea F", its configuration differing greatly from the Pangaea of Wegener. This supercontinent slowly drifted relative to the pole but maintained its configuration (Piper 1982a).

The second stage was in the early Proterozoic (2000–1600 m.y.). The motion of Pangaea F relative to the pole accelerates. Certain constructional movements within the supercontinent were possible, resulting in the Svecokarelian-Hudsonian orogeny (1900 m.y.), (Piper 1982a).

The third stage was in the early and middle Riphean (1600–1000 m.y.) and included the formation and subsequent closing of narrow oceans. Rifting was most intensive during the magmatic interval of Jotnian-Mackenzian (1300–1250 m.y.), which was also possibly the first global-scale rifting event (Pesonen and Neuvonen 1981). This stage terminated with the closing of oceans (since 1200 m.y.) and with the Grenville orogeny, which resulted in the formation of Pangaea E (Morel and Irving 1978; Poorter 1981).

The fourth stage was in the late Riphean-Silurian (1000–400 m.y.). During the first half of this stage (the 4a stage, 1000–600 m.y.), the existing Pangaea E differed quite distinctly from the configuration of the previous Pangaea F. A new stage in the global rifting at the Vendian-Cambrian boundary resulted in the formation of the proto-Atlantic and paleo-Asiatic Oceans (the 4b stage). The Gondwana megacontinent existed at that time in the Southern Hemisphere and some isolated Northern Hemisphere continents were mainly situated in the lower latitudes. The ancient Pacific Ocean was situated entirely in the Northern Hemisphere and occupied most of its area. Thus, the reverse situation of the continental and oceanic hemispheres existed then compared with now (Gorodnitsky and Zonenshain 1979).

At the end of this stage (the 4b stage), during the Caledonian orogeny, the proto-Atlantic and paleo-Asiatic Oceans closed and Pangaea D appeared (Morel and Irving 1978; Gorodnitsky and Zonenshain 1979).

The fifth stage was in the post-Silurian time (< 400 m.y.). In the time span of 400–150 m.y. (the 5a stage), Pangaea existed, but its configuration changed from time to time as a result of the opening and closing of small oceans: paleo-Atlantic and paleo-Tethys. These Hercynian tectonic movements resulted in the consequent formation of Pangaea C, B and A, which drifted relative to the pole (Morel and Irving 1978; Khramov 1981). The 5b stage started with Pangaea drifting away from the southern pole and then dispersal was initiated. The first break-up of Gondwana possibly occurred along the eastern coast of Africa, then the Indo-Australian-Antarctic block appears to be subjected to dispersal. The Atlantic Ocean was being formed, whereby its first phase was the separation of Eurasia and Africa. All fragments of Pangaea (except Antarctica) generally drifted northward, Tethys closed and the Eurasian megacontinent was being formed in the Northern Hemisphere. This dispersal of Pangaea has been studied in detail according to plate tectonics (see, for instance, Gorodnitsky and Zonenshain 1979), thus the whole fifth stage could possibly terminate in 150–200 m.y. in the future with the formation of a new Pangaea (the 5b stage) (Khramov 1983).

This concept is characterized by the evolution of significant plate tectonic processes (Khramov 1983): at the first and second stages (2700–1600 m.y.), the process of global drifting of the supercontinent relative to the pole begins and accelerates and the process of rifting is just initiated. This rifting then, increases from stage to stage resulting in the formation, first, of narrow and, then, of wider secondary oceans; during plate motion, the differential component (continental drift) increases; a general displacement of the continental lithosphere relative to the pole becomes dominant at the beginning of Paleozoic, but disappears almost completely toward the end of Jurassic.

It is clear that the proposed model is not necessarily supported by the paleomagnetic data now available, as their interpretation is not unambiguous, as was mentioned above, especially with respect to data on the Precambrian (see Sect. 7.3)

Horizontal movements in fold areas, their origin and present structure are governed by the position (past or present) of these areas in the collision zones of lithospheric plates. Thus, the generation of the Alpine-Himalayan belt is the result of the collision of Africa and India with the Eurasian continent. Many fold belts and island arcs were initially formed as linear or slightly bent structures; only later deformations complicated their shape.

Paleomagnetic data aids in the reconstruction of the lithospheric plate position relative to the geographic grid in past epochs. They clarify the older, structural-tectonic relations and the predeformational structure of fold belts, and, thus, they provide data on the origin of sedimentary basins and the distribution of mineral deposits.

CHAPTER 5

Paleomagnetism and Tectogenetic Problems

5.1 Paleomagnetism in Fold Belts

5.1.1 Components of Tectonic Movements and Their Paleomagnetic Studies

It can be clearly seen on the tectonic map of the world that the fold systems occur in particular belts and are extensive (1000 km and more). The axial lines of the fold belts extend as straight lines or arcs. This arclike shape of the axial lines is particularly evident in many portions of the Alpine-Himalayan fold belt, in the northern Urals and in the eastern margins of the Siberian platform (the Verkhoyansk system).

Thus the ever present question for geologists: Is the shape of the axial lines in fold systems only primary or is their curvature a result of horizontal tectonic displacements? If the shape of an axial line is secondary, then the shape of an ancient sedimentary basin, reconstructed according to sediments forming these belts, also appears to be secondary, i.e. the initial outlines are disturbed by horizontal tectonic displacements. What is the true shape of an ancient sedimentary basin? Paleomagnetic data partially answer these questions.

The interpretation of paleomagnetic data from fold belts poses several problems related to the task, which must clearly be necessary solved (Van der Voo and Channel 1980; Khramov and Pechersky 1982).

The first task: how to find the resultant amplitude of a tectonic movement (displacement, rotation, deformation) according to the paleomagnetic elements (values of declination D and inclination I of the paleomagnetic vector) for a body and how to estimate the kinematic elements of the movement (direction and amount of displacement, situation of a rotational axis and its angle).

The second task consists in the determination of the limits to which the obtained data can be extrapolated, i.e. to establish the boundaries of a coherent (rigid) bed, block, region and province, and need to be outlined in plan, as well as in depth.

Generally, the rock massifs are obviously subjected to tectonic movements on different scales: local, regional and global. Thus, while interpreting a tectonic structure and the evolution of an area, we face the necessity of solving a third task: the definition of the movements and deformations in both space (i.e. the recognition of local, regional and global types), and time (i.e. the recognition of the movements of different orogenic cycles, phases).

All these tasks can be classified as indirect geophysical tasks of a multiple nature. Thus, only in some cases can these tasks be accomplished by the use of only paleomagnetic data. In general, structural, geohistorical and additional data are required.

5.1.2 The Determination of Local Movements and Correction for Their Effects; General Solution

Evidently, on a local scale, the solution of this task does not require taking the sphericity of the Earth's surface into consideration. In this case, the composite tectonic motions to which the rocks are subjected can be divided into translational movements and rotations. Both types of movement generally affect the distributions of directions in predeformational rock magnetization, and only the displacements along the paleolatitude are exceptions. Then, any rotation can be separated into two components: a rotation around a horizontal axis and a rotation around a vertical axis.

If often happens in geologic observations that one of these components is known, e.g. the elements of a bed occurrence (i.e. the orientation of a horizontal rotational axis and its angle) or the rotational angle around the vertical axis (e.g. if it is set at zero according to structural data). The usefulness of the division of a rotation into two components follows from the fact that one difficulty is overcome which arises in the general solution of the task: if the coordinates D and I are known prior to and after the rotation, it is not possible to determine the situation of the rotational axis and the rotational angle of a body: identical results might be obtained by an infinite set of different rotations (Le Pichon et al. 1973).

First, let us examine a general method of solving this task (Bazhenov 1980).

Any rotation in three-dimensional space can be described by a square matrix of the third order [A] with a determinant equal to unity; the vector magnitude and angles between the vectors are meanwhile maintained constant. With a knowledge of the matrix [A] we find the direction of the rotational axis in space and the value of the rotational angle. In contrast, with a knowledge of the axis direction and rotational angle, we can determine the matrix [A]. It is thus possible to calculate the coordinates of any vector after rotation has been performed if its initial coordinates and rotation matrix [A] are known:

$$(x)\,[A] = (y), \tag{5.1}$$

where (x), (y) are rows of vector coordinates prior to and after the performed rotation, respectively; multiplication is carried out according to the rules of matrix algebra.

However, if the coordinates of one vector, prior to and after the rotation, are known it is not possible to determine unambiguously the rotation matrix and, hence, the axis direction and rotational angle; to perform this operation the coordinates of two noncolinear vectors, prior to and after the rotation, must be known. When calculating the rotation matrix it is possible, for convenience, to introduce a third vector, e.g. the vector product of the first two. Then, the rotation of the three-vector system can be written as:

$$[x]\,[A] = [y], \tag{5.2}$$

where [x], [y] are the matrices of the vector coordinates prior to and after the rotation.

We find the rotation matrix by solving Eq. (5.2) relative to [A] according to the known matrices [x] and [y]:

$$[A] = [x^{-1}] [y], \quad (5.3)$$

where $[x^{-1}]$ is an inverse matrix of [x].

Thus, to evaluate the effects of rotations around an arbitrary axis from paleomagnetic determinations, one must solve Eq. (5.1); and to find the rotational angle and the axis position according to paleomagnetic data, one must solve Eq. (5.3). A method of solving these tasks by plotting on a Wulff net was proposed by Bazhenov (1979); below, the sequence of operations is briefly examined.

Assume two blocks of rocks with known directions of the ancient (prefolding) magnetization J^a_{n1} and J^a_{n2} and the elements of rock attitude (dip angle and azimuth). One must find the position of the rotational axis and the rotational angle of one block relative to the other. Vectors J^a_{n1} and J^a_{n2}, and normals to the beds, are plotted on a stereoprojection in the recent system of coordinates. Normal azimuths are equal to the azimuths of bedding dips, i.e. "inclinations" of normals are equal to the angles of bed dips minus 90°. Great circles (meridians of the equatorial grid) are drawn through the vectors and through normals, and the pole position for each circle is determined. Then we draw again great circles through the centre of the arcs that connect vectors and normals and the corresponding pole. The cross-point of these arcs would then give the position of the rotational axis. Next, we choose a system of coordinates for the vectors and normals in which the rotational axis is in the centre. The difference between the declinations of the vectors and the normals would then be the value of the rotational angle. Proceeding from a geologic situation, we then decide which block is to be regarded as immobile and which has undergone rotation. It is possible that both blocks were displaced relative to the stable blocks and that during these movements, they appeared to be displaced relative each other. This method might also be used to solve such a task.

5.1.3 The Determination of Local Movements and Correction for Their Effects; A Special Case

In the practice of paleomagnetic studies a particular case often occurs, in which the elements of the rock attitude are known and a bed dip is caused by a simple rotation around the strike line that did not change its orientation. The effect of such a rotation can be easily excluded using corresponding methods of sampling and data processing. This method applies the ancient system of coordinates: the surface of the rock bedding is recognized as the xOy plane and the Oz axis is directed along the normal to it. Rotation of a rock around the vertical axis by a certain angle would cause a change in the ancient (prefolding) magnetization orientation in the ancient system of coordinates for the same angle (Rzhevsky 1966). It should be noted that this rule is absolutely valid for any elements of a rock attitude.

So, if the component of movement, resulting from rotation around the horizontal axis, can be excluded by choosing a corresponding system of coordinates (only the elements of rock occurrence are known for performing this task), then the component resulting from the rotation of rocks around the vertical axis cannot be excluded using geologic data. This component can be measured, however, according to a change of the paleomagnetic direction in the ancient system of coordinates. This provides a good opportunity for studying the nature and amplitudes of horizontal tectonic displacements, as exemplified below.

Let a fold axis initially represent a straight line. As a result of tectonic movement, a certain part of this fold is then displaced in a horizontal plane and the axial line of the fold becomes bent. The line of the bed dip in the same limb now displays different azimuths A, but the declination D' of the primary magnetization relative to the line of dip remains the same. As the declination of a vector of primary magnetization in the ancient system of coordinates is determined according to the formula $D^\circ = D' + A$, then a change of azimuth of the dip line to a specific angle causes a change in this vector declination in the ancient system of coordinates by the same angle. Therefore, this bend of an axial line of a fold might be regarded as a result of the rotation of rocks around a vertical axis. It might also quantify this rotation by a change in the declination of a vector of primary magnetization in the ancient system of coordinates. If the angle of rotation α and the situation of the rotational axis is known, then the amplitude of horizontal displacement is calculated from

$$L = 2l \sin(\alpha/2), \tag{5.4}$$

where l is the distance between the examined site and the rotational axis.

This method allows the determination of the presence of a rotational component in a certain part of a fold, i.e. it becomes possible to recognize the nature of fold deformation.

However, a situation can occur in which a fold does not change the shape of an axial line, but rotates in a horizontal plane as a rigid body. In this case, the directions of primary magnetization in different parts of a fold converge, but they differ from the directions of primary magnetization of parts of other folds that did not rotate, thus permitting the recognition of the nature of rotation and to an evaluation of the amplitude of the horizontal displacement of a certain part of a fold using the above formula.

It should be recalled that the declination of a vector of primary magnetization changes in the ancient coordinate system not only when a fold rotates around a vertical axis, but also when a fold is inclined, i.e. it rotates around any horizontal axis. If the rotational axis is perpendicular to the fold axis, then the increment in declination ΔD is approximately in direct proportion to the angle of fold axis inclination γ; the coefficient of proportionality is linearly dependent from the angle of bed dip

$$\Delta D \approx (\beta/90)\gamma. \tag{5.5}$$

The increment ΔD might be positive as well as negative, depending upon which fold limb was sampled and which direction of slope the fold had relative to this limb. For determining the sign of ΔD, the following rule can be applied. If the system of right-angle coordinates is oriented in such a way that the axis Ox extends along the line of bed dip, the axis Oy is directed to the right, the axis Oz is down and the fold slopes towards -y, then the increment ΔD is positive (the vector rotates clockwise). When the fold has a slope towards +y, then the increment ΔD is negative (the vector rotates counter clockwise).

Thus, if a fold axis has tilted, the vectors of the different limbs would rotate in opposite directions and the rotational angle would be the larger, the larger the angle of the bed dip and the steeper the fold tilting. If a fold has undergone rotation around the vertical axis, then the vectors of primary magnetization in the ancient system of coordinates would give the same direction and the same angle, irrespective of the angles of bed dip. Taking into consideration the rules described above and also the data on the character of the rotation of vector J_n on different fold limbs, a value of the tilt angle of the fold axis, then, it appears possible to evaluate which part of the rotational angle of vector J_n in a horizontal plane is due to rock mass rotation around the vertical axis and which part is due to rotation around the horizontal axis.

5.1.4 Regional and Local Components of Tectonic Movements

By definition, a regional component of tectonic movement consists only of rotations around vertical axes: the poles of rotation (see Sect. 4.1.1). Therefore, the definition of this component is reduced to the search for a rotational pole and a rotational angle from the data on the horizontal components of displacement and rotations in certain localities.

Detailed paleomagnetic studies in fold areas have revealed that deformations within separate regions, belts and zones, are distributed in such a way that the poles of relative rotations are situated within these zones or nearby (Bazhenov et al. 1978; Rzhevsky 1977; Grubbs and Van der Voo 1976; Van der Voo and Channel 1980, Shevlyagin 1982).

This means that the paleomagnetic poles obtained using local data are distributed on the Earth's surface along the arc of a small circle with its centre close to the site of observation and with a radius $r = 90° - \varphi_c$, where φ_c is the mean value of the paleomagnetic latitude of the mean (regional) pole of rotation (φ_p, λ_p).

The kinematic parameters of regional movement are defined by finding a small circle that approximates to a mean line along the band of paleomagnetic poles and then, the pole of this circle is found. The angles ψ are easily determined between the modern meridian of the rotational pole site and the radii for n local paleomagnetic poles and their mean value $\psi = \Sigma\psi_i / n$. The obtained coordinates of the rotational pole (φ_p, λ_p) and values of r and ψ define the situation of a "regional" paleomagnetic pole. The angular distance ψ_p between this pole and a paleomagnetic pole of the same age for an adjacent stable block (craton) would be the value of angle ψ_p of a regional relative rotation, and the differences $\psi_i - \psi_p$ would be the angles of relative local rotations. All these angles are counted off along the arc of the small circle with the centre in a common pole of the regional and local rotations.

Two different situations originating from this interpretation are shown in Fig. 5.1. In the first situation (left) the rotational pole lies practically in the centre of the observation site and the regional component of rotation is small. In the second situation (right) the rotation pole lies away from the observation site and corresponds to great angle ψ_p of a regional rotation. This difference can be explained by a different geologic environment and the difference in the bodies studied: in the first case, the paleomagnetic directions were distributed along the tectonic sheet that overthrust the North American craton and in the second case, the observations were conducted in the Caribbean microplates.

The determination of the kinematic parameters of the global (see Sect. 4.4.2), regional and local movements permits the construction of palinspastic maps (schemes) for a given region and for a given period of time.

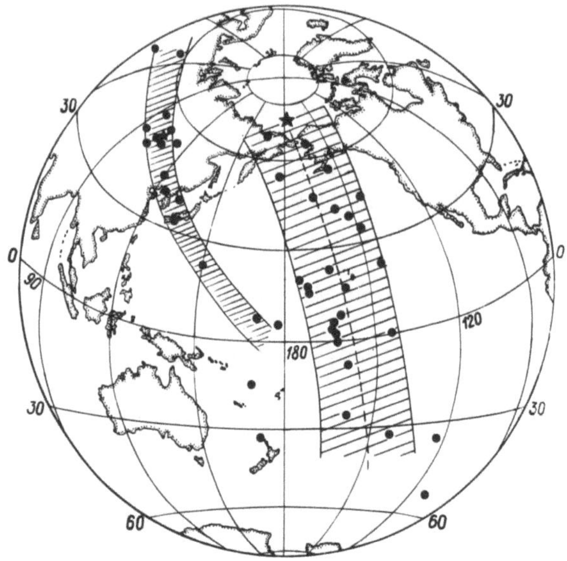

Fig. 5.1. Paleomagnetic poles for active tectonic zones. *Left* Triassic poles from the Idaho-Wyoming thrust belt. *Right* Cretaceous poles from the Caribbean region. (After Grubbs and Van der Voo 1976; Van der Voo and Channel 1980)

5.2 Palinspastic Reconstructions According to Paleomagnetic Data

5.2.1 Local Tectonic Movements in the Tajik Depression and in Southern Tien Shan

Reconstructions of the older sedimentary basins in the eastern part of the Alpine-Himalayan belt according to paleomagnetic data are of great interest from the viewpoint of its development as well as for exploration prospects. The Tajik depression and Southern Tien Shan are the areas best studied paleomagnetically of those areas examined.

The whole tectonic evolution of the Southern Tien Shan can be divided into several stages. During the first stage that embraced the entire Paleozoic Era, this region was part of an extensive geosynclinal belt where downwarping recurrently changed into mountain-building with contemporaneous magmatic activity. This stage terminated only at the end of Paleozoic, when the Paleozoic fold structures entered a new stage of evolution that was very similar in nature to the regime of younger platforms. During the second stage, that lasted for the entire Mesozoic and a major part of the Paleogene, the intermontane depressions were formed with arched uplifts and basement rises separating them. The Tajik depression is one of those intermontane depressions, where the platform formations are of considerable thickness (up to 4 km). In the area examined the tectonic movements started in the Oligocene and resulted in the creation of an actual mountain relief. The anticlinal folds are seen in relief as ridges and uplifts, while the synclinal folds form valleys and depressions separating them. The folds are mainly of a linear type plunging southward. In some places a lateral effect of one structure on another can be observed, expressed as a bending of the axial lines, or by overthrust structures (Zakharov 1964).

Because of these morphological peculiarities some workers concluded that the Tajik depression structures evolved under horizontal compression. The opinion of other workers, however, is that the folds in this depression are the result of differential vertical movements of basement blocks. The conventional geologic methods do not provide a definite answer as to whether the strike and shape of axial lines is primary or secondary. For solving this task, the Lower Cretaceous sediments were studied from the viewpoint of their paleomagnetic characteristics.

The Lower Cretaceous sediments in the Tajik depression are of great thickness (800–1500 m) and are mainly composed of red sandstones, siltstones and argillites. In most cases oriented samples are available, i.e. are ideal for paleomagnetic studies. Samples were taken along profiles oriented across the strike. Two to three profiles were sampled for each structure; if possible, samples were taken from both limbs to correct for a vector rotation due to axis tilting, in order to obtain a more precise angle for mass rotations around vertical axes in combination with method using the crossing of remagnetization circles.

We shall elucidate the concept of a local rotation by an example of the Nuranch anticline situated in the northeastern extremity of the Tajik depression (the western extremity of the Peter the First Range). This is a linear fold oriented north north eastward. Beds in the northwestern limb are of normal occurrence, the mean azimuth of the dip is 310° and the mean dip angle is 45°. Beds in the southeastern limb exhibit a mean azimuth of the dip of 151° and dip angles vary from 85° to 135° (most beds in this case are overturned). Along profile II in the southeastern limb, 36 samples of Albian age were taken. The following mean values characterize the distribution of magnetization vectors in the samples of this collection (Fig. 5.2a): $D = 57°$, $I = 72°$, $\alpha_{95} = 7°$, $K = 15$. At the same site 17 samples were taken from the Valanginian-Hauterivian stage (Fig. 5.2b); $D = 51°$, $I = 68°$, $\alpha_{95} = 12°$, $K = 8$. Along profile III, 53 samples of Valanginian-Hauterivian sediments were taken in the northwestern limb (Fig. 5.2c): $D = 341°$, $I = 49°$, $\alpha_{95} = 6°$, $K = 13$.

Remagnetization circles of the Albian and Valanginian-Hauterivian rocks (Fig. 5.2d) for the southeastern limb cross quite close to a point with coordinates $D = 352°$, $I = 56°$, which characterizes the

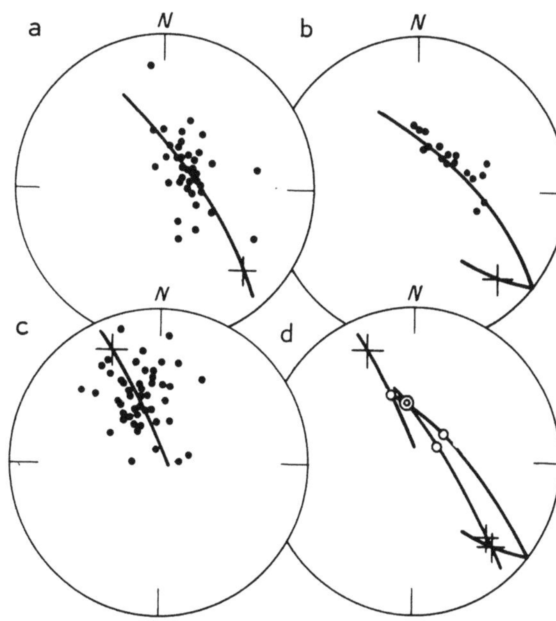

Fig. 5.2a-d. NRM directions of the lower Cretaceous red sediments forming the Nuranch anticline, Tajik depression. (Rzhevsky 1973). **a,b** South-east slope of the fold; **c** north-west slope of the fold; **d** resultant projection of remagnetization planes (a *double-circled* point denotes the direction of the prefolding NRM component)

direction of ancient magnetization of the Lower Cretaceous rocks in the Tajik depression (Rzhevsky 1977). Thus, it may be concluded that the southeastern limb did not rotate around a vertical axis and the fold was not subjected to any appreciable tilt. The remagnetization circle of the Valanginian-Hauterivian rocks in the northwestern limb lies away from the cross-point of the two above-mentioned circles. This can be explained by the fact that the rocks of the northwestern limb rotated counterclockwise in the horizontal plane. By rotation along the remagnetization circle, we find that the angle or rock rotation amounts to 10°. It is quite possible that the fold was subjected to additional horizontal compression, which resulted in this counterclockwise rotation of its northwestern limb. If it is considered that the rotational axis of the rocks extends along the northern pericline of the fold and that a distance of 10 km is acceptable as the distance from profile to the axis, then the width across the fold decreases for 1-1.5 km. Similarly, the rotation of vector J_n for other sections studied in the Tajik depression can be determined; their vectors J_n also form a rather tightly clustered group; the direction of the mean vector can be determined with high precision $\alpha = 3° - 12°$.

5.2.2 Construction of the Palinspastic Scheme of the Tajik Depression in the Early Cretaceous

If the corrections for the postfolding tilt of structures are removed from the directions of the horizontal component of the prefolding magnetization of the Tajik depression rocks, then the residual vector orientation differences are a result of the rotation of structures around a vertical axis (Fig. 5.3a). Directions of J_n^a to the north and west of the Major Hissar fault, and on the southern slope of the Alay Range, exhibit a northeastern declination. Southward of this fault, i.e. in the western extremity of the Tajik depression, on the southern slope of the Hissar Range and in the Peter the First Range, J_n^a vectors exhibit a northwestern declination and only slight differences. In the centre of the depression and in its eastern part, J_n^a vectors exhibit clear clockwise or counterclockwise rotations. It is thus reasonable to conclude that part of the Earth's crust, where the Tajik depression is situated, rotated as a single unit counterclockwise for 10°-20° and simultaneously, separate structures underwent local rotation, clockwise or counterclockwise. The Babatag structure was subjected to an additional, clockwise rotation by 15°-20° and the Sarsaryak structure, counterclockwise by 50°; the Darvaz Range structure underwent an additional, counterclockwise rotation by 60°. The boundary of the mobile part of the Southern Tien Shan approximately coincides with the Major Hissar fault and represents an arc of a circle with a diameter of 200-220 km and a centre in the area of the Vakhsh River mouth.

Using the above data, an attempt was made to construct a palinspastic scheme of the Tajik depression in which different parts of the depression are returned to their initial position. First, the structures are reversely rotated around the vertical axes situated in the southern terminations of ranges of the same names. The sites of the rotational axes were chosen on the basis that the amplitudes of the thrusts complicating these structures usually decrease southward. Furthermore, this principle in choosing the sites gives the best consistency in paleomagnetic inclinations for the different structures. The local rotations of the depression structures were excluded for this operation.

Before placing the entire Tajik depression into its initial position by rotation around a vertical axis, we need to consider the fact that the Peter the First Range is situated farther northward of the line forming the eastern continuation of the

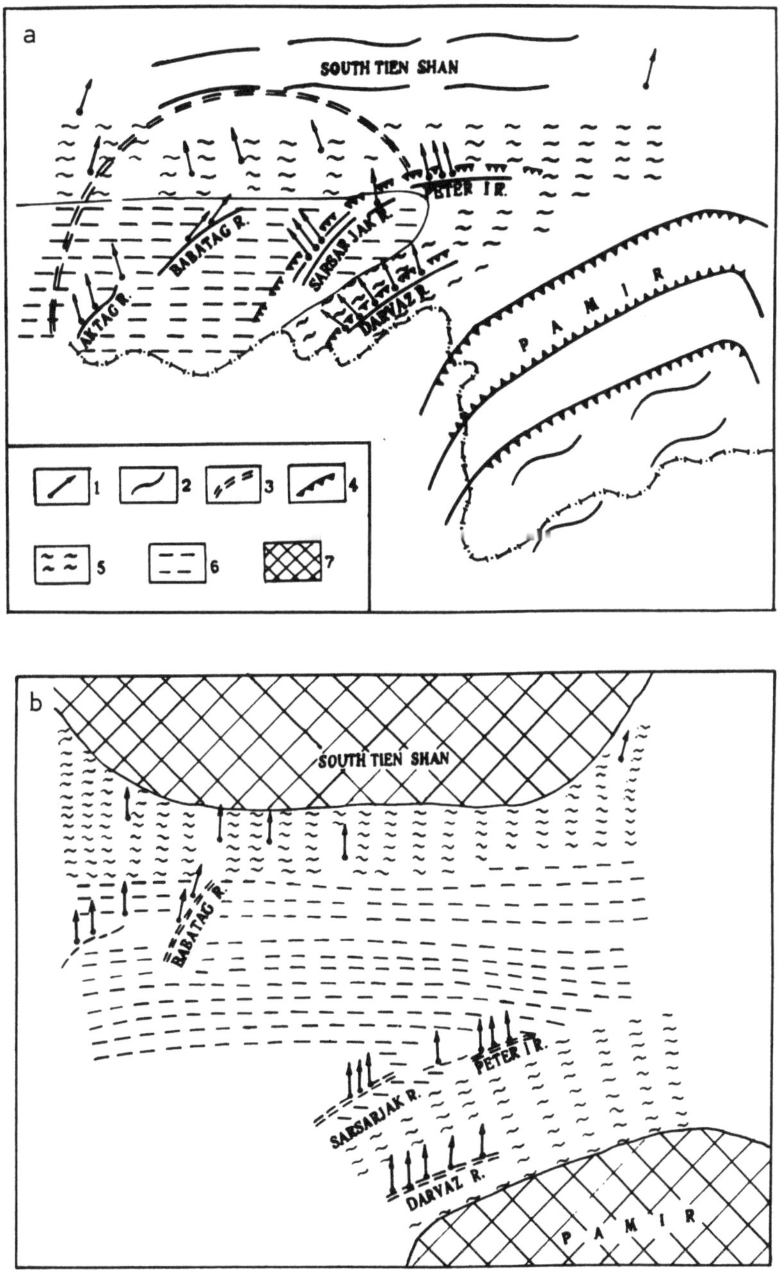

Fig. 5.3a,b. Lower Cretaceous paleomagnetic data and the distribution of Paleogene sedimentary facies in the Tajik depression. **a** Relative to the present geographic grid; **b** in the palinspastic restoration. (Rzhevsky and Khramov 1985)

boundary of the Southern Tien Shan mobile part. The Peter the First Range is likely to exhibit a translational northward motion of 160–180 km after it is rotated as a part of the Tajik depression. Thus, we move the Peter the First Range and Darvaz Range southward for the this distance in order to place the sequences studied in these ranges farther south of the line which forms continuation of the boundary of the Southern Tien Shan mobile part. As the sequences in the Sarsaryak Range are similar to those in the Peter the First Range (Davidson and Rzhevsky 1979), we also move these ranges southward by 160–180 km. A portion of the scheme to the south of the boundary of the Southern Tien Shan mobile part is rotated clockwise around a vertical axis by about 20° to obtain an azimuth of the horizontal component of the prefolding rock magnetization in the mobile part of the Tajik depression equal to that of the stable parts of the Southern Tien Shan. This rotation displaces the Peter the First and Sarsaryak Ranges southward by a further 50–60 km.

In the palinspastic scheme construction it was assumed that all of the displacements concerned in the Pamirs area so that its resultant displacement southward appeared to be 330–340 km. It should be noted, however, that in the palinspastic scheme construction we used only data on the rotational angles of this or that range about the geometrical size of the ranges proper. Consequently, in the scheme the minimal possible values of the horizontal displacements are shown; it is not excluded that the rocks composing the Peter the First and Darvaz Ranges were deposited even further southwards.

Using the obtained data we attempted to outline the sequence of the tectonic movements which resulted in the formation of the Tajik depression structure. At the same time, the following fact must be considered. When the resetting of those parts of the Tajik depression, which underwent rotation, was completed and they had acquired their original position, the azimuths of lines of rock dips appeared to be approximately equal to 165°–175° in almost every sequence. It is of interest that in the Southern Tien Shan (Konchoch, Darautkurgan), azimuths of the Cretaceous rock dips are quite similar. Thus, at the first stage of orogenesis the folds generated in bedded strata and these folds had an approximately latitudinal strike. Afterwards, the area south of the Major Hissar fault rotated as a single unit counterclockwise by 15°–20°. Then, an enormous northward displacement (up to 400 km) of the Pamirs occurred. During this motion, a strong horizontal compression, initiated in the sedimentary cover of the Tajik depression, resulted in a complication of folds by lateral interaction and thus in the formation of ranges and intermontane depressions. At that time, the rotation of rock masses around a vertical axis was taking place, accompanied by overthrusting.

5.2.3 Reconstruction of the Paleogene Sedimentary Basin of the Tajik Depression

In the scheme given here it can be seen that the sedimentary basin of the Tajik depression in the period of its initiation had an approximately latitudinal strike and the width of that ancient basin was 2–2.5 times greater than of the present basin. Reorientation of the tectonic zones from this latitudinal to a meridional direction mainly occurred in the Neogene-Anthropogene period (Zakharov 1964). Horizontal displacements in the Tajik depression should have changed the regularities of the original distribution of facies. The analysis of facies and the thickness of continental strata, however, revealed that this comparison is problematic due to inhomogeneity in the lithologic composition of the Lower Cretaceous rocks in the Tajik depression, which obscures any general regularities in sedimentation. Thus, it is preferable to analyze the Paleogene marine deposits which have a more obvious change in the general plan of their facies zonation.

However, the question arises, whether it is possible to use the palinspastic scheme, based on paleomagnetic data for the Lower Cretaceous sediments, in reconstructing the process of sedimentation in the Paleogene. A positive answer can be given, because the paleomagnetic data for the Paleogene red beds, exposed in the Outer Arc of the Pamirs, are consistent with the paleomagnetic data for the Lower Cretaceous sediments of these sections (within the precision of the method). Thus the nature and values of displacement of the Paleogene and Lower Cretaceous rocks in the Tajik depression are similar (Bazhenov 1980; Rzhevsky 1977).

Results of the Paleogene sediment studies in the Tajik depression that were conducted for many years revealed a submeridional (relative to the present geographic grid), lithologic facies zonation due to the effects of the source areas: in the north, the Tien Shan area and in the Pamirs in the east (Fig. 5.3a). To the north of the Tajik depression (the Southern Tien Shan extension) and to the east (the Pamirs extension), zones of shallow-water deposits are known to occur. South of the Southern Tien Shan and west of the Pamirs, they pass into zones of relatively deep-water sediments. In the shallow-water zones the Paleogene strata are made up of lagoon-continental rocks and in the relatively deep-water zones, the rocks are of lagoon-marine and marine origin. Thus, a normal submeridional plan in the distribution of facies appears to exist from the continental and littoral to sublittoral types.

In the system of coordinates of the palinspastic scheme, the facies distribution appears to be latitudinal. Starting from the Pamirs towards the Darvaz Range and then towards the Peter the First Range, a gradual deepening of the sedimentary basin is observed within the shallow-water zone (from conglomerates in the Darvaz Range to red sandstones and clays in the Peter the First Range). Further northward (along the Southern Tien Shan) the basin of sedimentation becomes shallower again and a band of shallow-water deposits is recorded there. In the present map (Fig. 5.3a) the shallow-water sediments of the Peter the First Range occupy the same latitude as the relatively deep-water Paleogene sediments of the Sarsaryak Range, i.e. there is an inconsistency in facies distribution in the latitudinal plan. In the palinspastic scheme (Fig. 5.3b) the area of sedimentation in the Peter the First Range is displaced southward. It was stated earlier that the results of paleomagnetic studies are concordant with a displacement of this type and size.

Thus, in the palinspastic scheme, the sedimentary basin during the Paleogene also displays a latitudinal orientation: the Southern Tien Shan served as the northern margin of the basin and the Pamirs served as its southern margin. Relative to the present geographic grid, the Paleogene sedimentary basin of the Tajik depression appears to be submeridional. If the palinspastic scheme of the Tajik depression is considered as correct, then it can be concluded that the northward displacement of the Peter the First and Darvaz Ranges occurred as a large-scale overthrust and the Paleogene shallow-water deposits underlie the overthrust folds of the Mesozoic and Cenozoic which comprise the Peter the First Range. It is hoped that the results of deep drilling will confirm the validity of this conclusion.

5.2.4 Palinspastic Scheme for Central Asia During the Early Permian Epoch

Paleomagnetic data on the Upper Paleozoic rocks of fold areas in the southern part of the USSR are available from published works: e.g. on the Mangyshlak Peninsula, Chatkal-Kurama, Central Kazakhstan and Tajikistan. In Mangyshlak and Tuarkyr Rzhevsky, Slautsitais (1976) studied the Permian and Triassic red beds and concluded that their remanent magnetization was synchronous with sedimentation and that it might be used for paleomagnetic reconstructions. Akhmatov (1966) studied volcanic formations of the late Paleozoic in the Chatkal-Kurama region. Tursunov (1976) followed with his studies in the Zailiysky Alatau and in the Southern Dzungaria and both established a high paleomagnetic stability for the rocks studied. Kumpan et al. (1968) also underlined the high paleomagnetic stability of the Carboniferous and Permian deposits in the Tengiz, Dzhezkazgan-Sarysu and Chu depressions.

In 1975–1980 Rzhevsky conducted paleomagnetic studies of the Paleozoic formations in different parts of the Hissar Range. The rocks of the Lower and Middle Paleozoic are known to have late Paleozoic magnetization and the extensive block tectonics in the Hissar Range, during Alpine folding, resulted in significant differences in the direction of the mean magnetization of the blocks studied. If all local rotations of the blocks are considered, then, for the early Permian rocks in the Hissar Range, the mean direction of a stable component of remanent magnetization would have coordinates $D = 176°$, $I = -36°$ (Rzhevsky and Khramov 1985).

Table 5.1 contains the paleomagnetic data from the works mentioned above. Unfortunately, data on the Lower Permian for Mangyshlak and Tuarkyr are not as yet available, therefore, we used information from the Upper Permian of these regions which seemed tenable since the major phase of folding has been registered as occurring at the end of the Triassic and beginning of the Jurassic. It is also quite important that during the Permian period, the displacement of Eurasia was insignificant (see Chap. 4). In Fig. 5.4 the sites of the paleomagnetic determina-

Table 5.1 Permian paleomagnetic directions for Central Asia and Kazakhstan

Area		Age	Paleomagnetic direction D	I	Paleolatitude φ_m	Reference[a]
Mangyshlak	44°N, 58°E	P_2	222°	–40°	23°	1
Tuarkyr	41°N, 55°E	P_2	225°	–34°	19°	1
Chatkal-Kurama region	41°N, 70°E	P_1	140°	–40°	23°	2
Zailiysky Alatau and S. Dzungaria	45°N, 77°E	C_3-P_1	141°	–52°	33°	3
SE Chu-Ili interfluve	44°N, 75°E	C_3-P_1	209°	–46°	28°	3
Central Kazakhstan	52°N, 67.5°E	P_1	233°	–56°	36.5°	4
Southern Tien Shan	39°N, 69°E	P_1	176°	–36°	20°	5

[a] *1* Khramov et al. 1974; *2* Khramov 1979; *3* Tursunov 1976; *4* Kumpan et al. 1968; *5* Rzhevsky, this work.

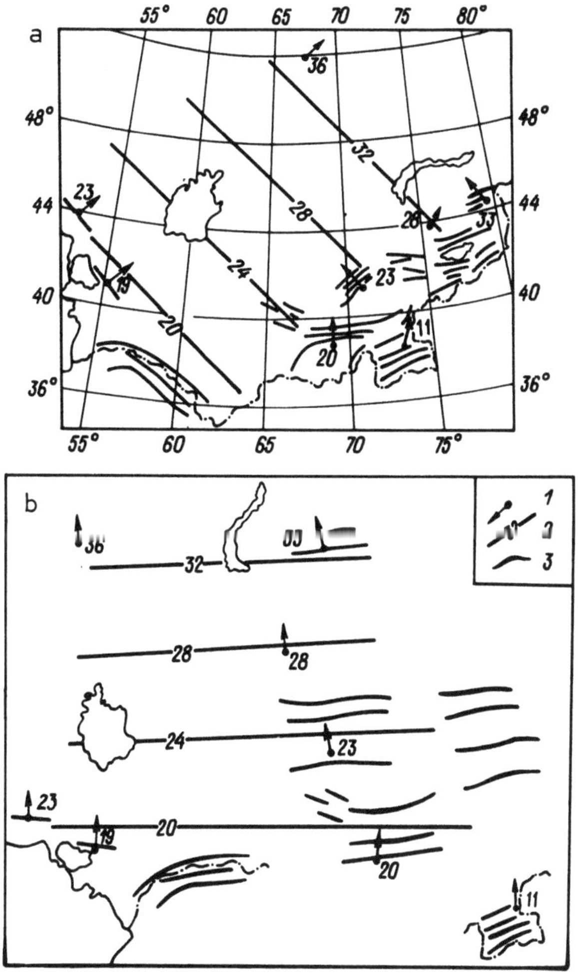

Fig. 5.4a,b. Palinspastic restora tion for Soviet Central Asia in th Permian according to paleomag netic data; **a** present position o the sites and orogenic units; **b** thei Permian position. (Rzhevsky anc Khramov 1985). *1* Paleomagneti direction and paleolatitude in the sampling locality; *2* lines of pa- leolatitudes according to paleo- magnetic data from Russian plate; *3* main mountain ridges and an- cient positions of the rocks from which they are composed

tions, the orientations of ancient meridians and the paleolatitude values are presented. It is notable that paleomeridian directions and paleolatitude values in Tuarkyr, Mangyshlak and in the Central Kazakhstan show good consistency.

Thus, it may be concluded that Tuarkyr, Mangyshlak, the Central Kazakhstan and the regions separating them were displaced northward after Permian times and rotated clockwise as a single unit. For Central Kazakhstan the amplitude of this displacement (14°) and the rotational angle (24°–27° clockwise) were determined by Kumpan et al. (1968). Folded areas of the Southern Dzungaria, Zailiysky Alatau, Chatkal-Kurama and southern Tien Shan are likely to have undergone additional northward displacement and counterclockwise rotation as the paleolatitudinal values in these regions differ greatly from the extrapolated values. Furthermore, the paleomeridians are rotated counterclockwise by significant angles relative to the extrapolated values.

Thus, it has been attempted to eliminate these additional displacements and rotations of the fold areas by using the geometrical sizes of the displaced areas, the values of the rotational angles and the differences between the paleolatitudes φ_m and the extrapolated paleolatitudes.

In Zailiysky Alatau and southern Dzungaria the paleomeridian azimuth is 140°, the azimuth of the extrapolated meridian is 225°, i.e. these areas underwent additional counterclockwise rotation by approximately 85°. Proceeding from the supposition that the rotational axis was situated in the western ends of the ranges, we return these areas back to their original position by a clockwise rotation of 85°. This means that the paleolatitude of the reconstructed areas now coincide with the extrapolated paleolatitude. However, these areas underwent only this additional rotation, and the additional northward displacement was insignificant. The eastern parts of the ranges, however, moved northward for 350–500 km as a result of rotation. It must be emphasized again that these rotations could result in the generation of mountain ranges. Therefore, generally, with respect to the palinspastic scheme, we do not operate the mountain ranges, but with the axial lines of future ranges.

In the area of Chu-Ili the paleomeridian azimuth is 209°, the paleolatitude is +28°, the azimuth of the extrapolated paleomeridian is 225° and the extrapolated paleolatitude is +32°. In order to eliminate the additional rotation and displacement, this area must be rotated clockwise by 16° and displaced southward by 4°.

In the Chatkal-Kurama region the paleomeridian azimuth is 141°, the paleolatitude is +23°, the azimuth of the extrapolated paleomeridian is 225° and the extrapolated latitude is 28°. Consequently, this area underwent additional counterclockwise rotation by approximately 84° and was displaced northward by 5°. After reconstruction different parts of the Chatkal-Kurama region appeared to move southward by varying distances. The southwestern ends of the ranges were displaced southward for only 300 km, whereas their northeastern ends, due to rotation, were moved a further 350–500 km.

In southern Tien Shan the paleomeridian azimuth is 176°, the paleolatitude is +20°, the azimuth of the extrapolated paleomeridian is 225° and the extrapolated latitude is +24°. Consequently, this area sustained additional counterclockwise rotation of approximately 49° and moved northward by 4°. During the construction of the palinspastic scheme, certain parts also appeared to have been displaced by varying distances: i.e. western terminations of the ranges were displaced southward by only 3°, while the eastern terminations were displaced a further 400–500 km due to rotations.

Thus, when the additional displacements and rotations are excluded, the territory examined can be returned to its initial position as a single unit relative to the geographic grid. In order to carry out this operation, we rotate it counterclockwise for 25° around an axis situated in central Kazakhstan and move it southward by 14°. At the moment paleomagnetic data are available for the Upper Permian strata of the southeastern Pamirs (see Table 5.2) and in the palinspastic scheme this is placed according to these data. The palinspastic scheme then shows that, in the early Permian, the tectonic zones were latitudinal lineations and their total width was about 1600 km. The axial lines of the future ranges were approximately parallel. When comparing this scheme with the present situation of the ranges, it may be concluded that since the early Permian and up to the present, the width of the territory examined diminished by a factor of two. Certain areas, separated then for 350–500 km, now appear to be closely situated (mainly the northern terminations of the ranges).

5.3 Deformations in Certain Fold Belts of Eurasia According to Paleomagnetic Data

5.3.1 On the Nature of Tectonic Deformations in the Alpine-Himalayan Fold Belt

Paleomagnetic studies with the goal of solving tectonic problems have revealed that rotations of the crustal blocks and their displacement for great distances are a common phenomenon in the Alpine-Himalayan fold belt (Asanidze and Pechersky 1979; Van der Voo and Channell 1980; Khramov and Pechersky 1982).

Table 5.2 Cretaceous and Permian paleomagnetic data from the eastern segment of the Alpine-Himalayan fold belt

No. of entry	Sampling Area		Age	Pole position[a]		Kinematics[b]		Reference[c]
				Φ	λ	Δφ	Δψ	
1	Caucasus, Lok-Karabakh	41.0°N, 45.0°E	K_1	61°	158°	16°	30°	7
2	Caucasus, Kafan	39.0°N, 46.0°E	K_1	66°	169°	15°	22°	7
3	Alborz Range	36.3°N, 51.8°E	K_1	61°	148°	7°	33°	10
4	Lut Block	33.2°N, 57.5°E	K_1	58°	315°	11°	−34°	9
5	Lut Block	31.6°N, 55.8°E	K_1	68	306°	9°	−22°	4
6	South Tien Shan	40.0°N, 69.0°E	K_1	78°	205°	9°	10°	8
7	Tajik Depression	39.0°N, 69.0°E	K_1	83°	320°	3°	−8°	2
8	Afghanian Microcontinent	31.6°N, 65.8°E	K	66°	312°	11°	−23°	4
9	NW Himalayas	34.4°N, 76.9°E	K_1	56°	271°	32°	−8°	5
10	NE Baluchistan	30.1°N, 66.6°E	K	11°	295°	59°	−54°	5
11	Sulaiman Range	30.2°N, 68.2°E	K	18°	295°	54°	−50°	5
12	NE Himalayas	28.5°N, 83.5°E	K_1	15°	288°	68°	−31°	6
13	North Eurasian Plate		K_1	77°	180°	6°	13°	
14	Indian Plate		K_1	19°	298°	56°	−46°	
15	Greater Caucasus	44.0°N, 42.0°E	P_1	30°	146°	33°	60°	3
16	Lut Block	31.6°N, 55.8°E	P_2	24°	328°	18°	−70°	4
17	SE Pamirs	38.0°N, 74.0°E	P_2-T_1	39°	183°	26°	48°	1
18	Afghan Microcontinent	34.1°N, 68.5°E	P_2	8°	310°	46°	[illegible]	1
19	Salt Range	32.5°N, 72.5°E	P_2	−13°	317°	60°	−88°	5
20	NW Himalayas	33.9°N, 75.0°E	P_1	−28°	310°	76°	−99°	5
21	SE Himalayas	28.5°N, 83.5°E	P_1	−5°	290°	74°	−68°	6
22	North Eurasian Plate		P_1	40°	167°	35°	40°	
23	Indian Plate		P-T_1	−5°	307°	59°	−79°	

[a]Φ, Λ Latitude and longitude of the pole corresponding to the prefolding magnetization.

[b]Δ φ is an angular northward displacement of the sampling areas; Δ ψ is the rotation about the vertical axis (clockwise is positive); confidence radii for angular values and poles are not shown, they fall within a range of 2°–11°

[c]*1* Davydov et al. 1982; *2* Khramov 1971; *3* Khramov 1982; *4* Klootwijk 1979a; *5* Klootwijk 1979b; *6* Klootwijk and Bingham 1980; *7* Lordkipanidze et al. 1984; *8* Rzhevsky, this work; *9* Wensink 1982; *10* Wensink and Warekamp 1980.

This was registered for the first time in the study of the Mesozoic and Paleogene formations of the Iberian Peninsula, which had rotated counterclockwise relative to the North-European plate since the end of Mesozoic by 35° (McElhinny 1973; Van der Voo Channell 1980). The Sardinian and Corsican blocks, and parts of Italy were subjected to even larger rotations, up to 40°–50° counterclockwise. Those rotations occurred mainly in the Paleogene and at the very beginning of Neogene (Van der Voo and Channell 1980). The Argos Peninsula in post-Jurassic time rotated clockwise by 100°–120° relative to the Greek continental block which was itself subjected to a counterclockwise rotation. The same rotation has been registered for Jurassic rocks of the Transdanubian region according to paleomagnetic data. Paleomagnetic poles of all these regions, regarded as median massifs and microplates, are very similar to those of Africa (see Appendix). Probably, this indicates that these structural units belonged to the African plate and that, in the Mesozoic and the beginning of Cenozoic, they moved together.

Paleomagnetic data from the central part of the belt also reveal the dominance of counterclockwise rotations (see Table 5.2). Northern Turkey up to middle Cretaceous, and central and northern Iran up to the end of Paleozoic, used to be a part of Gondwanaland (the African-Arabian plate) and moved with it as a single unit. Local rotations are also known within this megablock. Thus, the Lut block in post-middle Paleogene time rotated counterclockwise and the central Alborz area rotated clockwise by 30°. On analyzing these data it was concluded that, in Paleozoic, the region was part of Gondwana; in the Triassic, it drifted as a separate microplate; and since the Jurassic, it drifted as part of the North Eurasian plate (Asanidze and Pechersky 1979; Van der Voo and Channell 1980). The Lesser Caucasus area, in the middle Jurassic, rotated counterclockwise by 40°–45° relative to the Greater Caucasus and afterwards, both areas acquired a low latitudinal position. At the end of Cretaceous and later, an archlike bending of the Lesser Caucasus area occurred as a result of the collision of the African-Arabian and East European continents with the Arabian wedge impressed in the Caucasus mobile belt (Asanidze and Pechersky 1979).

Thus, the lithospheric blocks in the western and central parts of the Alpine-Himalayan belt underwent mainly counterclockwise rotations. Deformations there appear to be a result of the collision of the Eurasian and African-Arabian continents and their pattern can be explained by the different rate of colliding in different areas. In the eastern part of the Alpine-Himalayan belt, the deformation pattern is produced mainly by the northward movement of the Indian plate and its collision with northern Eurasia. Paleomagnetic data show that the Iranian plate separated from Gondwana in pre-Permian time, drifting as a single microcontinent. It was only in the early Cretaceous that it joined the North Eurasian plate (Asanidze and Pechersky 1979; Van der Voo and Channell 1980). Later, deformations in its folded surrounding areas occurred at the end of the Paleogene and Neogene and were the result of the pressing in of the Indian (Punjab) wedge, as observed earlier in the Turkey-Lesser Caucasus area.

Now we will examine in more detail the pattern of displacement of some continental blocks in the eastern part of the Alpine-Himalayan fold belt. The mean paleomagnetic poles (see Table 5.2) show that structures, such as the Southern Tien Shan, Tajik depression, and Afghan microcontinent in early Cretaceous time, were part of the North Eurasian plate. The structures, such as the northeastern Baluchistan, Sulaiman Mts. and the southeastern Himalayas in early Cretaceous time were situated more southward when compared to their present position (Fig. 5.5) and were part of the Indian plate.

The northwestern Himalayas were an island arc (Valdiya 1984) situated on the equator, near the middle part of the Mesotethys Ocean. The northern coast of this ocean was the continental margin of the North Eurasian plate and its southern coast was the Indian subcontinent. The Mesotethys Ocean, according to paleomagnetic data, had a width of about 4000–5000 km.

The mean paleomagnetic poles from Table 5.2 show that in the late Paleozoic, the southeastern Pamirs, and also the structural units of the central part of the belt (the Greater Caucasus, Lut block, etc.), were part of the North Eurasian plate. The Afghan microcontinent, as can be seen from Fig. 5.5b, was situated in the middle part of the Paleotethys Ocean which had a width of about 4000 km. The Salt Range, the northwestern and southeastern Himalayas were not far away from the Indian subcontinent[5], which was situated in the southern latitudes and was oriented northward with its present western margin.

[5]It should be noted that in the late Paleozoic the above mentioned structural units did not exist, i.e. those late Paleozoic structures that later composed part of these structural units.

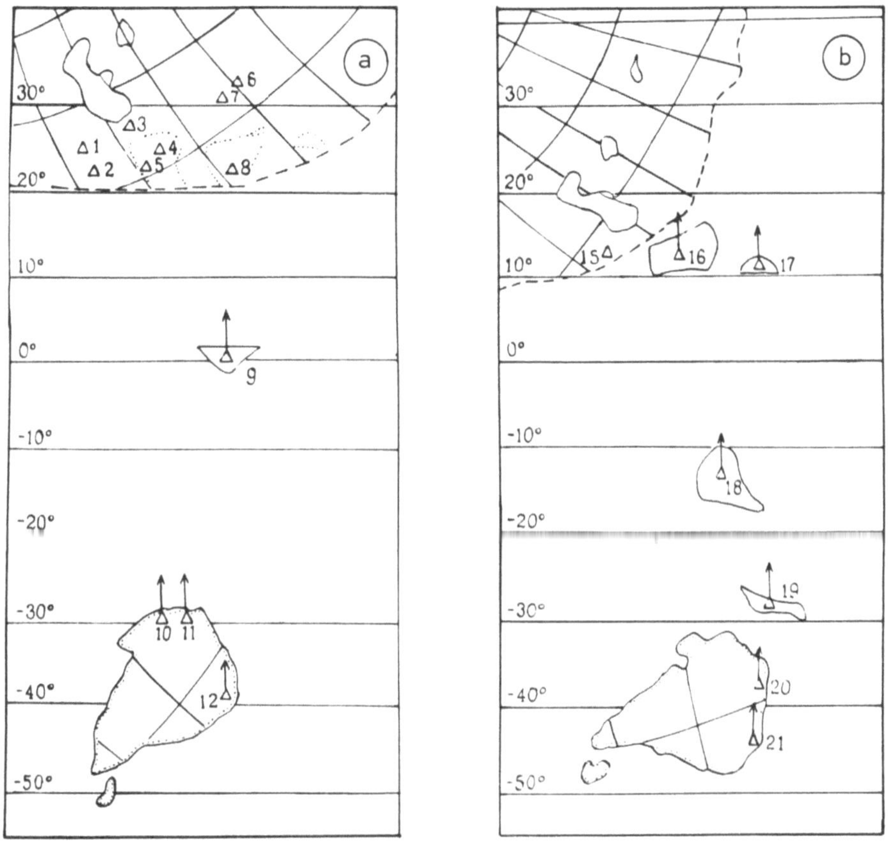

Fig. 5.5a,b. The global position of the Soviet Central Asia tectonic units for **a** the early Cretaceous and **b** the late Paleozoic. *Numbers* correspond to those given in Table 5.2. *Dotted line* shows the approximate edge of the northern Eurasian plate

Using global reconstructions for the late Permian and early Cretaceous, presented in Chap. 4, and also the methods of determination of the kinematic parameters of lithospheric plate displacement described in the same chapter, we can conclude that since the late Permian until the early Cretaceous, the North Eurasian plate rotated clockwise by 30° around the rotational pole with the coordinates: $\varphi = 0°$, $\lambda = 57°E$. For the same time interval the Indian plate rotated clockwise by 22° around a rotational pole with coordinates: $\varphi = 8°S$, $\lambda = 116°E$. These movements widened the oceanic basin to a certain extent (Fig. 5.5). At the same time, the Afghan microcontinent drifted considerably (33°) northward and joined onto the North Eurasian continental plate (Fig. 5.5a). Therefore, since the late Permian, a zone of subduction existed along the southern margin of the North Eurasian plate, which ceased to exist in the early Cretaceous.

In the inner part of the Afghan microcontinent of that time, a zone of accretion of the lithospheric plate has also been recorded.

Using the method described above, it was also possible to show that since the early Cretaceous until now, the North Eurasian plate rotated clockwise by 25°

around a rotational pole with the coordinates: $\varphi = 35°$N, $\lambda = 80°$E; and the Indian plate, for the same interval, rotated clockwise by 70° around a rotational pole with coordinates: $\varphi = 25°$S, $\lambda = 187°$E. These movements resulted in the closure of the oceanic basin that existed between the North Eurasian and Indian continents and resulted in the collision of these continents.

Comparison of the displacement values ($\Lambda\varphi°$) and rotations ($\Lambda\psi°$) of different blocks with those of the Eurasian and Indian plates (see Table 5.2) led to the conclusion that the kinematic parameters of global displacements of these blocks, before joining, were controlled by the kinematic parameters of the Indian plate displacement and afterwards by the kinematic parameters of the North Eurasian plate displacement, with respect to local and regional rotations. It must be emphasized that the great scattering of the $\Delta\varphi°$ and $\Delta\psi°$ values for different parts of the Indian plate is due not only to the fact that these regions occupy a different position relative to the rotational pole, but also to certain, additional displacements and rotations to which they were subjected.

If we compare the $\Delta\varphi°$ and $\Delta\psi°$ values estimated at one or another part and those calculated proceeding from the kinematic parameters of the plate displacement, we can estimate the amplitudes of the additional rotations and displacements of these parts. Thus, it has been shown that the northeastern Baluchistan and Sulaiman Mts. rotated counterclockwise by an extra 8°; the northwestern Himalayas rotated clockwise by an extra 40°; and the southeastern Himalayas, by an extra 16° clockwise.

Thus, these data led to the conclusion that the continental massif of Eurasia accreted at the expense of subsequent joinment of sialic blocks that separated from Gondwana. This attachment of blocks resulted in the closure of the oceanic basin in their frontal parts and the accretion of the oceanic crust in their back parts. The Indian subcontinent was the last large block that attached to Eurasia: its frontal collision with Eurasia was the cause of mountain-building in this part of the belt.

In this collision the relics of the earlier, closed oceanic depressions became fragmented and overthrust and drifted far northward. This conclusion is supported by the relics of the Paleozoic Paleotethys Ocean recorded from the Chatkal-Kurama region, Turkestan and Hissar Ridges (Burtman 1976; Kravchenko 1979) and in Tuarkyr (Panasenko 1981). Relics of the Mesotethys Ocean have been registered in the Central and Southern Pamirs, in Kara-Korum, the Himalayas, southern Hindu-Kush, Central Iran and in southwestern Afghanistan (Gansser 1964; 1979; Kravchenko 1979; Asrarullah et al. 1979; Ahmad and Abbas 1979; Sarwar and DeJong 1979; Berberian 1984).

In different parts of this region zones of subduction and obduction have been recorded (Ruzhentsev and Sokolov 1983; Khristov and Mikolaychuk 1983; Tahirkheli et al. 1979).

5.3.2 Deformations in the Urals Belt

At the present time the evolution of the Urals geosyncline is still being debated. This can be explained by the problems we face and by the existing difficulties in solving them. The real problem lies in the quantitative evolution of the horizontal movements in the Urals proper and in the adjacent areas. Unfortunately, the age determinations of the Urals metamorphic complexes are not abundant at the moment and these are badly needed for defining the timing of metamorphism and its distribution of the Urals eugeosyncline. However, problems of local tectonics have been satisfactorily tested. Therefore, the application of the paleomagnetic method for tectonic studies within the Urals fold system is of special interest. It must be noted that paleomagnetic coverage in the Urals is quite irregular. Major data have been obtained from the southern Urals (Danukalov 1975) and less data from the Central and Northern Urals.

Here, paleomagnetic determinations are analyzed (see Appendix). It was shown in Sect. 4.3.1 that there are two groups of paleomagnetic poles occurring in the Western and Eastern Urals separated by the Major Urals fault. Paleomagnetic poles from the Western Urals are very similar to those of the Russian platform, at least since the middle Paleozoic. It is difficult to make any definite conclusions for the early Paleozoic because of sparcity of paleomagnetic determinations. Paleomagnetic poles from the Western and Eastern Urals differ in the Ordovician time in particular. This variance is preserved in the middle and late Paleozoic: the paleomagnetic poles of the Eastern Urals are situated somewhat more to the east in comparison with the Western Urals poles (see Fig. 4.7). This divergence can be explained by two causes: (1) late Paleozoic remagnetization being favoured by the conditions in the Eastern Urals; (2) important tectonic movements that could occur in the Eastern Urals during its geosynclinal evolution.

A number of workers (Danukalov et al. 1977; Danukalov et al. 1983; Khramov et al. 1974; Komissarova 1969) have pointed out the late Paleozoic remagnetization of the Urals rocks. Metachronous (ancient secondary) magnetization of the older rocks is sometimes characterized by high stability; locally, it may be much higher than for primary magnetization. For destroying metachronous magnetization, high temperature cleaning is the most effective. Certain difficulties appear in paleomagnetic data interpretation due to this high stability of this later magnetization. It is not useful, however, in explaining the existence of two groups of paleopoles for the following reasons. Paleomagnetic determinations for the Ordovician of the Russian platform, Western and Eastern Urals differ greatly in pole positions, but the variations are quite consistent. This indicates that the differences in the pole positions are due to local or regional tectonic rotations. If the Western Urals area is rotated clockwise by 35° and the Eastern Urals by 55° counterclockwise, their paleomagnetic poles coincide within the error in determination (see Sect. 4.3). The more eastern position of the magnetic poles of the Eastern Urals in the middle and late Paleozoic, relative to the paleopoles of the Western Urals, might also evidence a counterclockwise rotation of many blocks in the Eastern Urals. All of the above factors indicate large-scale, horizontal movements in the Urals fold belt.

According to the geologic data (Perfil'yev 1979), the continental margin in the Northern Urals was characterized by quiet subsidence in the middle-late Paleozoic that sedimentation did not compensate. The oceanic area, situated east of the continental slope, exhibits quite different geologic development: intensive deformation of the transitional stage of evolution beginning in the Silurian, which was much earlier than in the adjacent continental slope. In the lowermost eugeosynclinal groups, an irregular band of metamorphic rocks of eclogiteglaucophane-schist grade, with an absolute age of 400±20 m.y., is observed close to the boundary with the structures surrounding the area in the west. This metamorphism might record high-pressure conditions and trace the surfaces of large-scale overthrusts. These events, however, did not affect the paleocontinental margin, which might be explained by the fact that those structures were widely separated at that time and only approached each other in the late Paleozoic. Consequently, paleomagnetic and geologic data appear to be quite consistent.

Paleomagnetic studies of Carboniferous rocks and older Paleozoic formations revealed a trend of counterclockwise rotation of the Urals structures by 20°–30° relative to the Russian platform (Ivanov 1968). The coverage of the area by paleomagnetic determinations is not sufficiently regular and, therefore, it is difficult to say whether all these structures rotated as a single unit or whether each separate part rotated as a result of deformation.

The Eastern Urals are typified by intensive folding, tectonic faults and block structures. Paleomagnetic studies of Triassic rocks on the eastern slope of the Urals (the Pyzhma River) record the existence of small blocks rotated relative to each other by up to 43°. Comparative studies of the Triassic poles of these blocks and of similar poles of the Russian plate led to the conclusion that separate blocks preserved their orientation relative to the Russian platform.

Thus, paleomagnetic data evidence global, regional and local tectonic movements within the Urals fold belt. These data also permit the definition of characteristic stages in the movement of the Earth's crust in the Urals. Large blocks appear to have moved in the early Paleozoic: e.g. the Russian platform, Western and Eastern Urals. The next stage resulted in the consolidation of the Russian platform and Western Urals. The sinistral movement of the Eastern Urals continued during the middle-late Paleozoic, relative to this block, along the Major Urals fault. Formation of the Urals fold belt terminated at the end of Permian. At the beginning of Mesozoic, however, the continuation of dislocations of the principal structures of the Urals have been recorded. The regional and local deformations may be regarded as differentiated manifestations of global tectonics.

Thus, it is quite evident that the paleomagnetic method is valid for tackling global, regional and local tectonic tasks in the Urals. The relations of the Urals geosyncline and the Russian platform in the geohistorical sequence of events is a key problem in understanding the regularities of crustal evolution in these areas as well as in the Eurasian continent as a whole. With the aim to solve this problem, it is necessary to conduct paleomagnetic studies in the following problems.

1. Metachronous magnetization of the Urals rocks, its diagnosis and methods of elimination or correction of its effects.
2. Studies of the regions adjoining the Urals: the Scythian-Turan plate, Kazakhstan block, Altai-Sayan fold area. Paleomagnetic reconstructions of these areas will permit the quantitative evaluation of their movement and the reconstruction of a pattern of the tectonic development of the Urals.
3. Studies of regional and local tectonics in the Urals as a basis for deciphering its global tectonics.
4. Paleomagnetic studies must be more extensive in the Central and Northern Urals.

5.3.3 Deformations in the Areas Surrounding the Siberian Platform

Studies of the geologic structure and tectonic evolution of the rigid Siberian platform need to be combined with studies of its relations with the adjoining areas: the Pacific fold belt in the east and the Arctic basin in the north. Geologic evolution of these crustal blocks is commonly studied proceeding from their stable position in space; their possible horizontal displacements and dislocations are neglected. However, some workers have shown that tectonic processes similar to the present processes in oceanic basins and continental margins are recorded in the evolution of the intracontinental depressions of the Arctic basin as well as of structures of the Verkhoyansk fold belt (Gusev 1974).

The paleomagnetic method assumes an ever greater role in studies of local and regional tectonics within the Siberian platform and adjoining areas. Gusev, in his first paleomagnetic work in the northern Siberian platform and Arctic basin, showed that this method permitted the establishment not only of the nature of tectonic movements (displacements, rotations), but also of the quantitative evaluation of rotations of certain crustal blocks and the directions of their rotation.

Thus, the inconsistency of the early Triassic paleomagnetic directions of the Western Taimyr and Siberian platform (Fig. 5.6) might be explained by horizontal displacement, i.e. Taimyr moved as a single block (since Triassic), in a northwestward direction along abyssal faults. Horizontal tension resulted in the rifting and separation of the Taimyr and Siberian crustal blocks. The rate of movement

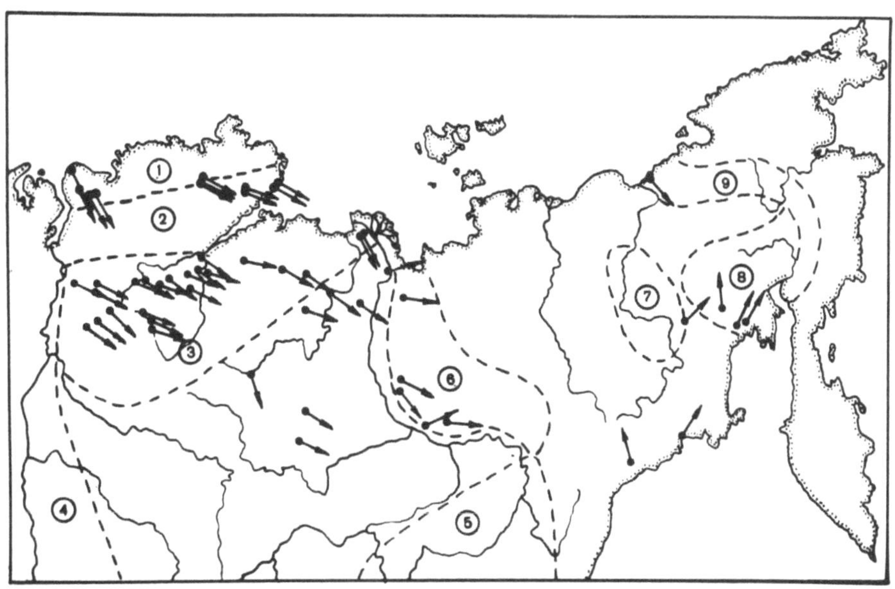

Fig. 5.6. Paleomagnetic data on the deformations of the tectonic units in the northern and eastern surroundings of the Siberian platform. Data from Gusev 1974, Rodionov et al. 1977 and from Tables 5.3 and 5.4 *1* Taimyr fold system; *2* Khatanga depression; *3* Yenisei-Anabar block; *4* western slope of Siberian platform; *5* Aldan Shield; *6* Verkhoyansk fold belt; *7* Pri-Kolyma block; *8* Omolon massif; *9* Okhotsk-Chukotsk zone

seems to have varied in different sites and was associated with block deformation, therefore, the western part of the Taimyr appears to have rotated clockwise by 30° relative to its eastern part. The probability of these displacements is confirmed by paleomagnetic mapping of the lava sheets in Western Taimyr. Within the fault zone, local rotations of certain blocks have been recorded and quantified (Gusev 1974).

From paleomagnetic studies in the areas adjoining the Siberian platform in the east and northeast (the Western Verkhoyansk and Lena River mouth regions), the usefulness of the paleomagnetic method in regional and local tectonics are evident. These results are listed in Table 5.3. Paleomagnetic data for the Western Verkhoyansk region show that in the northern (from 70°N and above) and southern (from 64.5°N and below) parts of the Verkhoyansk fold structure, the rotation that occured was counterclockwise 50°–60° as the middle part of this structure did not undergo any rotations (Rodionov et al. 1977).

These data are quite important for deciphering the tectonic movements in the area surrounding the Siberian platform. Thus, paleomagnetic determinations in the Western Verkhoyansk region suggest that the Verkhoyansk fold area was originally of meridianal orientation (relative to the present grid). The thrust structures and the structures overturned towards the Siberian platform, typical of the northern part of this area, are indicative of crustal mass movement in a westward direction, i.e. from the Verkhoyansk region. Rotation of the Verkhoyansk structure, which acquired an S-shaped form, seems to have occurred when it collided with the Siberian platform (Rodionov et al. 1977).

Table 5.3 Paleomagnetic poles for the northeastern and eastern fold surroundings of the Siberian platform (Rodionov et al. 1977; Pisarevsky and Iosifidi 1981)

Age	Sampling Area	Direction of the ancient magnetization J[a]		Pole for the Siberian platform		Calculated ancient field direction		Difference ΔD between calculated and observed declinations[b]
		D	I	Φ	Λ	D*	I*	
T_3-J_1	Olenek Channel 72.6°N, 124.7°E	174°	76°	59°	137°	156°	83°	18°
P_2	Olenek Channel 72.6°N, 124.7°E	164°	75°	46°	147°	147°	75°	17°
C_1	Bykov Channel 72.4°N, 127.0°E	201°	70°	30°	150°	151°	64°	50°
D_3	Krest Khomo 72.4°N, 127.0°E	211°	66°	20°	140°	164°	56°	47°
T-J	Chekurovka vil., Lena River 71.0°N, 128.0°E	83°	79°	59°	137°	159°	83°	−76°
P_1	Besyuke River 69.8°N, 128.5°E	290°	−76°	45°	141°	325°	−76°	−35°
T_1	Khorunku R. and Djolbuun R. 66.5°N, 126.5°E	131°	73°	49°	149°	138°	79°	−7°
T_{2-3}	Tirekhtyakh R. and Dyanyshka R. 66.0°N, 126.0°E	148°	80°	59°	137°	140°	85°	8°
P_2	Kelter River 64.5°N, 130.0°E	109°	81°	46°	147°	146°	79°	−37°
T_1	Kelter River 64.5°N, 127.5°E	82°	83°	49°	149°	137°	80°	−55°

[a] D* and I* are calculated for the sampling areas according to poles for the Siberian platform.
[b] Δ D>0 denotes clockwise rotation of the J_n^a and hence, of the sampling area; Δ D<0 denotes counterclockwise rotation.

It was supposed earlier that in the northern end of the Verkhoyansk fold structure its virgation resulted in the formation of three branches: Lena-Olenek, Kharaulakh and Kular. Paleomagnetic data, however, prove that the Lena-Olenek structure did not change its position relative to the Siberian platform since the Permian through Jurassic, and it rotated only at the end of the late Jurassic-early Cretaceous. Thus, the Lena-Olenek structure may be regarded as a single block that evolved separately from the Verkhoyansk fold area, otherwise this structure, being a part of the Verkhoyansk unit, had to sustain a counterclockwise rotation of approximately 60° at exactly same time.

Paleomagnetic studies of the rocks in the Chekurov anticline permitted the establishment of a rotation of this structure (or of its southern part at least) which is counterclockwise by 76° (Rodionov et al. 1977). Generation of the Chekurov fold was associated with trappean magmatism in two phases, according to paleomagnetic data: pre-dating and post-dating the folding. The Chekurov traps are then assigned to the middle Jurassic-early Cretaceous. Paleomagnetic determinations in the Upper Devonian and Lower Carboniferous of the Bykov channel of the Lena River (Table 5.3) suggest a clockwise rotation of this portion relative to the Siberian platform of 48°. The similarity of paleomagnetic data in the Upper Permian and Upper Triassic-Lower Jurassic of the Olenek channel of the Lena River show a stable orientation of the Chekanovsky Range, relative to the Siberian platform, but at the end of Jurassic and at the beginning of Cretaceous, it rotated clockwise by about 18°.

Deformations similar to the Verkhoyansk type were also recorded on the opposite side of the Pacific fold belt according to paleomagnetic studies in the North American Cordilleras (Grubbs and Van der Voo 1976). The evident consistency of changes in paleomagnetic directions and changes in structural orientations permitted the determination of the character (displacements, rotations) and quantification of deformations (the rotational angle of the southern end of the Rocky Mts. amounted to 80° clockwise).

Recently, new paleomagnetic data from the Northeastern USSR were obtained, which are to be used in studies of the tectonic evolution of the region. In 1982 a model of tectonic evolution of northeastern Siberia and adjacent regions was proposed (Fujita and Newberry 1983). According to this model, the eastern boundary of the Siberian plate extended, in the early and middle Paleozoic, approximately along the Novosibirsk Islands-Alazeya Uplift-Omulev Uplift. The Prikolyma block of that time was situated slightly to the south (relative to the Siberian plate) and was turned clockwise by about 90° compared to its present position. In Triassic, Jurassic and early Cretaceous times, these blocks took their present position relative to the Siberian plate as a result of the northwestward drift of the Omolon block and its collision with the Prikolyma block. Then, in the late Cretaceous the Alaskan-Chukotkan block became attached.

New paleomagnetic data (Table 5.4) confirm the validity of this model (Khramov et al. 1985). The paleomagnetic poles for the Ordovician, Silurian, Devonian and Jurassic rocks of the Omulev Uplift cluster near the APWP of the Siberian platform and the mean direction of NRM for the Devonian rocks of the Prikolyma Uplift forms an angle of about 90° with the corresponding direction for the Siberian plate.

Table 5.4. New paleomagnetic data for the northeastern USSR[a]

Age	Rock type	Sampling area	Paleomagnetic direction				Number of vectors		Pole position		Reference[b]
			D	I	k	α_{95}^{o}	N	R	Φ	Λ	
J_3	Volcanics	Omulev uplift, 65.4°N, 148.5°E	211°	74°	4	14°	20	17	38°	130°	1
D_3	Sediments	Prikolyma uplift, 63.0°N, 151.0°E	258°	−63°	65	15°	–	3	44°	226°	2
D_3	Sediments	Prikolyma uplift, 63.0°N, 151.0°E	263°	−73°	71	7°	–	8	52°	208°	2
D_3	Sediments	Prikolyma uplift, 63.0°N, 151.0°E	268°	−72°	23	13°	–	4	49°	207°	2
D_3	Sediments	Prikolyma uplift, 63.0°N, 151.0°E	77°	61°	38	9°	9	–	42°	229°	2
D_2	Sediments	Prikolyma uplift, 63.0°N, 151.0°E	280°	−72°	33	14°	–	5	44°	201°	2
D_{2-3}	Volcanics and seds	Omulev uplift, 65.5°N, 148.5°E	223°	−2°	3	18°	9	18	−19°	103°	1
S_2	Sediments	Omulev uplift, 65.3°N, 148.5°E	70°	2°	30	14°	–	5	−9°	77°	1
O_2	Basaltic lavas	Omulev uplift, 65.3°N, 148.5°E	221°	−18°	4	47°	4	1	−27°	102°	1
O_2	Basaltic lavas	Omulev uplift, 65.3°N, 148.5°E	154°	26°	30	23°	1	2	−9°	174°	1

[a] Mean values and statistics are calculated at the sample level within data for cleaning (h-and t-cleaning) specimens only.

[b] *1* Pisarevsky, Pogarskaya, Slautsitais, new data; *2* Gagiev et al. 1982.

Explanations of the paleomagnetic data on the Verkhoyansk region, which were mentioned at the beginning of this section, however, present difficulties within the limits of this model. The above interpretation of these data can be explained by a west-north-westward drift of the Kula plate and its collision with the Siberian plate. An alternative explanation of these paleomagnetic data is also given by the supposition that the bending of the orginally linear Verkhoyansk structure was as a result of the collision of the Siberian and the Alaskan-Chukotkan plates, (Khramov et al. 1985). In this case, however, it is necessary to assume that the block, including the Omulev Uplift, was subjected only to a latitudinal drift relative to the Siberian plate. It is also quite realistic that the general S-shaped plan of the Verkhoyansk fold area is of an original nature and the drift of the Pacific plates, mentioned above, resulted only in a sinistral shift in the basement of the eastern margin of the Siberian plate, which was recorded in the sedimentary strata as local, counterclockwise rotations of allochthonous sheets.

The deformations, such as horizontal displacements and rotations, appear to be typical of the fold areas adjoining the Siberian platform not only in the north and east, but also in the south. Comparative paleomagnetic determinations for the Siberian platform and the Altai-Sayan fold area in the early-middle Paleozoic (see Appendix) show that the paleopoles of the Altai-Sayan area evidently tend to be displaced southward in the middle Paleozoic and westward in the early Paleozoic, relative to contemporaneous poles of the Siberian platform. The Altai-Sayan area is likely to have existed as a single block in the early-middle Paleozoic and at that time underwent a horizontal displacement and clockwise rotation. It is only in the early Carboniferous that it became a single unit with the Siberian platform.

Thus, paleomagnetic data provide additional information on the time sequence, scale and pattern of tectonic movements of the Earth's crust in the area of junction of the Siberian platform and fold structures. Future paleomagnetic studies should be concentrated in the Altai-Sayan fold area and in the northeastern part of the USSR, in particular to explain their tectonic evolution.

5.4 Tectogenesis and Rhythmicity of Geomagnetic Reversals

Certain features in the geomagnetic field evolution for the last 600 m.y. have been defined by analyzing the Phanerozoic paleomagnetic scale (Fig. 3.4). From the viewpoint of geotectonics the most important features are as follows (Molostovsky et al. 1976):

1. During the Phanerozoic, three regimes of the geomagnetic field existed, each affecting each other: a relatively stable field of reversed polarity, a relatively stable field of normal polarity, a disturbed field with frequent reversals.
2. According to these regimes, three principal stages of evolution of the geomagnetic field are recognized: the Paleozoic, Mesozoic and Cenozoic to which three major magnetostratigraphic units (megazones) correspond in the paleomagnetic scale. The geomagnetic field during these intervals was dominated by reverse, normal and mixed polarity, respectively. The beginning of every stage of the field evolution occurred before the start of the corre-

sponding geological era by several millions of years. Each megazone is characterized by irreversible peculiarities of structure, thus, reflecting the geomagnetic field evolution.
3. The paleomagnetic scale includes 12 large units: i.e. hyperzones that are comparable in their stratigraphic range to geological systems (see Fig. 3.4).
4. Progressive-cyclic development of the geomagnetic field is seen from the recurrence of large intervals of frequent reversals approximately every 200 m.y. These intervals are commonly registered at the end stages of geotectonic cycles (the Caledonian, Hercynian and Alpine) or at the principal folding phases during these cycles.

The regime of geomagnetic reversals and geotectonic events is related not only in the aspects mentioned above. During the initial period of paleomagnetic studies, it was observed that the displacement of paleomagnetic poles along APWP curves was of an exceptionally irregular character. For every platform long periods of minor displacements alternate with short periods of major displacements. Sheinman (1975) and Molostovsky et al. (1976) noted that the stages of fast pole displacements are coincident with epochs of maximum tectono-magmatic activity, at the end of Caledonian, Hercynian and Alpine cycles (Fig. 5.7) and these mark the termination of the most intensive orogenic epochs.

It follows from the concept of mobile lithospheric plates that the oscillations of sea level of the World Ocean depend mostly on the intensity of the global tectonic processes. These oscillations are recorded in transgressions and regressions and their scale is estimated according to the area of the present continents that were once occupied by epicontinental seas during some geologic epochs. Diagrams of changes in this area were plotted many times for the whole planet as well as for separate continents. Notable is the fact that the periods of frequent reversals are comparable with the world regressions and the start of these periods commonly predates regressions by approximately 10 m.y. It was noted above (Khramov 1978) that the changes in the pattern of movement (in rate and direction) of lithospheric plates have been recorded in the same periods.

Thus, these correlations also support the well-known assumption about the origin of tectonic movements and about the unity of the whole endogenetic process of the Earth. The exact mechanisms of the relations of geomagnetic and geotectonic phenomena still remain problematic. It may be suggested, however, that gravitational differentiation at the lower mantle boundary provides energy for the mechanism of the geomagnetic dynamo, which in turn produces the mantle convection which governs the movement of lithospheric plates. Thus, the retardation of geotectonic events in comparison with the geomagnetic ones then appears obvious; this retardation provides a means of evaluating the rate of translation of disturbances through the mantle. It is necessary to make a detailed correlation of the reversal epochs (critical geomagnetic intervals) and the tectonic activity epochs that are characterized by uplifting of the crust and folding (Fig. 5.8).

Dating of such critical geomagnetic intervals has been made according to the magnetostratigraphic scale of the USSR; the epochs of general uplifting and folding phases have been recorded according to Bogdanov (1969). The correlation

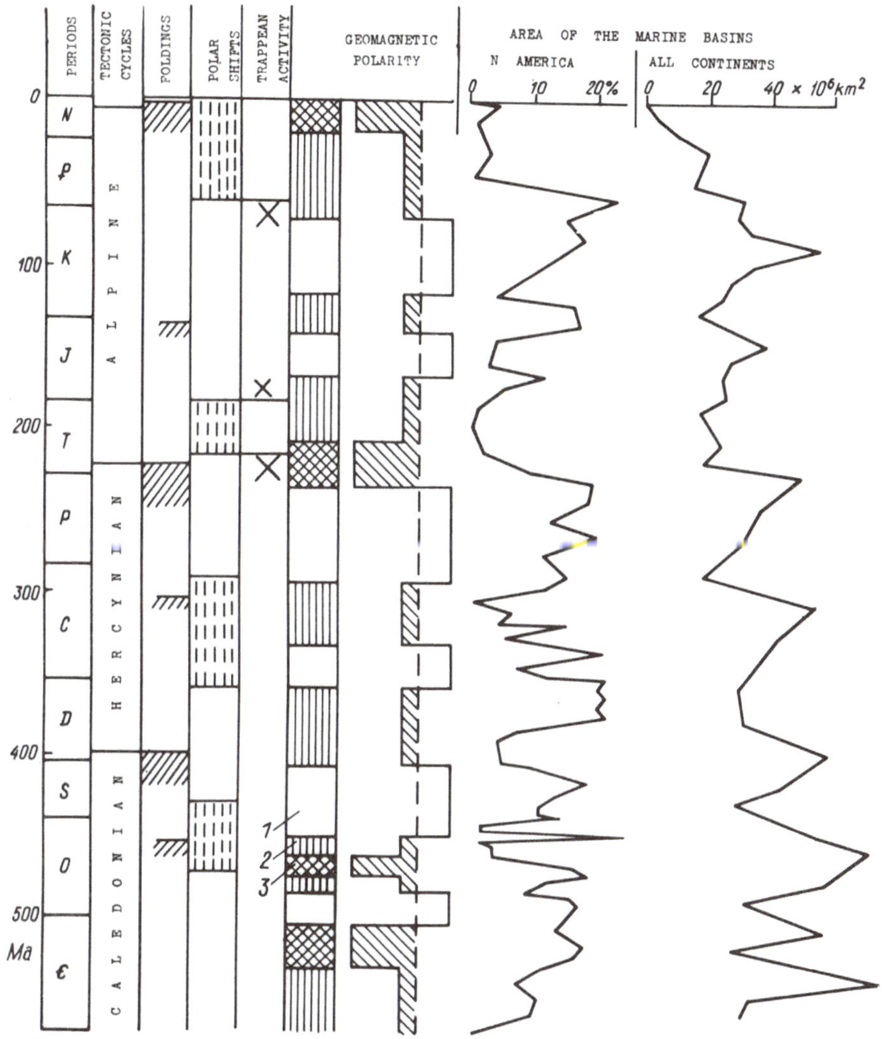

Fig. 5.7. Correlation between global tectonic phenomena and geomagnetic reversals. (After Khramov 1978, slightly modified). Geomagnetic reversal regimes: *1* domination of a single polarity with very rare reversals; *2* rare reversals; *3* frequent reversals

of the epochs of these processes is obvious; its peculiarity lies in the retardation of the uplifting and folding epochs relative to the critical geomagnetic intervals. This difference in time between the beginning of the paleomagnetic field critical interval and the beginning of the tectonic event varies from 15 to 70 m.y. Three maxima and four minima have been recorded for retardation and the general trend of the curve can be seen: the time difference decreases when approaching the Recent epoch. Maxima for retardation have been registered for 120; 315; 430 m.y. ago; their minima, 48, 210, 350; 515 m.y. ago.

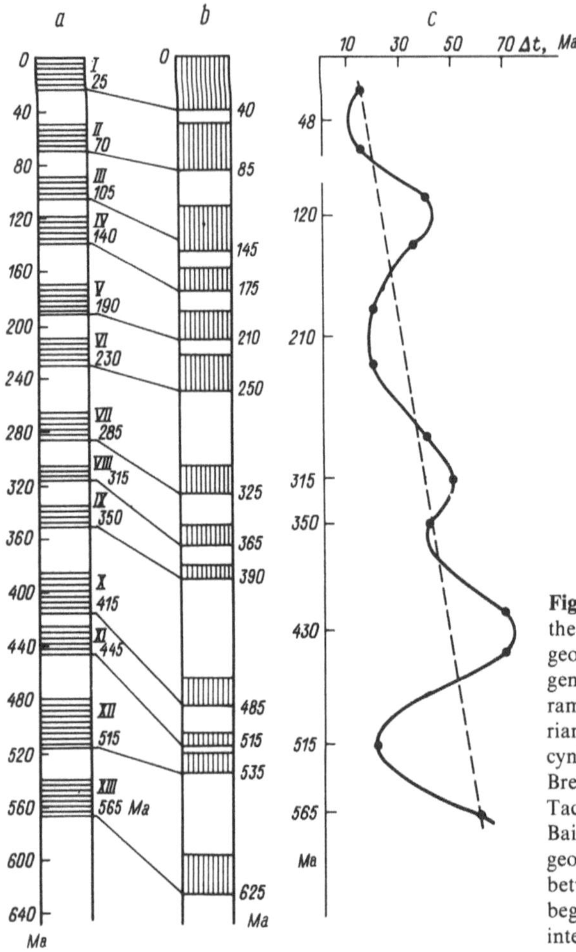

Fig. 5.8a-c. Temporary sequences of the active tectonic epochs and geomagnetic critical intervals. **a** Orogenic phases (*I* late Alpine; *II* Laramide; *III* Austrian; *IV* late Kimmerian; *V* early Kimmerian; *VI* late Hercynian; *VII* Uralian; *VIII* Sudetian; *IX* Bretonian; *X* late Caledonian; *XI* Taconian; *XII* Salairian; *XIII* late Baikalian. **b** Intervals of frequent geomagnetic reversals. **c** Time lag between the orogenic phase and the beginning of the geomagnetic critical interval

If we proceed from the assumption that the gravitational differentiation of mantle matter at the core-mantle boundary provides energy for the geomagnetic dynamo as well as for tectogenesis, then we can reasonably explain the retardation of tectonic activity epochs relative to the epochs of an unstable geomagnetic field. Any significant redistribution of energy in the core produces an immediate reaction by the geomagnetic dynamo. As to heat energy, a certain period of time is needed for it to be convected and so accounts for the retardation of the tectonic activity epochs following the epochs of an unstable state of the geomagnetic field (Khramov 1978). The minima of the retardation curves correlate easily with the most important stages of the Earth's transformation. According to the general tectonic scheme of Bogdanov (1969), the Hercynian folding terminated at the time boundary of 210 m.y.; the 350 m.y. time boundary fell within the range of termination of the Caledonian folding (395 m.y.) in the European and African continents and of the Paleozoic folding (340 m.y.) in Southern and Eastern Asia and in Australia. The 515 m.y. time boundary corresponds to the terminal stage of the Baikal folding (500 m.y.) for Europe, Africa, Southern and Eastern Asia. And it is only the 48 m.y. time boundary which does not correspond with the folding time boundaries in the scheme of Bogdanov; this inconsistency cannot be explained as yet. The maxima in the retardation curve agree with the epochs of the least intensive tectonic activity.

The quantity of energy emitted during gravitational differentiation of the matter at the core-mantle boundary probably determines the mode of this processes. The more energy emitted, the higher the rate of convective matter transportation in the mantle, and consequently, the shorter is the time for the disturbance to reach the Earth's surface. If the quantity of energy emitted is insignificant, the time necessary for the disturbance to continue would be longer because a certain amount of energy must be accumulated for convection to start.

Thus, with a large quantity of energy emitted, tectonic activity acquires a global scale and initiates some essential transformations of the Earth. A trend of the retardation curve is likely to be related to the growing emission of energy in the core, which makes the convection rate in the mantle higher and, thus, reduces the duration of tectonic cycles and increases their frequency from Precambrian to Recent. This conclusion has been confirmed by the reversal frequency that has increased from the Vendian to Cenozoic. If the correlation of tectonic epochs and unstable field intervals reveals their relationship to a single source of energy situated at the core-mantle boundary, then the global transformations probably associate with the maxima in the emission of this energy from this source.

In this schematic model the geomagnetic cyclicity is determined by the rhythms in the Earth's core growth. Thus, it is suggested here not only as the cause of the relationship between the geomagnetic and geotectonic phenomena, but also the endogenetic results of the geomagnetic cyclicity. At the same time, new analyses have confirmed the harmonics with periods close to the duration of the Galactic year ($\sim 220 \times 10^6$ years) in phenomena such as the change in the geomagnetic field intensity or in the geomagnetic reversal frequency (Bol'shakov and Solodovnikov 1981; Khramov and Kravchinsky 1984). Thus, it is imperative to elaborate the models, which examine the exogenetic nature of geomagnetic cyclicity. The basic problem in this case would be a mechanism that transfers the energy of the cosmic body movement into the energy of tectonic movements through the magnetic field (Khramov and Kravchinsky 1984).

Naturally, these correlations are indirect and cannot be considered certain at the present state of knowledge. They may be accepted as correct only for the larger stages in the geomagnetic and geotectonic evolution. To an even greater extent it concerns the correlation of paleomagnetic phenomena with the stages in biosphere evolution. Certain epochs of great changes in the Phanerozoic fauna and flora are known to occur close to changes of the principal stages of geomagnetic field. However, it is hardly possible that some cases of correlation between changes in faunal species and specific geomagnetic reversals are regarded as an effect of the geomagnetic field on Life on the Earth.

5.5 A General Hypothesis of Tectogenesis

5.5.1 Hypothesis of Planetary Equilibrium

This hypothesis was first proposed and elucidated by Forsh in 1969 in order to further develop Wegener's ideas on the importance of rotational forces in the drift of the crustal plates (see Khramov et al. 1974). The hypothesis of global equilibrium assumes that the cause of tectogenesis of the lithosphere lies in a constant disturbance and restoration of the global equilibrium as a result of the rapid rotation of the Earth. A disturbance of global equilibrium may be produced by mass transfer in the abyssal zones of the Earth as well as in its upper parts. The problem of the scale and the rate of such abyssal processes and their implication in lithospheric tectogenesis is still at the hypothetical stage. Mass transfer in the upper parts of the Earth evidently disturbs global equilibrium and is one of the most important causes of tectonic processes in the lithosphere. Accumulation of excess mass, irrespective of location, displaces the centre of the Earth mass from the rotational axis and disturbs the planetary equilibrium. Considerable stresses then appear that stimulate a subsequent restoration of equilibrium by means of rebuilding certain spheres of the Earth. To restore a stable equilibrium it is necessary for the excess mass to approach the rotational axis of the Earth as closely as possible, i.e. near the pole. Redistribution of masses in the upper parts of the Earth is negligible compared to the Earth's total mass and cannot produce any significant displacement of the rotational axis. Since the lithosphere is separated from the Earth's main mass by a sufficiently viscous asthenosphere, the planetary equilibrium might easily be restored by a displacement of the lithosphere as a single shell, along the asthenosphere, relative to the pole of the Earth, displacing the excess mass to the nearest pole.

It is known that even an insignificant mass transfer produced by a seasonal formation and distribution of the Siberian atmospheric maximum results in a displacement of poles up to 15 m (Sidorenkov 1982). As this is a seasonal phenomenon, the pole returns annually exactly to its previous site. A more intensive effect is expected from mountain-building, the transfer of large masses of terrigenous sediments, and from the appearance and disappearance of extensive masses of continental ice.

These surficial transfer of masses is quite important compared to the mass of lithosphere, which represents only a thin shell of the Earth and yet this produces a significant rotation relative to the Earth's rotational axis. Thus, in this manner the global equilibrium can be restored.

5.5.2 Drift of the Lithospheric Plates and the Coriolis Force

Displacement of the lithospheric shell relative to the Earth's rotational axis inevitably produces a redistribution of the rotation rate in its different zones, which, due to the inertia of the rapid rotation movement of the Earth, results in fractures of the lithosphere and a relative movement of the lithospheric plates. This, probably, is the principal cause of tectogenesis of the Earth's crust.

When the position of the lithospheric shell changes relative to the Earth's rotational axis, the rate of rotation in different zones of the Earth's surface varies. Therefore, the lithospheric surface can be divided into four quarters with the equator and the great circle perpendicular to the direction of the pole displacement, which serves as the boundaries of these quarters. In adjacent quarters the rates of change are of opposite sign. In those two quarters to which the poles are moved, the rotation movement becomes slower; and in those two quarters from which the poles move, the rotation rate accelerates. In the zones with the decreased rotation rate, the lithospheric plates, due to inertia, have to get an increase in motion in the direction of the Earth's rotation and in the zones with the accelerated rate, the stimulated movement would be against the direction of the Earth's rotation.

The highest change in the rate of rotation occurs in the zone of the great circle which coincides with the direction of the pole shift. In this zone, the velocity gradient approaches zero perpendicular to this direction, and, thus, due to a difference in the rate change, meridional fractures are expected to appear between these zones. As a result, in both hemispheres (northern and southern) at least four to six fractures of the meridional direction might originate. In this case, at the boundary of zones of the backward inertia movement and at the boundary of zones of more and less intensive inertia movement where the first zone moves quicker, the meridional zones of collision and compression appear. At the boundary of the zones of diverging inertial movement and at the boundary of the zones of less, but more intensive, diverging movement, where the first zone lags after the second, meridional fractures are expected to be generated as this zone in characterized by a separation of lithospheric plates. All these meridional fractures extend from the poles towards and across the equator. Where the inertia movement is in the opposite direction, the zones of plate collision and separation are strongly moved.

The forces of inertia movement also change substantially along the meridional direction. From the equator to 10° latitude where the Earth's circumference changes slightly latitudinally, the rotation rate only changes insignificantly with a change in latitudinal position. However, between 10° and 45° latitude the Earth's circumference starts to decrease in latitude quite rapidly. A sharp change in the rotation rate here occurs results in a change in the latitudinal position. More than 45° toward the poles, the change in the circumference remains almost constant at different latitudes. The change in the rotation rate as a result of a movement in latitudinal position also remains constant, together with the forces of inertia movement.

As a result of the lithospheric shift relative to the poles, fractures of the latitudinal direction perpendicular to the above-mentioned meridional fractures appear: (1) in the equatorial zone where the forces of inertia movement acquire an opposite direction; (2) in the zone between 10° and 45° latitude where the forces of inertia movement subsequently increase to 45° latitude. (3) The zones of the Earth from 45° up to the poles in which a change of latitudinal position is characterized by the highest force of inertia movement and in plate drift, they overtake the rate of the zones situated closer to the equator, thus producing a latitudinal shift.

This hypothesis explains well the whole concept of geodynamic motions and plate drift. To support it we will examine the history of horizontal movements of lithospheric plates in the Phanerozoic as seen from global paleomagnetic reconstructions (see Sect. 4.5). In the late Cambrian the southern pole was close to the western coast of Morocco (Fig. 4.26). In this case three quarters of the near-polar area were situated over the ocean and only one quarter was over northwestern Africa. It was most likely at this time that glaciation started there. The centre of the near-polar continental area, which presumably coincided with the centre of the glaciation, was in the western Sahara. This thick ice cover disturbed the Earth's gravitational equilibrium. To restore it, the pole had to shifted to the centre of glaciation and to move southward[6]. It produced a displacement of both the near-polar continental area and of the glaciation centre to the south. The pole shifted there as well and in the late Ordovician, it appeared to be near the coast of the Gulf of Guinea (Fig. 4.25). In this process the near-polar continental area continued overlying Gondwana. The glacier expanded, its mass centre kept its southward motion. The pole shifted after it and , in the late Silurian, it reached the Rio de Janeiro region (Fig. 4.24).

Due to the pole shift in the early Paleozoic the enormously large continent of Gondwana moved more and more towards higher latitudes and, due to inertia, had to drift in the direction of the Earth's rotation, overthrusting the Pacific Ocean. As a result, an extensive fold belt started to develop along this entire boundary to the west of South America, at the margin of Antarctica, and to the east of Australia. North America, in the early Paleozoic, was situated in the near-equatorial zone and its mobility was low. However, Europe, at the beginning of Paleozoic, was in the Southern Hemisphere and moved into the equatorial zone only in the Devonian. Due to inertia, it had to drift against the direction of the Earth's rotation and thus collided with North America with the resultant extensive zone of the Caledonian folding.

The glacier continued to extend, embracing the continental parts of Gondwana and so its centre, with the pole, became displaced into South Africa in the Devonian and Carboniferous (Fig. 4.23). As this took place, the near-polar continental area reached the Pacific Ocean along the coast of Chile and western Antarctica and so the glacier could not extend further in that direction (Fig. 4.21). So it then started to move eastward to occupy extensive continental areas of Antarctica, India and Australia. The pole continued its shifting, following the glacier centre, which changed its path at an acute angle and reached Victoria Land in the early Permian (Fig. 4.20).

This acceleration in the polar shift in a new direction resulted in a significant change of the rotation rate in the western and eastern parts of Gondwana. The pole moved off from Africa and South America, whereby their rotation rate had to increase. Due to inertia, they had to shift westward, i.e. opposite to the direction of the Earth's rotation. The pole was displaced into eastern Antarctica. The rate of rotation of the Indo-Australian, eastern Antarctica part of Gondwana had to decrease, and due to inertia, it had to move in the direction of the Earth's rotation, i.e. eastward. As a result, a fracture originated along the eastern coast of Africa and a large aulacogen appeared that gradually narrowed from the north to the south toward the southern extremity of Africa. However, the opposite situation occurred

[6]All directions of pole shifting are given relative to an actual geographic grid.

to the south. As the pole shifted into eastern Antarctica, due to inertia of rotation, it acquired a spur in the direction of the Earth's rotation and so the pole moved off western Antarctica, South America and Africa. Due to inertia, these were impelled to move in a direction opposite to the Earth's rotation and to collide with eastern Antarctica. An area of compression and folding appeared between them as a result, embraced the Trans-Antarctic Mts, the Cape Mts, and the mountains in southern South America. That was the first break-up of Gondwana into two plates. Their limit became a large aulacogen extending along the East African coast that appeared in the zone of dispersal of these plates, and extended to the south in Antarctica. The Trans-Antarctic Mts. fold system that developed at the same time in the zone of compression served as this boundary. The aulacogen at the point of intersection of the path of the pole naturally passes into a fold area.

As Africa and South America moved westward, in the middle Devonian they collided with North America and Europe and thus an extensive belt of Hercynian folding was formed. At the end of Carboniferous and in the early Permian, owing to the displacement of the Gondwana glacier into Antarctica, the southern pole also shifted there. A general rotation of the lithosphere occurred and the northern pole rapidly approached Europe. When it appeared in higher latitudes, it had to move eastward in the direction of the Earth's rotation due to inertia. However, the Kazakhstan block did not change its latitudinal position, but kept its low mobility. As a result, it collided with the European platform, thus forming the Urals Mts. (Figs. 4.19 and 4.20).

In the early Permian the Gondwana glacier reached the eastern coast of Australia and Antarctica and emerged into the Pacific Ocean. The glacier could no longer extend eastward and so lost its role as driving force of the lithosphere. At this time, however, the lithospheric rotation produced by the "pursuing" of the southern pole after the mass centre of the Gondwana glacier, brought out the northeastern part of the Asiatic continent into the northern near-polar area. There, in Permian time a considerably large glacier seems to have been formed. The planetary equilibrium became disturbed and, again, a motive appeared for the lithosphere to move to draw the glacier centre towards the north pole. Since that time, the glacier has played the role of a driving force on the lithosphere moving it in the same direction, i.e. from the southern to the northern pole. The lithosphere also moved towards the northern pole under the effect of large structures of the Hercynian folding that appeared in the Northern Hemisphere at that time as a result of the collision of Gondwana and Laurasia. As a result, Antarctica and Australia left the polar area and the Gondwana glaciation ceased. Before the formation of the Gondwana glacier, Gondwana was situated mainly in low latitudes. This situation probably corresponded to a nonglacial, global equilibrium of the lithosphere. During the Gondwana glacier "activity", the lithosphere became displaced for more than 90°. Thus, after this glacier disappeared, the lithosphere did not move backward, but continued its former motion northward for some time in order to restore the previous position of Gondwana at low latitudes. The disappearance of the enormous Gondwana glacier most likely produced a sharp increase of the Earth's temperature on a global scale, and was therefore, probably the cause of the melting of the Permian East Asiatic glacier, and so the Triassic warm, nonglacial period was initiated.

In the Middle Jurassic the northern pole had a slightly more northern situation relative to the Siberian platform, and the southern pole was in the Pacific Ocean, 20° from Antarctica. However, at the end of late Jurassic northern Eurasia moved away from the northern pole, according to paleomagnetic data, and Antarctica approached the southern pole. This has been suggested to be caused by a major rotation of the lithosphere relative to the poles. With this approach of Antarctica towards the southern pole, the second glaciation of Antarctica started at the end of the late Jurassic. It seems that due to this motion of the lithosphere, Africa moved southward, away from Europe and North America, and a wide oceanic basin was formed between them. The onset of glaciation on Antarctica promoted a further displacement of the southern pole towards the centre of Antarctica. This event probably produced a sharp disturbance in the global equilibrium and an acceleration of lithospheric shell rotation relative to the Earth's poles with a resultant manifestation of an inertial force. In the zone of the great circle perpendicular to the direction of pole displacement, a meridional fracture was generated. The pole shifted to Africa, which had to reduce its motion. Due to inertia, it moved eastward in the direction of the Earth's rotation. The pole receded from South America, which had to accelerate its motion and, due to inertia, had to move westward, opposite to the direction of the Earth's rotation. This is assumed to be the case in the Southern Hemisphere; north of the equator, the opposite trends are expected. According to paleomagnetic data, the equator in the Jurassic was situated along the northern coast of the Gulf of Guinea. As a consequence, North Africa and the eastern part of the central Atlantic Ocean, which had already been formed at that time, were displaced eastward together with southern Africa. Thus, the hypothesis of global equilibrium permits the formulation of a simple and clear explanation of the greatest geotectonic puzzle: the appearance of a fracture between Africa and South America that forms a right angle. It is formed by the meridian of the great circle perpendicular to the direction of the pole displacement and the equator.

At the end of Mesozoic the site of the northern pole is near the Alaskan coast and it later moved to its present position. If we consider this motion mainly as a result of displacement of the lithosphere relative to the rotational axis of the Earth, then the quarter-sphere where the North American continent is situated is driven westward: the pole moves away from the North American continent, the rotation rate then increased and due to inertia it moves against the Earth's rotation, i.e. westward. The quarter-sphere, containing Eurasia, is driven eastward: the pole approaches Eurasia, it then slows down, and due to inertia its movement corresponds with the direction of the Earth's rotation, i.e. eastward.

South of the equator a movement of opposite direction realized. However, the mass centre of the South American continent lies further to the east than that of North America, which lies to the east, outside the great circle perpendicular to the direction of pole displacement. This displacement of poles will draw the equator nearer to the South American continent, the rotation rate will increase and, due to inertia, this continent will be driven westward, just as it was in the case of the North American continent. The major northern massif of the African continent is most likely to have been situated close to the equator in the Mesozoic. Because of the pole displacement, the equator extended southward from the centre of its mass. The continental motion slows down and due to inertia, it moves eastward. Thus, a general pattern of the continental drift and of the Atlantic Ocean formation can be seen, which is recognized by mobilists.

The eastern part of Gondwana separated into three large parts: Antarctica, Australia and India. Antarctica had to diverge farther southward because it was situated nearer to the pole and rotated around a small circle. Australia moved southeastward, being farther from the pole, and rotated around a great circle, whereas India drifted mainly eastward. The separation of these parts of eastern Gondwana was also affected by the subsequent changes in the pole positions during the Mesozoic and Cenozoic. At that time, the Indian Ocean was mainly formed.

With the Indian Ocean growing the African-South American continent moved far westward. However, when its centre reached the line where the Mid-Atlantic Ridge is now situated, the general geotectonic situation had already changed. As the pole was significantly displaced during the Alpine cycle, the Laurasian and African-South American continents divided along the line of the present Mid-Atlantic Ridge. Both Americas moved off to the west and Africa to the east.

The African and Indian blocks in their movement eastward had a northern component and drifted E-N-E with respect to the West African coast and Mid-Atlantic Ridge where this movement is most likely to have begun formation. At the same time, Northern Eurasia had a southern component and drifted E-S-E due to inertia from its previous rotational movement around the pole near Chukotka during its eastward divergence. As a result, both of these huge blocks had to collide and the Alpine-Himalayan fold belt originated at the location of their collision. Australia, in the quarter-sphere subjected to deceleration of rotation, had to move eastward and the Indonesian massif, W-S-W with the equator approaching it. This type of interpretation provides a good explanation of the relationships of the Alpine-Himalayan and circum-Pacific fold belts. The western part of this circum-belt is formed by the thrusting of two separate vast areas of the crust: the North Eurasian and Australian areas over the Pacific Ocean.

An evidently reasonable interpretation of the principal pattern of large-scale tectonics in the Phanerozoic follows from the hypothesis on planetary equilibrium. All of the continents are situated in the expected manner, according to this concept. This is impressive evidence in its favour.

It is commonly accepted that the continental drift is caused by the convective flow of the subcrustal mass. Data on the physical state of the mantle and on its layered structure do not quite support the possibility of free convection. The causes of these motions, as well as the stability of the character of the motion, are not known. The hypothesis of planetary equilibrium explains the driving forces as well as the regularities in the horizontal displacements of the Earth's crust. The driving force of the processes examined results from the interference of two strong, true factors: the potential energy of gravitation and the kinetic energy of the Earth's rotation. This hypothesis does not rule out some other force which would affect the planet from within or from space. This influence, however, is not assumed to contradict the general system of planetary equilibrium. Divergence from the rotational axis from the centre of the Earth's mass is likely to produce a significant stress that will stimulate the later shifting of the lithospheric shell relative to the rotational axis and redistribution of the motion rate.

High-power exogenetic processes on the Earth's surface produce a centre displacement of its mass exceeding a thousand times that resulting from the Siberian maximum of atmospheric pressure. The influence of the latter on the position of the rotational axis is a known fact. However, even in this case the pole divergences are expected to cover only some kilometers or some tens of kilometers. Probably, the processes acting in the inner parts of the Earth produce a more appreciable effect. Besides, one more powerful factor is quite likely to act independently. The proper rotation of the Earth is possibly also included in the process of centre displacement of the Earth's mass and so causes a recurrence of the disequilibrium state.

The continental drift only occurs due to inertia of motion and this occurs irrespective of the consequences caused by this motion in planetary equilibrium. If this drifting of the continental massifs favour equilibrium, then it consists of a single act. This is likely to have occurred in the early Triassic on the Siberian platform. Despite deep fractures and intensive lava outflows, the separate parts of the platform did not move far apart and an ocean did not originate in their place. The motion stopped just after starting. A different situation has been observed in the Atlantic Ocean. There, the separation of the continents probably assisted disequilibrium and produced a recurrent displacement of the rotational axis, which was then responsible for a new stage in drifting. This new stage continued until the moment the drifting continents were situated in the zones where their motion contributed not to disequilibrium, but to the reestablishment of planetary equilibrium.

In the light of the planetary equilibrium hypothesis, the crustal motion mechanism is as follows: during the separation of the crustal plates produced by a divergence of the rotational axis, a fracture originates between the plates that reaches the asthenosphere. A sharp decrease of pressure causes the melting of its material and the filling in of this fracture with lava.

The pole displacement is not supposed to follow a straightforward line, but constantly forms loops. At the initial stage of the pole displacement along the loop, the crustal plates move apart. The opposite trend occurs in the final stage, but the plates cannot return to their original position as the fracture separating them is filled with the congealed lava. As a result, the oceanic crust, which had formed, is accreted away from the spreading axis. The continental massifs, making up the central parts of the crustal plates and possessing the highest inertia and kinetic energy, gradually glide in the direction of the dominating movement and are thrust over the oceanic crust.

As the poles are continuously making loops the fracture zones, that serve as the boundaries of the crustal plates, show separation alternating with compression, and this plays an important role in the formation of the mid-oceanic ridges. At present we know that in the zone of the Mid-Atlantic Ridge, and in many more ridges that significant deformations of the sedimentary cover and the basalt basement and angular nonconformities in sediment occur. This process appears very picturesque in the transverse profile of the Red Sea; a number of rift valleys are seen, enclosed one inside another, with their slopes uplifted over the bottom of the inner valley as well as over the bottom of a more vast rift depression enclosing it. It gives an evident pattern of periodical alternating of separation with the formation of a rift valley and of compression with uplifting in the central part of a valley.

Under rhythmical horizontal displacements of the crustal plates, the compression and separation at their opposite sides might be asynchronous. Thus, it becomes clear why a rift valley is situated along the axis of the Mid-Atlantic Ridge, indicating an expansion, whereas to the west of South America there is no rift valley along the axis of the East-Pacific Ridge, but there is a crest, suggesting compression. The mid-oceanic ridges seem to grow intensely in the time intervals when the poles are recurrently making loops in approximately the same place. When the poles display a quick and directed movement, mid-oceanic ridges are not necessarily formed as the reverse movements of the plates are quite insignificant under these conditions. Rhythmically recurrent reverse movements of plates that produce gaps in the formation of the oceanic crust are supposed to be much shorter than the intervals between the geomagnetic reversals, thus permitting the reversals to register in the magnetic properties of the oceanic crust.

Thus, the essence of tectogenesis is the constant disturbance and restoration of the planetary equilibrium under the rapid rotation of the Earth. The causes of disequilibrium are of different nature: exogenetic, endogenetic and cosmic with the additional of an intensive inertial drift of an extensive portion of the crust. The main factor in the restoration of planetary equilibrium is most likely to be the displacement of the lithospheric shell relative to the rotational axis and the change in the Earth's shape. This process is of a directed nature as the rotational energy is continuously and irrevocably consumed in the redistribution of rotation rates, in continental drift and in associated extensive zones of strong folding and compression. Thus, if there is no force to accelerate the rate of rotation, it will gradually decrease.

The decrease in the rotation rate gradually lessens the tectonic activity that is directly related to the rotation of the Earth according to the proposed hypothesis. This specific character of tectogenesis of the Earth is similar to a quickly rotating body. Consequently, it can be concluded that planets and satellites, having a very low rotation, e.g. the Moon, tectonics is of a quite different nature in comparison with that of the Earth. The tectonic processes typical of the Earth seem to be quite impossible there, where only intraplanetary and cosmic forces are expected to act In contrast, planets having a higher rotation rate, e.g. the Earth, tectonic activity is suggested to be much more intense.

CHAPTER 6

Paleomagnetology and Mineral Deposits

6.1 Introductory Remarks

Systematic paleomagnetic studies in their initial stages were mostly dedicated to theoretical considerations, but now paleomagnetology deals with the practical tasks of geology with increasing efficiency. Paleomagnetic studies revived the fundamental conceptions of Wegener (1926, 1928) and provided a radically new concept from which the new theory of global tectonics or plate tectonics, finally emerged. This theory provides the key for solving many tectonic problems directly related to the origin and distribution of mineral deposits. As a result of the many uses of paleomagnetic data, a detailed concept of plate drift and of the pattern of plate changes in the past can be established as far back as the Proterozoic. We now know that we live not only on a planet that rotates, but also on continents that drift. As a result, a number of older hypotheses have been revised and new ideas and concepts began appearing in theoretical and practical geology, as well as in the field of forecasting mineral deposits (Irving 1964).

This chapter includes a discussion of structures in relation to paleolatitudes, types of sedimentary basins, their genesis and distribution, regularities in paleolatitudinal distribution of mineral deposits, evidence for global transgressions in relation to paleolatitude and paleomagnetic-tectonic criteria in the prospecting of mineral deposits. All conclusions drawn here are based on the paleomagnetic data collected and generated by the workers of the Paleomagnetic Laboratory of VNIGRI and shown on the global Phanerozoic reconstructions in Chap. 4.

6.2 Paleolatitudinal Constraint on Crustal Deformation

Darwin (1911) appears to have been the first to mention the acceleration and deceleration of the Earth's rotation and of its effect on the lithosphere. Veronnet (1912) and Leybenzon (1955) mathematically calculated the possibility of the crustal layer breaking at certain latitudes. Krasovsky (1953) reported on "significant tectonic dislocations" during the times when the shape of the Earth's ellipsoid changed. This problem was discussed in detail in the works of Tillo (1887), Bondarchuk (1970), Lichkov (1965), Tsaregradsky (1963), Katterfeld (1962) and others. The well-known geologists Shatsky (1955), Khain (1957), Yanshin (1963) and Peive (1956) argued that the rotational force must, to a certain extent, influence tectonic dislocations. Voronov (1968) stated that the rotational force has a major role in the formation of the planetary fracture pattern and in the structure of the entire lithosphere.

Stovas (1975) mathematically analyzed the tension of the Earth's ellipsoid during rotation and reported general regularities in the geographic zonation of crustal deformations. As a result four latitudinal zones were recognized: (1) polar zones, 90°- 80°; (2) a zone centred on 62°; (3) a zone centred on 35° and (4) an equatorial zone. In that author's opinion these are the critical latitudes. Under alternating changes in the rotation rates, the crustal layer becomes alternately compressed and extended, and vertical displacements of the crust take place at the poles and in the equatorial area. At latitude 62°, vertical displacements, lithospheric fracturing and the initiation of faults occur. Tangential tensions, i.e. mountain-building processes, occur at 35°. The idea of critical parallels, though an attractive one, did not make further progress, for it became clear that not all of the intensive tectonic events of the past fall into modern latitudinal zones.

Goncharov (1970) analyzed the role of rotational force and the occurrence of diastrophic events at certain parallels from a different point of view, applying the theory of plate tectonics. It is generally understood that the structural pattern, faulting, folding and global fracturing at certain ages form an orthogonal system generally following the paleogeographic grid of that period, which changes consistently as lithospheric blocks move into new positions. These tectonic elements will be most evident within narrow paleolatitudinal intervals at critical paleoparallels only when the lithospheric areas are situated at those paleolatitudes for a sufficiently long period of time. Because of plate movement, active diastrophism migrates along the strike of lithosphere intersecting older fold systems, resulting in the reworking and changing of structural patterns on a global scale. Plate drift and rotation of the Earth, however, place constraints on all of these changes as they become superimposed on the global tectonic pattern created by plate tectonics. Studies of the relations of crustal structural elements and paleolatitudes, made recently for all Phanerozoic epochs and for almost all of the continents, support these conclusions (Goncharov 1973, 1974, 1976).

Initiation of planetary fracturing and its later transition into the class of faults is closely related to the localization of lithospheric blocks and their orientation relative to the geographic grid of parallels, especially the critical ones, and to the rotation of the Earth. This conclusion is extremely important because recently a leading role in the evolution of any particular area is assigned to the faults that pierce the entire crust and form global megafracturing (Voronov et al. 1962; Ivanov 1975; Katterfeld 1962; Kopystyansky 1978; Moody and Hill 1956; Permyakov 1949; Muratov 1977; Stovas 1975; Khain 1978; Chebanenko 1963; Shul'ts 1964). The abyssal faults are associated with the formation of geosynclines and fold structures, increasing magmatic activity and other tectonic processes. Commonly, oil and gas fields in sedimentary basins occur in the near-fault and supra-fault structures and in areas of intersecting supra-fault anticlinal zones (Fischer and Judson 1976). In connection with these, the orientation of the near-fault structures becomes of great importance in prospecting.

The regularity in the distribution of fracture and fault patterns, relative to the rotational axis, is acknowledged and the workers recognized two to eight (and more) systems of dislocations. In general, workers observed an orthogonal character for these dislocations. Some argued only for a longitudinal and latitudinal directions of dislocations (Aleksandrov and Zanynkov 1969; Goncharov 1970;

Stovas 1975), others considered mainly the diagonal ones (Chebanenko 1963; Moody and Hill 1956), and finally the third group acknowledged both types of directions in dislocations (Voronov 1968; Shul'ts 1979). It follows from the analysis (Goncharov 1970) that two systems of dislocations existed in every geologic epoch. They were conjugated into an orthogonal grid that reflected the rotational tensions in the lithosphere originating along the meridional and latitudinal azimuths. According to the concept of those advocating a rotational force (Stovas 1975), just this type of tension is generated in the Earth's crust.

In comparative tectonic studies we see, as a rule, sharply differing structural plans in the crust: variously oriented zones of faults at every stage, varying structural contours, fold systems, zones of volcanic activity, metallogenic belts, etc. All these zonal rebuildings of the lithosphere are caused by mobile paleolatitudes when the continental blocks were situated there. In a later redistribution of lithospheric blocks in noncritical paleolatitudes, the older dislocation zones become welded and turned into "masked" zones. At the critical paleolatitudes, new systems appeared that were rotated or displaced at a certain angle relative to the previous systems. If the blocks move along a latitude or longitude, without any change in their orientation, a certain inheritance of orogenic activity might be observed.

Bubnoff (1930) discussed the changes in the dominating structural strike through time, with reference to Europe. Those strike changes were counterclockwise and a certain orientation of zones, i.e. rising and submergence, corresponded to every phase of each folding cycle. The most reasonable explanation of this process might be given when examining the drift of the European plate through the critical paleolatitudes. According to paleomagnetic data, this continent was drifting during the Phanerozoic with a clockwise rotation. Consequently, the changes in the dominant strikes had to be counterclockwise.

The existence of about ten differently oriented systems of dislocations in the Earth's crust (Shablinskaya 1977) is most likely not just an accidental observation reported by researchers. These systems very likely do exist, but they seem to be of different ages. Each system had its own paleolatitudinal-paleolongitudinal grid (Goncharov 1974). The older the lithospheric block, the more systems of dislocations are registered, thus the definition of the global pattern of dislocations becomes less evident. On examining the global pattern of fracturing, it can be concluded that the orthogonal systems are exceptionally well defined, i.e. their distinction is clearer and easier, as the diagonal ones are much more numerous (Voronov 1968). For instance, in the newly formed oceanic crustal portions, only one latitudinal-longitudinal system of dislocations has been recorded. It is most distinct near the mid-oceanic ridges where the young crust has not been as yet subjected to significant dislocations and rotations. With increasing distance from the ridge axes or with displacements of the ridges proper, the orthogonality of the system is preserved relative to the geographic grid. However, the strike directions begin forming an angle and the greater it becomes, the more significant is the rotation of the lithospheric block. Eventually, the orthogonal system of dislocations in the newly formed crust becomes complicated by the present field of the planet's rotational tension, and a number of orthogonal dislocations grids are generated, which are variously oriented relative to each other. Just this type of dislocation set has been recorded in the older portions of the lithosphere (Figs. 6.1 and 6.2). The

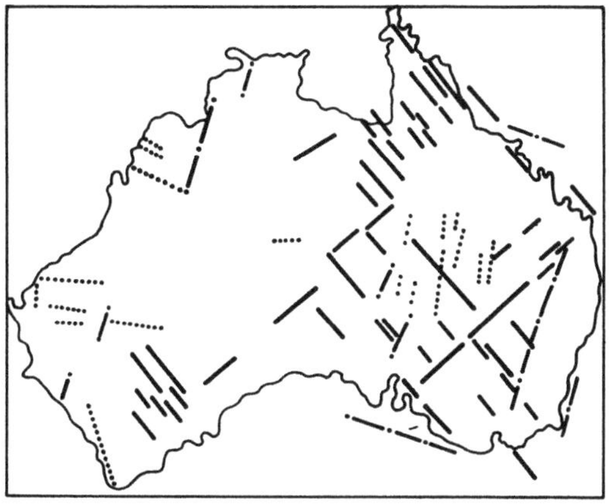

Fig. 6.1. Differently oriented fault systems in Australia. (Hills 1953)

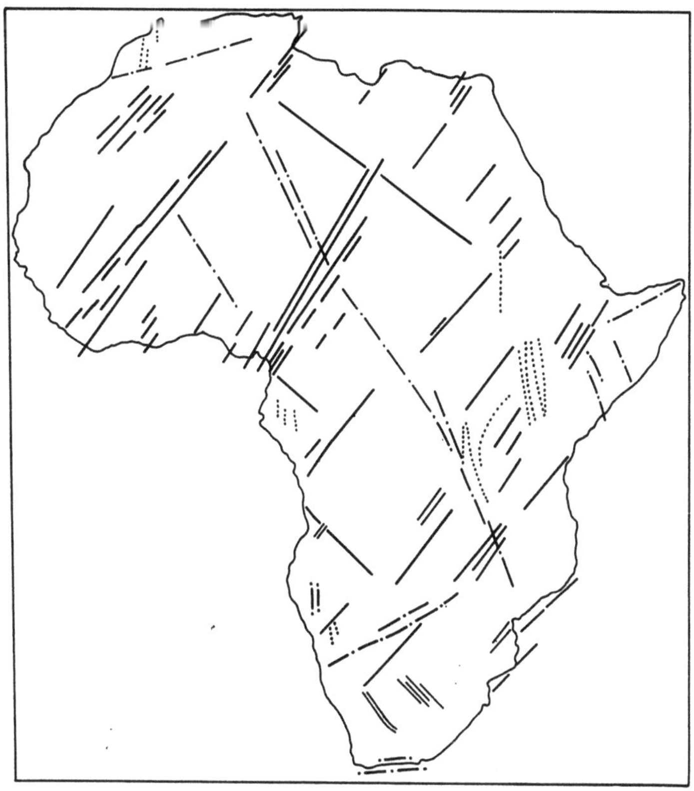

Fig. 6.2. Differently oriented fault systems in Africa. (Chebanenko 1963)

regularities now known permit the suggestion that the dislocation pattern of the present orthogonal directions might be registered in young lava plateaus, in the Antarctic ice sheets and in the ice cover of large lakes.

Comparative studies of the global fracturing of different ages and paleolatitudes do not contradict the regularities mentioned above (Table 6.1). Table 6.1 shows that the major strikes of the global fracturing, relative to the contemporaneous paleoparallels and paleomeridians, are of an orthogonal direction or similar to orthogonal directions, i.e. they correspond to those assumed to be generated in the lithosphere according to the hypothesis of rotational force.

The proposed conceptions might provide an explanation for the origin of global fracturing and fault pattern, consolidation of platforms and their reactivation, zonal distribution of metallogenic belts, their orientation and extension. The regularities recognized would thus yield the time of formation of various structural elements and their geographic coordinates. The known regular relationship of structural elements of the Earth's crust and the paleolatitudes helps a great deal in forecasting metallogenic belts, because the relationship of endogenetic deposits and fault tectonics is a generally accepted fact.

6.3 Types of Oil and Gas Basins, Their Genesis and Distribution

Palinspastic reconstructions must be conducted in order to more accurately determine the evolution of sedimentary basins, the type of oil- and gas-bearing basins and their geographic position. Eventually, sufficient paleomagnetic data had been accumulated and many researchers made such reconstructions.

The classification of oil- and gas-bearing basins can be based on their regional tectonic pattern and on the occurrence of these basins in the areas of the continental

Table 6.1 Comparison of the paleomagnetic latitudinal line orientation and that of rock fracturing

Area	Rock age	Direction of the paleolatitudinal lines[a]	Direction of rock fractures[b]	
			Latitudinal system	Meridional system
Estonia	O	323°	316°	339° + 90°
Northwest Russian platform	S	325°	320°	337° + 90°
Latvia	D	307°	285°	315° + 90°
Middle Volga region	C_1	337°	345°	333° + 90°
Middle Volga region	C_2	323°	340°	338° + 90°
Middle Volga region	C_3	320°	–	307° + 90°
Northwest Russian platform	C	313°	324°	318° + 90°
Middle Volga region	P	316°	345°	–
Latvia	P	300°	314°	324° + 90°
Volga – Kama region	P	318°	314°	315° + 90°
Estonia	N	270°	270°	270° + 90°

[a]Directions of the paleolatitudinal lines were calculated according to the poles listed in Sect. 4.3.1 and in the Appendix.
[b]Orientations of fracturing lines were taken from Shul'ts (1973).

or oceanic crust. The classification given here is based on tectonic criteria, according to Weeks and Lewis (1952) and Uspenskaya and Tauson (1972). These types of basins were then tested from the viewpoint of plate tectonics. Klemme (1975) recognized eight principal types of basins proceeding from this theory.

Type I. These are inner basins with a simple structure, moderate size and saucer-shape, filled with gently dipping rock piles that were formed during one cycle of sedimentation. These basins are observed in the inner continental regions near the Precambrian shields. They are commonly built up of Paleozoic sediments deposited in platform seas and oceanic environments. In most cases this type of basin is difficult to discover and the chance of finding industrial hydrocarbons within these areas is about 50% and less, the probability of finding a giant field is extremely low.

Type II. These are the intracontinental basins of composite structure, large (subcontinental miogeosynclinal) and small (intermountain) that generated during two or more sedimentary cycles. The basins are of a more composite structure compared to type I and are situated within the inner regions of a craton. Twenty-three basins of this type have been discovered, including 85 giant fields with 25% of the world oil and gas resources.

Type III. These are the rift, graben and semigraben, small and middle-sized basins which subsided along faults. These basins are likely to be generated at an initial stage of seafloor spreading (rifts within a stable craton) in post-Paleozoic time. In some cases they evolved within the limits of type II basins that were formed earlier. Basins of the third type are rich giant fields. The discovery coefficient of these basins is high: five of nine explored appeared to be of an economic type.

Type IV. These are extracontinental basins occurring in regions downwarped toward the middle-sized oceanic depressions. These basins are situated along the continental margins and continue into middle-sized oceanic depressions. The basins are commonly of post-Paleozoic age. Nine basins of this type have been discovered, including 86 giant fields with 50% of the world resources. The success of exploration in this case amounts to 50%.

Type V. These are fully separated, linearly-extended, unilateral basins of a stable coast bound by faults. These basins are situated along the continental margins, built up of Mesozoic and Cenozoic rocks cut by faults. They are probably those basins of the third type (rift) which were separated during seafloor spreading and moved apart for a width of an ocean (the "passive margin" of a continental block). Success in finding a significant industrial field amounts to 30%. Only three giant fields are known within these basins.

Types VI and VII. These are intermountain basins formed during the second stage of sedimentation. The basins are situated along the continental margins where the oceanic plate is likely to underthrust the continental plate. The sixth type of basin is situated across the strike of the continental margin, and the seventh type of basins are parallel to its strike. These basins are of Mesozoic and Cenozoic age. Large fields commonly occur in the basins of types VI and VII. Thirteen basins, including 37 giant fields with 10% of the world resources of hydrocarbons have been discovered.

Type VIII. These are the Neogene deltas, shaped like a "bird's foot", built up of Miocene to Recent deposits. They are situated along continental margins intersected by rivers that drain vast areas of the continent. These basins are filled with deposits of marine and nonmarine origin. One of the two deltas explored appears to be of an industrial type.

The Permian basins of western Texas and New Mexico, basins of the Rocky Mts., the Appalachians, Alberta and many more are examples of type I and II basins. The craton basins are usually situated in the continents, but some extend off shore (the Aquitanian basin, for instance). The basins of the Red Sea, Rhine graben, the Cambay basin of India, the Sirte basin in Africa and also the North Sea basin, all with giant fields, exemplify the type III basins. Thus, during the prospecting and exploration study of an area, it is necessary to pay special attention to the graben-rift structures. Basins of type IV include the Gulf basins, the Tethys zones, the China Sea, the Beaufort Sea, the northern slope of Alaska, the East Venezuela basin and also the basins of the Persian Gulf. Basins of type V are now being explored on the Atlantic Coast and in the coastal regions of India and Australia. These were developed within the passive margins of the separating plates, the opposite coasts of the Atlantic Ocean: in Greenland and Arctic Canada, eastern Africa and India and also in some regions of the western, southern and eastern coasts of Australia (Fischer and Judson 1976). The transversal and longitudinal basins of types VI and VII might also occur in island-arc subduction zones (for instance, the basins of Sumatra and the Java Sea) and in the development zones of transitional crust where the continents are underthrust or overthrust the oceanic plates (fields in the western part of the USA: Los Angeles, Ventura). Basins of type VIII (deltaic) were formed during the evolution of very large drainage systems, such as of the Mississippi, Mackenzie and Niger Rivers.

The tectonic situation of the basins may change with time and combinations of various prototypes may reappear. Basins of type VIII might exemplify this situation. They are superimposed on the basins of type IV (the Eastern Kalimantan); basins of type VIII overlying the type IV basin (the Mackenzie delta); basins of type V or VIII underlain by the type III basins (eastern Canada), etc. Thus, to recognize the basin type from the viewpoint of plate tectonics, all of the known geologic aspects must be considered for hydrocarbon exploration.

6.4 Paleolatitudinal Distribution of Mineral Deposits

6.4.1 Coal

Paleogeographic distribution of mineral deposits, and also of oil and gas fields, have been a subject of considerable attention. Some workers reported that there was an evident relationship between coal formation and humid zone latitudes; between salt genesis and arid zones; between oil and gas accumulation and low equatorial latitudes. The explanation of this assumption lies in the favourable physicochemical environment that provided optimal conditions for a high development of biomass, i.e. the initial production of hydrocarbons and coal. However, later studies in this field did not support either the lower latitudinal distribution of oil and gas basins or latitudinal-climatic distribution of coal and salt. The fixist concept does not reveal their occurrence in areas of lower latitudes, nor can it provide an explanation for the generation of coal and salt deposits and oil and gas fields at high polar latitudes. Thus, current practice does not support the theoretical concepts on latitudinal-climatic zonation of these deposits.

Similar analysis from the viewpoint of plate tectonics, in combination with statistical methods, permitted the establishment of certain rules in the paleolatitudinal occurrences of coal, salt, oil, gas and other deposits with a sufficient degree of reliability.

Three conditions are essential and absolutely indispensable for the origin of coal: humid climate, paleogeographic environment and tectonic regime. Climate plays the major role as the "principal geographical and geochronological regularities in coal accumulations (the belts and the epochs of coal accumulations) are governed by climate" (Sinitsyn 1967, p. 26).

We will now examine the distribution of initial coal reserves of the world deposits relative to the present latitudes (from the viewpoint of fixism) and then relative to the paleolatitudes (from the viewpoint of plate tectonics). Here, we use a paleomagnetic base and a map of the world's initial coal reserves (Matveyev 1972). Upon summation of the initial coal reserves of all ages in 10° latitudes intervals for South and North America, Africa, Australia and Eurasia and correlating these with the total areas of continental crust in the same intervals, we obtain a histogram of the so-called effective thickness of coal according to the ranges of Recent latitudes (Fig. 6.3a). In the case of continental drift, the average values of effective thickness are estimated by dividing the coal reserves by the areas of corresponding paleolatitudinal intervals defined for the coal-bearing sequences, i.e. the initial Devonian coal reserves were divided into by 10° interval areas of Devonian paleolatitudes, the Carboniferous reserves, by the Carboniferous areas, etc. To obtain sufficiently uniform statistical data, the areas of 10° latitudinal (paleolatitudinal) intervals considered were only in the rigid portions of the Earth's crust.

The distribution plot of the initial coal reserves relative to the present latitudes is unclear, i.e. it is not satisfactorily distinct; the increase of plant biomass is shown to extend from the equatorial to polar latitudes, thus disagreeing with realistic principles. The Antarctic coal is not considered because of lack of data, which also poorly supported this concept. This type of histogram prompts an improbable conclusion (under present climatic conditions) on the higher rate in plant biomass increase in the polar areas compared to the equatorial regions and on the absence of a distinct climatic zonation. This is the conclusion to be drawn from the paleoclimatic interpretation of the histogram when regarding the coal accumulation primarily as a climatic phenomenon.

The advocates of a stable position of continents suggested that this climatic type of coal accumulation evolved during geologic history (Sinitsyn 1967). The initial stage of this process in the middle Devonian is assumed to be related to the

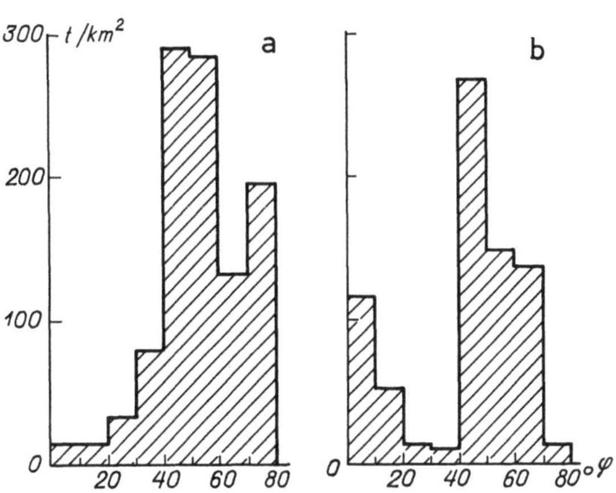

Fig. 6.3a,b. Recent latitudinal **(a)** and paleolatitudinal **(b)** distributions of the initial coal reserves. (After Khramov et al. 1974)

coastal-marine environments of a tropical arid climate. During the early-late Carboniferous, this process continued under humid climatic conditions in the maritime lowlands. In the late Carboniferous-Permian and Mesozoic time the coal accumulation process was of a continental nature associated with areas of boreal climate. Later, during the Neogene, coal accumulation was again distributed in boreal as well as in tropical areas.

We should not adhere to suppositions of this type, if we favour the analysis of coal accumulation from the concept of plate tectonics. The corresponding histogram (Fig. 6.3b) shows quite distinctly a relationship with climate. Two quite evident maxima are seen: one in the equatorial paleolatitudes, the other in the middle ones and almost a complete lack of coal accumulation in the 20°–40° paleolatitudinal interval, clearly indicating that the climatic environment, since the Devonian time at least, was of a distinctly zonal character, similar to the present zonation. Rather poor reserves in the 20°–40° interval and away from the 70° paleolatitude evidence a depressed type of vegetation at those latitudes. At the present time, arid and polar conditions are dominated by poor vegetation that cannot provide an initial product for any significant coal formation in these latitudes.

Dominantly tropical or temperate-humid environments of coal formation might be recognized from the analytical peculiarities, from the presence or absence of annual rings, from the frequency of transition and interbedding of coal and either carbonates or variegated red beds containing bauxite. The available data indicate that the plant composition of coal varies depending on the latitudes in which it was formed, and thus, the tropical and temperate paleolatitudinal areas of coal formation can be recognized. This observation is especially important, not only as an additional argument in favour of climatic zonation in the past, but also as an indicator of different humid conditions: i.e. tropical-humid and temperate-humid.

This example of paleolatitudinal analysis of coal accumulation is an impressive argument in favour of the plate tectonics concept; at the same time, it demonstrates perspectives of the paleomagnetic method.

6.4.2 Salt

In general, salt-bearing sequences, irrespective of their composition, thickness and distribution, are supposed to be a reliable indicator of an arid climate environment. In this case, we have no satisfactory explanation of salt occurrences in the Devonian strata of the islands in the Arctic Ocean, in the Permian and Carboniferous sediments of Spitzbergen, in eastern Greenland, etc. Certain difficulties arise when interpreting salt occurrences in the present areas of equatorial climate: Carboniferous and Permian coal-bearing basins of South America; Triassic, Jurassic and Cretaceous salt deposits of South America and Africa, etc. Similar difficulties appear when conducting a paleolatitudinal analysis of the distribution of halogenic sediments. If a grid of paleomagnetic parallels is superposed on the plot of distribution of the Paleozoic evaporite deposits and the number of basins is counted in 10° intervals, then it becomes obvious that the salt-bearing sequences were accumulated in practically all paleolatitude intervals (Fig. 6.4a).

Many workers have emphasized that the distribution of halogenic formations is connected with two factors: (1) a change in paleoclimatic conditions in time and space and the situation of belts of arid climate; (2) peculiar aspects in the tectonic development of the Earth. During folding, the areas of continental masses are growing as the orogenic systems are being formed, thus producing a change in the Earth's atmospheric circulation and in the appearance of arid zones (belts). Statistical analysis, however, did not prove any significant latitudinal or paleolatitudinal concentrations of salt in the arid belts. Halogenesis was probably not related to such factors.

At present it is accepted that deep-seated faults affect the formation of the crustal structures and that the basement and the subsalt beds commonly have a block structure. Factual data on basins of buried halogenic deposits show an activation of deep-seated faults during salt formation. As a rule, the halogenic strata are situated in the areas of crustal downwarping. These were the sites of fault and fracture zones that served as feeding channels for an easy circulation of internal waters and brines.

The mechanism of formation of downwarped areas in the crust is sometimes explained by the irregular rate of the Earth's rotation, which results in the redistribution of geodynamic stresses in the upper mantle (Voronov 1968; Stovas 1975; Shul'ts 1979). The rotation of the planet produces not only a redistribution of stresses, but also their localization within definite narrow intervals, i.e. the critical parallels. Then, postulating an association of salt formation and tectonically active regions (these regions occur at quite definite parallels), we expect to find the maximum concentration of halogenic rocks just at these latitudes. From the viewpoint of plate tectonics active salt accumulation takes place at the corresponding paleolatitudes and also in the regions of spreading and probably of subduction. This relation can be seen from the plots in Fig. 6.4 and also in other publications. Thus, if salt origin is connected not only with climate, but also with depth, we can reasonably explain the appearance of salt at latitudes that undebatably did not belong to zones of arid climate.

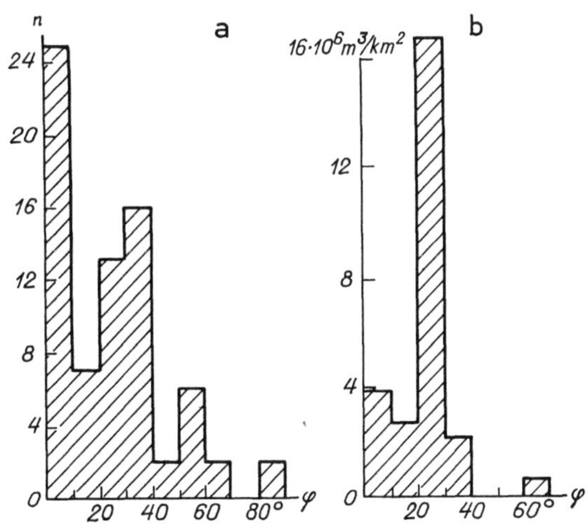

Fig. 6.4a,b. Paleolatitudinal distribution of the salt-bearing basins **(a)** and of the initial salt reserves **(b)**. (After Khramov et al. 1974)

Recently, data on the determination of salt reserves were assembled, mostly referring to Paleozoic times. They permitted the application of statistics, thus making a paleolatitudinal analysis of halogenic rocks more precise. Dividing the salt volume into the paleolatitudinal intervals by the corresponding areas we observe that, upon plotting the histogram (Fig. 6.4b), major salt reserves are associated with the paleolatitudinal intervals 20°–40° which are now characterized as the optimal environment for salt deposition. Removing from this histogram the salt volume that was accumulated in arid paleolatitudes, we obtain the maxima just at the paleolatitudes that are called critical. This histogram, however, shows only the trend because of scarcity of data. Nevertheless, in climatic reconstructions and in forecasting salt deposits, it is necessary to consider the deep origin for specific parts of the salt concentrations that might accumulate in the arid belts (climatic control) as well as in the intervals of mobile latitudes (tectonic control).

6.4.3 Oil and Gas

The concept of plate tectonics provides a new approach to the problem of the origin and distribution of oil and gas (Irving and Gaskell 1962; Irving et al. 1974). This concept also provides an explanation of the highly dynamic development of the Earth's crust in the older and recent continental margins. The major potential and reserves of oil and gas are associated with these regions (Fischer and Judson 1976). We know several satisfactory models of formation of oil and gas fields in zones of spreading and subduction based on the plate tectonics concept, such as the models proposed by Sorokhtin, Kovalev, etc. (Sorokhtin 1979b). These models do not oppose any other hypothesis of oil and gas accumulation, but at the same time, they do not explain the mere existence of intracontinental oil and gas basins and their association with the paleoequator and paleolatitudes 30°, 60°, 80°–90°. In some works (Goncharov 1973, 1976, 1981) the favourable conditions for hydrocarbon accumulation and redistribution in the mobile paleolatitudes are suggested to cause this preferred pattern.

More than 400 sedimentary basins are recognized in the world, and oil and gas fields are known in only half of them. Why do the oil and gas fields in the other half remain unknown, although the necessary preconditions for the accumulation of oil and gas exist? Probably, these basins were not located in the critical latitudes for an optimal time interval, i.e. they did not go through the necessary tectonic active stage, which is considered to be one of the most important requirements conditions for a productive basin. In the opinion of some workers (Boyko 1982; Gavrish 1978), geodynamic stress in the zones of intensive crustal extension-compression in fault regions affects the distribution of oil and gas fields. This variety of geodynamic stresses in the crust seems to have the probability of maximum development in the critical latitudes. Their role in the genesis of certain mineral deposits was shown in some works involving the plate tectonics concept (Goncharov 1970; Khramov et al. 1974).

The critical intervals of paleolatitudes are most favourable for penetration, redistribution and accumulation of gaseous, fluidal and plastic-fluidal formations due to high jointing of rocks, fault tectonics and structure generation. The initial hydrocarbon reserves examined in the paleolatitude intervals also supported these arguments (Goncharov 1977). The method applied in the statistical analysis was the same as in the case of coal.

The total initial oil and gas reserves (Table 6.2) were divided by the total area of the continents along the 10° intervals of paleolatitudes of different age, the histograms were plotted according to the data thus obtained (Fig. 6.5). Four sufficiently distinct maxima in the distribution of oil reserves are revealed (Fig. 6.5a): at the equatorial paleolatitudes, at 30°–40°, at 60° and at polar latitudes. These are considered the most mobile and most favourable for penetration and formation of hydrocarbon deposits. In this histogram the distribution of the initial gas reserves, according to paleolatitudes, does not appear regular enough (Fig. 6.5b). Here, the maxima occur in the equatorial and polar latitudes. In middle paleoparallels this peak is insignificant and at latitudes of 30°–40° it is not observed at all. Gas is the most mobile component among hydrocarbons and the recognized

Table 6.2 Paleolatitudinal distribution of the world oil and gas resources[a]

Interval of paleolatitudes	Area of the continents within this interval $10^6 km^2$	Oil resources t/km^2	Gas resources m^3/km^2
0°–10°	99.8	614	93 800
10°–20°	107.5	62	55 250
20°–30°	94.1	18	32 510
30°–40°	91.0	252	28 770
40°–50°	82.2	8	17 250
50°–60°	65.9	22	24 600
60°–70°	45.8	960	49 650
70°–80°	18.6	3	47 390
80°–90°	4.1	220	260 900

[a]Data on the world resources were taken from Vysotsky (1976).

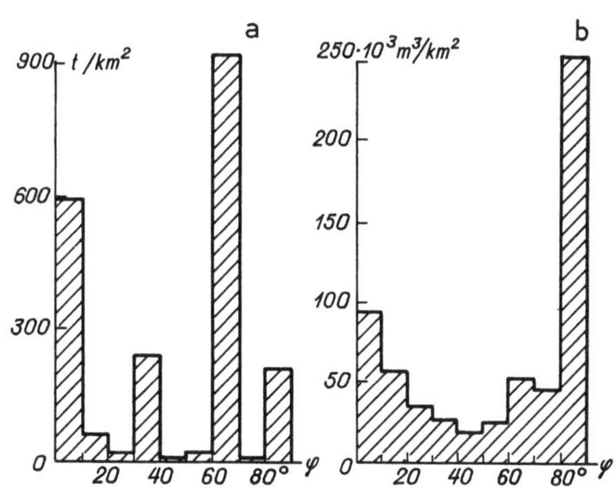

Fig. 6.5a,b. Paleolatitudinal distribution of the initial reserves of oil **(a)** and gas **(b)**. (After Khramov et al. 1974, modified according to new data)

regularities of oil distribution are probably, not realized in this case. The mobile critical paleolatitudes serve as the most favourable medium for gas penetration, but not for its migration or accumulation.

The established regularities in the paleolatitudinal distribution of oil and gas should be tested by practical work. Nevertheless, even now when the deep-seated strata are explored or when oil is prospected in poorly exposed areas and in water areas, it is possible to recommend certain paleolatitudinal belts where the sediments of a given age might be most productive or it is possible to forecast the potential stratigraphic intervals in certain regions.

6.4.4 Ore Deposits

A prevailing humid and, with that, hot or sufficiently warm climate is a necessary condition for the genesis of certain sedimentary ores, e.g. iron, manganese, bauxite, cupriferous sandstone and phosphate. In other words, ore deposits of iron, manganese and aluminium appear only in those parts of humid areas where the climate is most favourable for migration of these elements in the crust by weathering and natural water circulation. Phosphate deposits are known not only in humid areas, but also in arid zones. The most important factor for complete bauxitization is for heavy precipitation alternating with dry seasons (Strakhov 1962).

The paleomagnetic method described in the previous sections was applied for recognizing the paleolatitudinal regularities in the distribution of the ores mentioned. In the maps of paleomagnetic latitudes a particular ore deposit was indicated and estimated according to 10° intervals of paleolatitudes. These data were plotted in the graphs in Fig. 6.6.

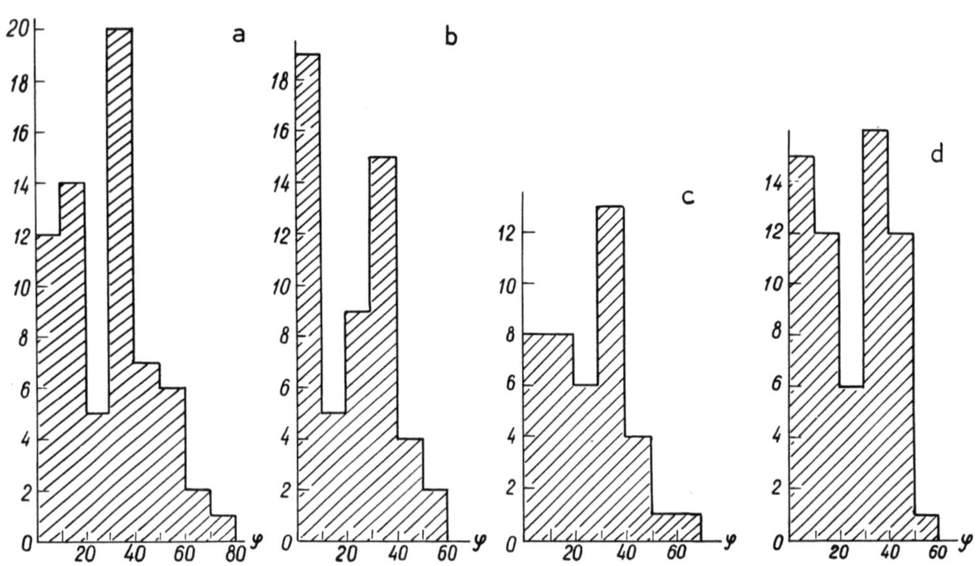

Fig. 6.6a-d. Paleolatitudinal distributions of iron **(a)**, manganese **(b)**, bauxite **(c)** and phosphorite **(d)** deposits

In all graphs a paleolatitudinal interval is quite clearly, favourable for the accumulation and genesis of the mineral deposits examined. These are mainly, concentrated in the intervals of hot and temperate climate, and just a little higher than 40°–50° paleolatitudes. In the arid 20°–30° paleolatitudes the concentration of iron, aluminium, phosphate and manganese deposits is much lower. The arid paleogeographic environment is considered unfavourable for the genesis of these deposits. As 95% of manganese deposits accumulate in a marine environment, the arid paleolatitudinal interval in Fig. 6.6 has been reduced to a certain extent. The cupriferous sandstone needs an extremely hot climate for its accumulation and it has not been recorded anywhere with a higher than 40° paleolatitude. These mineral deposits can serve not only as indicators of climate, but also of the tectonic activity of an area at the time of their genesis. Thus, the iron-manganese ores of the Caledonian epoch reveal a tendency to be localized in geosynclinal zones of sedimentary ore formation. In the Hercynian tectonic stage the platforms became the main areas of Al-Fe-Mn ore formation. In Alpine times this ore formation occurred in the stable, consolidated areas and became more evident than in the two previous tectonic stages.

The relation of the Al-Fe-Mn ores and the surface of erosion can be used as additional evidence of a paleogeographic environment in the regions of ore formation. Bauxites, for instance, are recognized to occur near transgressive marine basins. Iron and manganese ore formation is related to the onset of a transgression or a regression.

Thus, the consistent regularities in the distribution of certain mineral deposits according to paleolatitudes confirm the usefulness of the models of continental drift mentioned above. Besides, the localization of a particular deposit in an area can help in understanding its tectonic activity (see Sect. 6.2), marine environment (see Sect. 6.5) and paleogeography.

6.5 Association of Transgressions with Certain Paleolatitudes

As far as in the 19th century, many workers pointed out the wide development of many transgressions, but only in the fundamental studies of Bubnoff (1930) was a concept of global-scale transgressions and regressions presented. Since that time numerous publications have appeared, whereby the concept of global transgressions was either wholly accepted or refuted. The quantity and time of transgressions are conflicting, according to data of different authors. The total number of transgressions since the Cambrian amounts to 12–18 in the opinion of some (Nalivkin 1962) and 56 according to others (Strakhov 1949). It should be noted that these studies were based on the assumption of a stable position for the continents.

Global transgressions and regressions have also been discussed in many works from the viewpoint of plate tectonics. Kravchinsky (1977), for instance, suggested that transgressions and regressions are related to the rate of plate movement. When this rate increases, the continents "float" with a resultant regression, while a decreasing rate results in their being "submerged" and becoming covered by ocean water. Aparin (1979) is a propenent of a contrasting opinion. A witty hypothesis was proposed by Sorokhtin (1979a). He suggested that during fast spreading the volume of the oceanic ridges sharply increases, the water level becomes higher and a transgression starts. With a low rate of spreading, the ocean bed submerges, consequently resulting in a regression. All these mechanisms suggest a simultaneous distribution of transgressions and regressions in the continents. In reality, we observe an evident asynchroneity in the evolution of these processes. Thus, in the early Permian in the east of the USSR, a transgression has been recorded, while in Europe, a regression occurred. The maximum transgression in the early Carboniferous in Europe, corresponds to a maximal regression at the same time in South America. The late Cretaceous period is characterized by the maximum regression in Australia and Eastern Siberia and by one of the largest transgressions in Europe. Many other events of this kind could be enumerated for Phanerozoic times.

However the following should be remembered when transgression was dominant in the early Permian to the east of the USSR, this region was in high polar paleolatitudes. The European regression was recorded at lower paleolatitudes. In the early Carboniferous, the European part of the continent was situated at the equator and South America was at high polar paleolatitudes. In the late Cretaceous Australia and eastern Siberia were in the same high paleolatitudes and the major part of Europe, was in near-equatorial and lower paleolatitudes. Thus, a certain regularity appears in the distribution of transgressions and regressions that tend to group in the regions of polar and high paleolatitudes or in those of equatorial and lower paleolatitudes. This regularity becomes most evident in successive examinations of the Phanerozoic reconstructions. To show this regularity, we examine the distribution of marine and continental formations that show transgressive and regressive epochs with reference to the above discussed global reconstructions in the Carboniferous, Permian and Triassic.

At the end of Carboniferous and at the beginning of Permian the marine regime of sedimentation was dominant in Gondwanaland, which was situated at that time in the high polar paleolatitudes (Khain 1971). In South America, at the paleolatitudes of 50°–70°, marine sediments were registered. At the end of the late Carboniferous in Australia the earlier formed depressions were filled with marine sediments against a background of a general uplift. Continental sediments are distributed over the entire African continent, but thick fluvio-glacial deposits begin forming in its southern part, at higher and polar paleolatitudes. In the Siberian platform a transgression was registered during the early Permian at higher polar paleolatitudes, in contrast to the European block, where a regression dominated at that time, which was in an equatorial paleolatitudinal range. In the lower and near-equatorial paleolatitudes a break in marine sedimentation was registered over a major portion of the North American platform.

During the time interval that followed-approximately since the second half of early Permian until the first half of the late Permian – the depressions in South America closed, the sea area in the central part of the platform decreased in high and polar paleolatitudes, but the northern depressions on the platform again started to submerge in the lower paleolatitudes. In Africa the basins of that time are known from a range of equatorial paleolatitudes. Unconformities and breaks in sedimentation have been recorded in its southern regions, but these are also known at higher paleoparallels. At the end of the early Permian and during the major portion of the late Permian in Australia, the marine facies pass into continental facies at higher paleolatitudes. At the same time, at higher and polar paleolatitudes of Antarctica, the coal-bearing and glacial formations develop. In Siberia, at the end of the early and at the beginning of the late Permian, a continental environment dominated at 40°–60° paleolatitudes. During that time, in North America and Europe, the sea increased in the equatorial and lower paleolatitudinal range.

The time interval since the end of the late Permian until the beginning of the late Triassic is again characterized by the development of transgressions at high and polar paleolatitudes and by extensive regressions at low and equatorial paleolatitudes. In North America, in the range of 0°–50°, an abrupt decrease of marine sedimentation started in the late Permian and reached its maximum in Triassic. The Triassic marine sediments have been recorded only in the range of 50°–60° paleolatitudes. The South American platform was an area of uplift and erosion in the lower and middle paleolatitudes. Thick sedimentary piles accumulated only in the southern part of the continent at high paleolatitudes. Continental formations are known from the entire African continent in the range of equatorial and middle paleolatitudes. In the southern part of the continent, however, marine formations appeared at the end of the late Permian and in the early Triassic, again at higher paleolatitudes. A significant Permian transgression developed at high paleolatitudes in Australia. At that time, in the Russian platform, the marine and lagoonal facies pass into the alluvial-boggy type. The Siberian platform underwent a general submergence at higher paleolatitudes. This analysis was done for the entire Phanerozoic since Cambrian until Recent time.

As a result of this analysis, 19 alternating equatorial-polar transgressions and regressions have been established for the Phanerozoic period (Figs. 6.7 and 6.8). Along the ordinate below the abscissa, occurrences in continental regions with marine sediment environments are shown in a given 10° paleolatitude range. Above the abscissa the environments with continental sediments are presented. For example, if in a given geological time the marine basins in the range of 0°-10° are stated (fixed) on the six continents in the range 0°-10°, six conventional divisions are plotted, on the four continents four divisions and so on. Of course the principle of histogram plotting is simplified and rather rough, but is admissible in this analysis. It can be seen in Figs. 6.7 and 6.8 that the transgressions in the polar and equatorial parts of the Earth successively alternate with the regressions. The interval of each regime lasts about 30 m.y. This regularity is of exceptional importance as the problem of periodicity of geologic processes has been debated for a long time. However, a generally accepted opinion has not yet been reached.

This periodicity, although not as reliable, was established earlier for the Paleozoic and, with a higher reliability, it was determined for the Mesozoic (Vereshchagin and Nalivkin (eds), 1972). In the Mesozoic the continents were situated at latitudes close to the present ones. These cycles appeared more perceptible for that period as, in the Paleozoic time, the distribution of transgressions is blurred due to a significantly different situation of the continents in comparison with that of the present. This periodicity for the entire Phanerozoic can be reliably traced only in the reconstructions made according to the mobilistic theory. However, in this period the true picture of periodicity is also locally blurred because the available stratigraphic and paleomagnetic data are insufficient.

The regularities thus established seem to be governed by a fundamental mechanism. It is most likely a change in the rate of the Earth's rotation. This change is expected to produce an instant response by the hydrosphere in terms of the redistribution of its mass, and the lithosphere would thus accomodate to the new regime with a delay. Then in a region characterized by a general transgressive background, a pattern of alternating transgressive-regressive regimes would be registered with the transgressive type dominating. The regressive pattern in these regions would be more evident, for the withdrawal of oceanic water would preceed the downwarping of the lithospheric blocks. Displacement of the lithospheric blocks might also be a cause of an obviously distinct expression of the regressive epochs. This is probably just the situation that explains the Permian-Triassic regressive cycle, i.e. when the continents became grouped into a single supercontinent, Pangaea, it prevented the lateral movement of transgressions inside the continents and a sea appeared to be distributed only along the Pangaean perimeter.

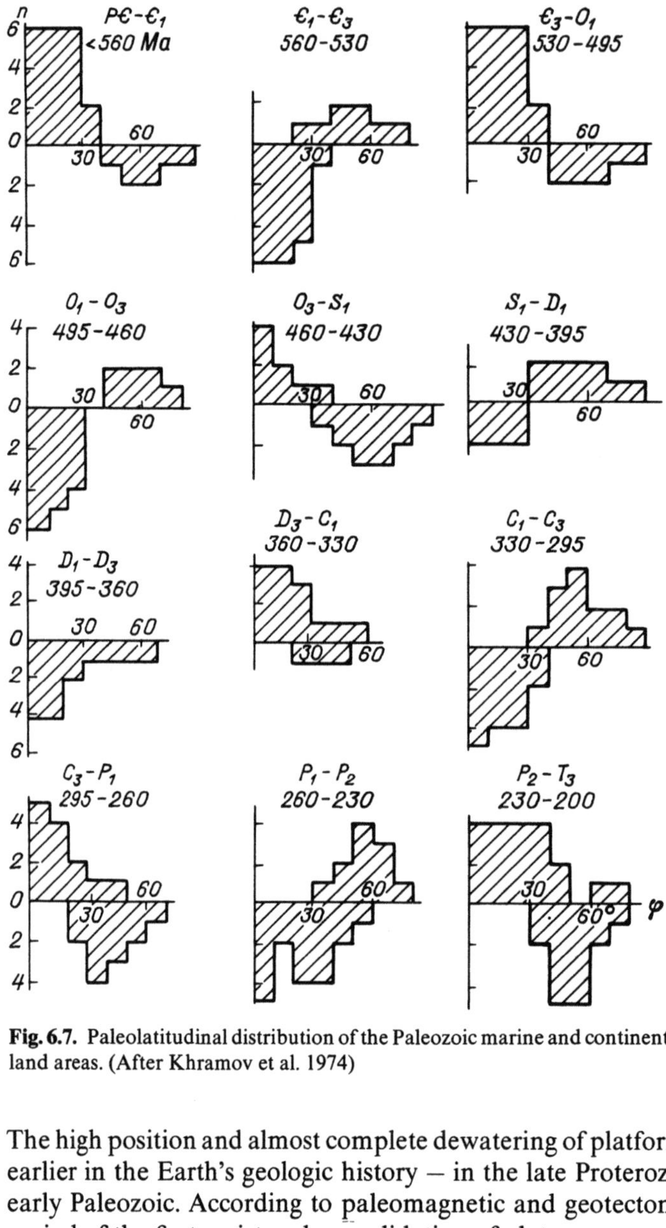

Fig. 6.7. Paleolatitudinal distribution of the Paleozoic marine and continental sediments in the present land areas. (After Khramov et al. 1974)

The high position and almost complete dewatering of platforms was also recorded earlier in the Earth's geologic history – in the late Proterozoic-beginning of the early Paleozoic. According to paleomagnetic and geotectonic data, this was the period of the first registered consolidation of plates.

It appears evident from the established regularities that it is not valid to discuss only the time of global transgressions and regressions in the continents. We must also introduce another facts to obtain a higher validity for our statements, i.e. the location of a lithospheric block in the time interval discussed. Paleomagnetic methods also help in establishing the range of paleolatitudes (with a different grade of precision, but generally it is quite appropriate). Probably, the periodicity of transgressions should not be discussed in terms of time, but rather in terms of space and time.

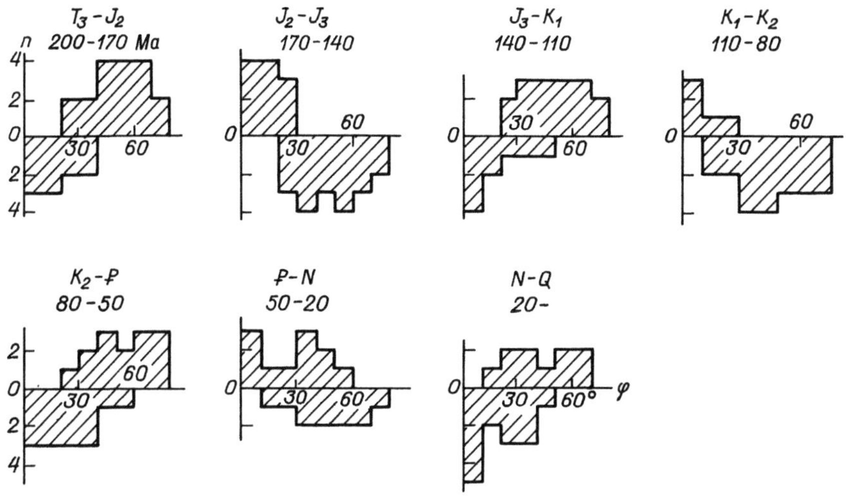

Fig. 6.8. Paleolatitudinal distribution of the Mesozoic and Cenozoic marine and continental sediments in the present land areas. (After Khramov et al. 1974)

The distributions thus obtained (however, they need refinement) permit the forecasting of the general trend of transgressions in any part of the planet. Naturally, the ruggedness of the relief, the situation of the lithosperic blocks and their high or low position require certain corrections to the scale of transgressions. Thus, the marine transgressions with their general directions from the poles or from the equator might penetrate into a region not strictly along antipodal azimuths, but rather into regions of ocean water in the equatorial (warm) or the Arctic (cold) latitudes. This would certainly result in significant changes in the physicochemical, facial-lithologic, ecologic and other types of environments in a region.

Thus the global scale of transgressions, which is most likely to be explained by the rotation of the Earth, permits the establishment of a general trend in the evolution of transgressions, the forecasting of changes in the physicochemical conditions, the determination of the possible (marine or continental) genesis of sediments in the area studied and finally, the application of all these data in practical geology.

6.6 Paleomagnetic-Tectonic Criteria in Outlining Oil and Gas Fields

According to the established regularities using the paleomagnetic method, the criteria for the perspective areas in mineral deposits, and especially in oil and gas can be elaborated.

The prehistory of continental block movement can be reconstructed according to paleomagnetic data supported by geologic information, which helps in reconstructing the zones of spreading and subduction that recurrently appeared during the geologic evolution of the Earth. The type of oil- and gas-bearing basins in the continents can be approximately determined when the sizes and boundaries of the blocks, their relative situation and the contours of the paleooceans are recognized (see Sect. 6.3). The evaluation stage of the regional prospects is followed by the recognition of those parts of the basins which would appear of the highest productivity according to the regularities discussed in Sect. 6.4.3. To reach this goal, it is necessary to plot the paleolatitudinal grid on the map of a given area for every stratigraphic range that covers the structural evolution of the basin. It is known, from the paleolatitudinal analysis, that the initial reserves of hydrocarbons tend to be located in narrow paleolatitudinal intervals: the critical paleolatitudes, and that parts of the basins cut by these paleoparallels are considered to be the most promising. Thus, these two procedures sufficiently narrow the areal and stratigraphic intervals that might be of high potential for large accumulations of hydrocarbons.

Hydrocarbons are known to occur mostly in sequences of marine origin. The paleolatitudinal analysis of the distribution of marine and continental strata (see Sect. 6.5) makes it possible to determine the type of genesis of buried sediments. Thus, among the most potential areas we distinguish the ones with probable marine formations. For this purpose we use the diagrams of the distribution of transgressions for the time interval concerned. The structural strike of the potential deposits under evaluation is established following a general orientation of the paleoparallels and considering the relations of structure formation and paleolatitudes (see Sect. 6.2).

The same criteria are applicable in determining the potential areas and strata for other types of mineral deposits. In any particular case of mineral reserve evaluation in any region, it is necessary to consider the basic parameters that govern the most favourable conditions for the origin of a deposit. Still, in any case the paleomagnetic data on the latitudinal situation of the crustal areas, which also includes the basins, are the most important, for the paleolatitudinal constraints are well known in the origin of all deposits.

Naturally, all these criteria on potential areas and strata should be based on high precision determinations of paleolatitudes and on their reliable stratigraphic allocation and, furthermore, all of the conventional geologic methods of mineral prospecting should also be properly applied.

CHAPTER 7
Certain Problems of Paleomagnetism

7.1 Synchronous and Metachronous Components of the Ancient Magnetization: The Problem of Late Paleozoic Remagnetization

Paleomagnetic reconstructions and conclusions are mainly based on synchronous magnetization, i.e. coeval with the age of the rock. The laboratory and field methods of recognizing synchronous magnetization were discussed in Chap. 2. A component of natural remanent magnetization established by any of the methods, or by a combination of methods, is not always synchronous with the sedimentary or magmatic rock. During evolution, rocks might be affected by different factors: e.g. subsidence, dynamic loads, heating by intrusions. They could also undergo different physicochemical changes that might bring partial or complete destruction of the synchronous magnetization and then, a new-metachronous magnetization appears i.e. an ancient magnetization that is younger than the rock.

This problem of synchroneity of the ancient magnetization occurred for the first time when Devonian rocks were studied. The first two determinations made on the Old Red Sandstone and lavas in England gave discrepancies in the coordinates of the paleomagnetic poles. The states of the geomagnetic field with these polar coordinates were designated DI and DII (Creer 1964; Creer and Embleton 1967). Heating of samples of the Old Red Sandstone to more than 600°C gave a component corresponding to the field DII (Chamalaun and Creer 1964). Further determinations of the main Devonian field on the Devonian rocks of the Urals, and the Dniester region showed a dominance of the geomagnetic poles of the DI field and only a few results gave the DII field (Khramov et al. 1974). The DII field is typified by lower latitudes of the pole and a westward longitudinal displacement compared to the DI field (see Appendix).

Comparative studies of paleolatitudinal grids of the DI and DII fields and of paleoclimatic belts yielded a good consistency for the DII field. The equator, corresponding to the DI field, is displaced westward for 10°–20° and is rotated counterclockwise relative to the paleoclimatic equator (Khramov et al. 1974 Nalivkin and Pozner 1975). It is doubtful whether the DI field is Devonian and its proximity to the late Carboniferous and Permian field supports this uncertainty. Creer explained the appearance of magnetization corresponding to the DI field by a regional remagnetization of the Devonian rocks in Europe in Hercynian times when the arid conditions favouring lateritization were dominant during the late Paleozoic (Creer 1968).

The wide distribution of the late Paleozoic magnetization in the pre-late Paleozoic rocks (since Riphean up to the middle Paleozoic) in western Europe, in the Russian platform and in the Urals resulted in the rocks of different agent having paleomagnetic directions that approximately correspond to the same situation of the paleomagnetic pole. According to data on west European rocks, the paleomagnetic poles concentrate at 45° latitude, which is close to the latitudes of the paleomagnetic poles estimated on the red bed formations of the Upper Carboniferous and Permian of the European part of the USSR. Magnetization of the Lower-Middle Paleozoic rocks of the Tungus syneclise, affected by the Triassic traps, serves as an example of metachronous magnetization resulting from a regional heating. In the zone of trap magmatism the latitudes of the Triassic paleomagnetic poles of 50°–60° are typical of rocks of different age (Khramov et al. 1974). The older strata of the Urals and the Old Red Sandstone of the Anglo-Welsh cuvette are likely to acquire thermoviscous and thermochemical magnetization as a result of subsidence and heating of rocks during Hercynian orogenesis.

Thus, the data support the possibility of a regional remagnetization of rocks, but its mechanism might be quite different. In the Russian platform, for instance, the late Paleozoic metachronous magnetization in the rocks of older age is very likely to appear due to lateritization in arid conditions. In the Urals the processes of thermoviscous and thermochemical magnetization during Hercynian orogenesis were quite important. In Siberia, during the Triassic maximal magmatic activity, remagnetization appeared upon regional heating of rocks, the origin of which might be thermoviscous as well as thermochemical.

The most general hypothesis assumes that the appearance of metachronous components of magnetization as a result of slow and continuous processes of chemical alteration of rocks (Danukalov et al. 1977). Under the gradual lithification of a rock, i.e. the crystalline phase, particles of low-water iron hydroxides (goethite) and hematite are precipitated from the colloidal nonmagnetic phase. These particles precipitate a different times, and pass through the super-paramagnetic boundary and so acquire a magnetic moment along the direction of the geomagnetic field at the time of this transition. If the geomagnetic field is subjected to reversals and the epochs of N- and R-polarities (normal and reversed) are approximately of equal duration, then the resultant magnetization of a rock would equal zero. During the evolution of the geomagnetic field, the intervals of a stable state of a single polarity with a duration of tens of millions of years have been registered in the late Paleozoic and Mesozoic. The Kiaman is one of these intervals of reversed polarity. A stable interval of a normal polarity, comparable with the Kiaman interval, falls in the second half of the Mesozoic. If a field of single polarity is persistent, then a noncompensated magnetization appears. This mechanism of the origin of remagnetization is shown by the presence in samples of mutually compensating phases that might be isolated under A.F. and thermal demagnetization (Khramov et al. 1974), and by a strong excess of the detrital magnetization value over the metachronous in the same rocks; this was established in the rocks of the Urals (Danukalov et al. 1977).

Cleaning of samples from the Upper Riphean Katav Formation exhibiting late Paleozoic (Hercynian) magnetization by alternating fields, showed that a successive effect of an alternating field from 4×10^3 to 5×10^4 A/m changed the magnetization direction from the normal to the reversed one and vice versa (Fig. 7.1). A major rotation of magnetization with a change of its value occurs in the

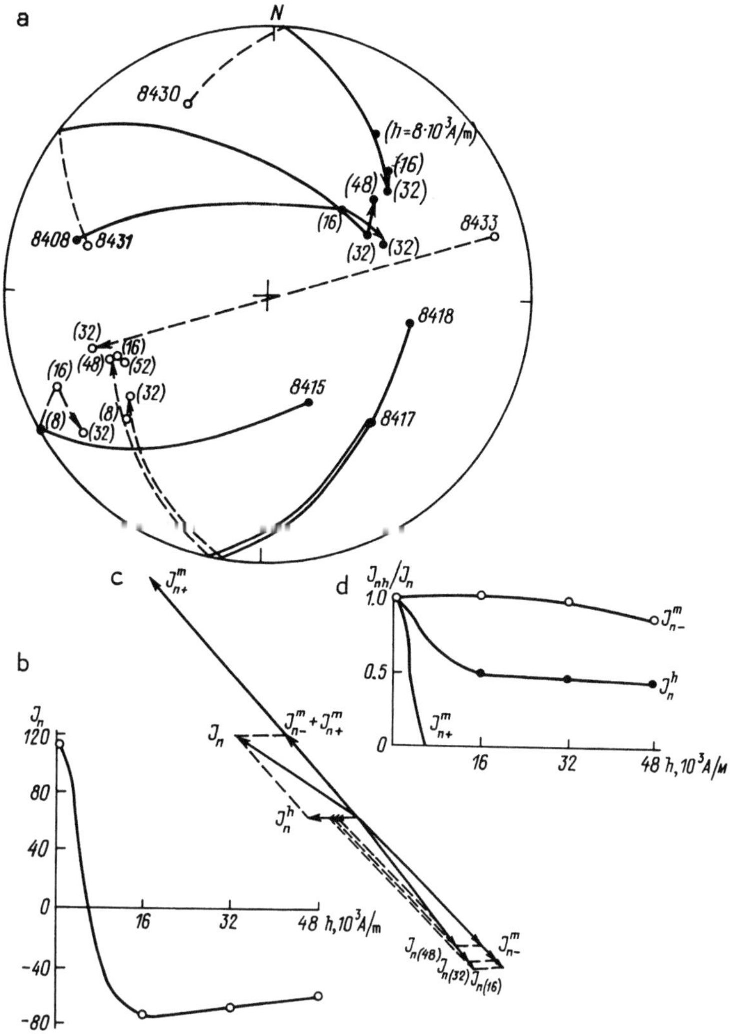

Fig. 7.1a-d. Determination of two components of the Hercynian remanent magnetization during A.C. demagnetization of the Upper Riphean rocks (Southern Urals). **a** Stereographic projection; **b** a change of J_n projection on the J_n^a vector axis; **c** vector diagram (specimen 8417); **d** a change of complex magnetization components with A.C. demagnetization (the curve was drawn in accordance with the component resolved in **c**)

interval up to 8×10^3 A/m and with a further increase of a.c. amplitude only slight changes of directions of J_n are registered in the area and corresponded to the mean direction of the late Paleozoic magnetization of the formation. This magnetization is stable relative to the Earth's magnetic field. The magnetization direction is changed to an opposite one under A.F. cleaning because two magnetizations are present in the rock, their directions differing by 180°. The normal J_{n+}^m and reversed J_{n-}^{m} magnetizations are almost of the same value but differ in stability; the rock is also characterized by the presence of a secondary magnetization J_n^h which resistant to A.F. cleaning.

The process of rock formation is of extremely long duration. During this time the crystallization of hydroxides, dewatering and the transformation of ferruginous, paramagnetic minerals may all occur, with the additional precipitation of ferromagnetic grains. This process is exceptionally long in the case of red rock formations, which results in the appearance of a complex magnetization (Roy and Park 1974). This magnetization might include detrital and chemical types and even represent a superposition of several components corresponding to different states of the geomagnetic field.

The studies of North American red beds showed that some of these components are much younger than the rocks (Roy and Morris 1983). Thus, the dominant or the most stable components of J_n of the red beds of the Pennsylvanian Casper and Dunkard Groups, of the Mississippian Mauch Chunk and Barnett Groups and of the late Devonian Catskill Group were formed almost at the same time: at the end of Carboniferous and in the early Permian (Roy and Morris 1983). These results confirm the above-mentioned mechanism of the appearance of the late Paleozoic magnetization.

Distributions of paleomagnetic directions and poles for major epochs and regions appear to show a satisfactory consistency with the hypothesis of a central dipole. Deviations from this regularity, when seen in a band of paleomagnetic poles, are probably indicative of metachronous magnetization. Certain methods enable the establishment of the presence of intermediate aged magnetization in a rock. For instance, there is the method of determination of coefficients of redeposition for sedimentary rocks (Bagina 1970), the method of Sholpo and Mamedov (1969) based on the dependence of the coercive spectra of the isothermal magnetization of rock on its thermal history, thus distinguishing between chemical and thermoremanent magnetizations. Thellier's method is used for recognition of the thermoremanent magnetization; the statistical method allows the evaluation of the importance of between-bed scattering and in combination with a separate analysis of physical components of J_n, recognized in laboratory demagnetization, it provides the possibility of distinguishing the nature of every component.

If a rock is characterized by a primary (synchronous) magnetization and the ancient secondary (metachronous) magnetization, the graphical method allows the resolution of a multicomponent vector into its components if their directions are known (Fig. 7.2). If we pass a plane through the vector J_{nh} (the result of cleaning in a field $h = 48 \times 10^3 A/m$) and the direction J_n^h (present field), then it will cross the plane of the "ancient" directions of the primary J_n^o and metachronous J_n^m magnetizations along the line J_{nI}. Vector J_{nI} shows the relationship between ancient prefolding magnetizations within a given magnetization J_{nI}. Resolving vector J_{nh} into components J_n^h and J_{nI} and then vector J_{nI} into J_n^o and J_n^m and with the known length of vector J_{nh}, we can determine the values of vectors J_n^h, J_n^o, J_n^m on the plot or according to formulae (7.1)–(7.3) and can determine their components according to the axes of coordinates x, y, z.

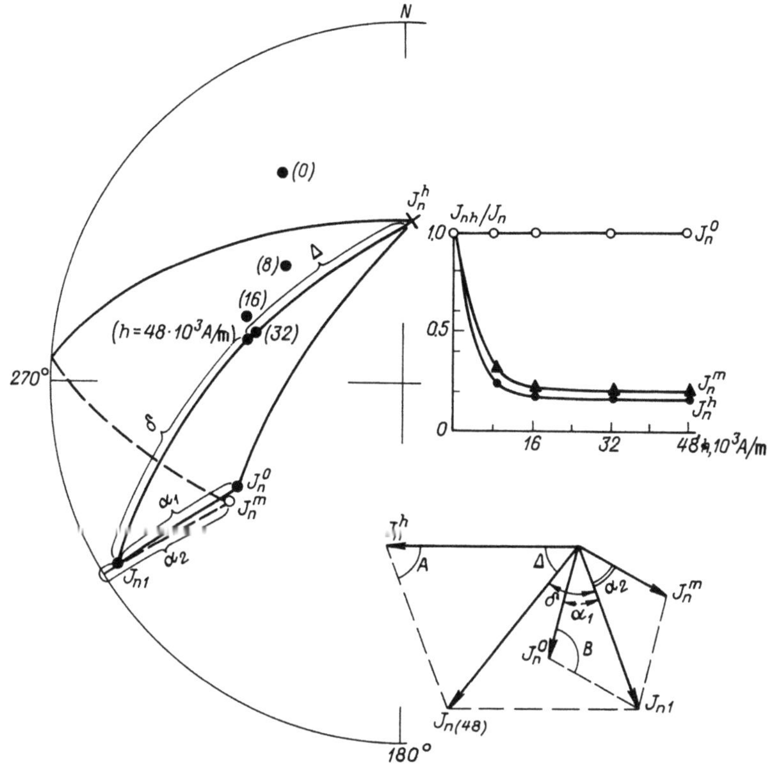

Fig. 7.2. The component analysis of J_n vector and synchronous, metachronous and recent remanent magnetization behaviour during the A.C. cleaning (Basa Formation, Southern Urals)

Angles A and B are obtained by

$$A = 180° - \Delta - \delta;\ B = 180° - \alpha_1 - \alpha_2. \tag{7.1}$$

With A and B angles known, the following calculation can be carried out:

$$J_{n1} = J_{nh} \sin\Delta / \sin A;\ J_n^h = J_{nh} \sin\delta / \sin A. \tag{7.2}$$

Therefore,

$$J_n^o = \frac{J_{nh} \sin\Delta \sin \alpha_2}{\sin A \sin B};\quad J_n^m = \frac{J_{nh} \sin\Delta \sin \alpha_1}{\sin A \sin B} \tag{7.3}$$

According to Eq. (7.3) the values of synchronous and metachronous magnetizations are determined. Thus, performing this resolution by stages of cleaning and analyzing the behaviour of components J_n^h, J_n^o, and J_n^m, it is possible to determine their nature. It can be seen in Fig. 7.2 that J_n^h and J_n^m change identically under the effect of the field h; these components seem to be due to magnetizations of the same kind and their carriers are grains of the same minerals. Probably, these magnetizations are viscous and thermoviscous magnetizations close to the super-paramagnetism boundary of finely dispersed hematite grains. Component J_n^o is of high resistivity, which does not change in fields up to 46×10^3 A/m, which is indicative of an association with hematite. Thermal tests confirmed that single-domain hematite is the real carrier of magnetization J_n^o.

The fold test might also be used for removing metachronous magnetization if the ages of the folding and of the rock are known. Information obtained by this method, however, can be erroneous if the origin and time of folding are not duly considered. A typical example might be the results of the examination of the Trenton limestone (McElhinny and Opdyke 1973). Directions of J_n appeared to be almost the same for horizontal and deformed beds. Special testing of the anisotropy of magnetic susceptibility established that the magnetization of these rocks was acquired after the formation of folds, but prior to the lithification of the sediment. In this particular case, magnetization J_n, though postfolding, is still almost synchronous as the folding was synsedimentary.

Thus, the possibility of metachronous magnetization should be tested by a combination of paleomagnetic methods. Certain cases of satisfactory recognition of synchronous and metachronous components of J_n are known, however, further refinements are needed in the methodology for distinguishing metachronous magnetizations.

The properties of the late Paleozoic metachronous magnetization were analyzed and can be formulated as follows:

1. Metachronous magnetization exhibits high resistance to thermal- and A.F.-cleaning. During cleaning, however, the magnetization direction might sometimes change for 180°, i.e. the pole coordinates remain the same.
2. If many samples are tested, it might appear that some of them lose metachronous magnetization under thermal cleaning and a still more ancient component is revealed.
3. Metachronous magnetization is weak compared to the primary magnetization of the lower Paleozoic rocks of the same lithologic composition, as it is in most cases the superposition of non-coeval components of opposite directions.
4. Metachronous magnetizations display an insignificant between-bed scattering, which indicates a proximity of the magnetization process in time.
5. The directions of metachronous magnetization do not depend on the age of a rock; reversed magnetization are dominant, probably due to the noncompensated reversed-magnetization that appeared during the Kiaman geomagnetic interval.
6. Metachronous magnetization can be of a chemical, thermoviscous nature, although it seldom seems to be partially thermal, as it is not removed under thermalcleaning in a temperature range of 200° to 600°C.

Now the following mode of recognition and removal of metachronous magnetization can be suggested: total, high-temperature (up to 600°C) stepwise demagnetization, because magnetization J_n^m of sedimentary rocks is commonly related to finely-dispersed hematite; chemical demagnetization and statistical analysis at every step of cleaning with a separate testing of removed and remaining components of J_n.

On the one hand, metachronous magnetization presents an interference in the examination of a geomagnetic field for the epoch of rock formation and in the elaboration of a paleomagnetic geochronologic scale; on the other hand, it might provide information of the geomagnetic field during the following epochs and produce extra evidence on the further geologic evolution of the rocks and of the entire region.

On the whole, the problem of the middle and early Paleozoic paleomagnetism cannot be unambiguously solved at the moment, because of the few available determinations, the wide distribution of late Paleozoic remagnetization, the ambiguous interpretation of this phenomenon and the difficulties in the recognition of synchronous components of magnetization in sedimentary as well as in volcanogenic rocks against this background.

7.2 Hypothesis of the Earth's Expansion

A hypothesis of the Earth's expansion was suggested prior to the concept of plate tectonics as an alternative to the continental drift hypothesis. At the moment, however, it remains as it is, for many workers think that many geologic phenomena (Carey 1976) cannot be explained in terms of plate tectonics. In particular, calculations have shown that the area of newly formed lithosphere since the beginning of Jurassic, formed by the process of ocean floor spreading, exceeds by one-third the area of the lithosphere which has been underthrust and reworked in the zones of subduction. This approximately corresponds to a 10% expansion of the Earth (Carey 1976; McElhinny et al. 1978b).

Two hypothetical versions of the Earth's expansion are known. According to the first one, the Earth's surface at any moment of pre-Mesozoic history slightly exceeded the resultant area of the continental crust of the corresponding age. This led to the assumption of a relatively high expansion rate of the Earth and of an exponential law of the expansion (Carey 1976). The second version, suggested by Egyed (1961), was also based on paleogeographic and tectonic data, which proposed moderate expansion of the Earth. Egyed was the first to use paleomagnetic determinations to calculate the Earth's paleoradii. Later, many evaluations of the Earth's paleoradii became available according to paleomagnetic data for different epochs. Two methods appear especially accurate among the numerous determination methods: i.e. the paleomeridian method suggested by Egyed and the minimal dispersion method proposed by Ward (1963).

The paleomeridian method is applicable for calculating the paleoradius of the Earth R_a, if the sites of paleomagnetic determinations lie along a single paleomeridian. Therefore:

$$R_a = 57.3l/(\varphi_1 - \varphi_2),$$

where l is the actual distance (km) between the sites of determinations with paleomagnetic latitudes φ_1 and φ_2 (degree). As it is unlikely that the sites fall exactly on a single paleomeridian, the distance l is the interval between the paleoparallels drawn through every site (Khramov and Sholpo 1967).

The procedure of minimal dispersion consists in the calculation of the precision parameter K of the paleomagnetic poles at different distances δ from the sites to the pole and in the determination of the maximum of a function K (δ). Thus, the angle δ is related to the ancient radius R_a by

$$\delta = (R_a/R)(90° - \varphi),$$

where R is the present radius of the Earth and φ is the paleomagnetic latitude at the site.

Both methods are applicable only when the area with the determinations to be compared are from the same rigid block that did not change its size or shape. The first estimates of R_a by these two methods appeared to be sufficiently discrepant and, as a rule, the first method gave underestimated values of the radius (Table 7.1).

A common disadvantage of these determinations is the limited extension of the areas along the paleomeridian in the first method and along the paleoparallel, in the second one. Besides, if the observations are conducted only in the Northern Hemisphere, it is difficult to distinguish between the effect of the northward displacement of the dipole approximating the paleomagnetic field and the effect of a decrease in the paleoradius compared to the present one. Thus, the data of Table 7.1 might also indicate that the Mesozoic and Late Paleozoic paleomagnetic field was a dipole field displaced northward, with while the Earth's radius remained unchanged. Data for the entire Earth must be analyzed using paleoreconstructions to solve this dilemma: radius change or dipole displacement. This work was done by McElhinny et al. (1978b) (Table 7.2). A slightly modified method of minimal dispersions revealed a good inner consistency of data (high values of precision parameter of poles K_{max}) as well as external consistency (for different supercontinents). According to these results, the Earth's radius could not have changed by more than 5% in both directions from its present value for the last 400 m.y. Thus, both hypotheses of exponential expansion and moderate expansion of the Earth at the expense of oceanic lithospheric accretion is contradicted by the paleomagnetic data.

Table 7.1 Values of the Earth's paleoradius R_a estimated from paleomagnetic data and expressed as a fraction of the present radius R. (Khramov and Sholpo 1967; McElhinny 1973)

Period	Plate	R_a/R	
		Minimum dispersion method	Other methods
K	North European	0.97	0.99
T	North European	0.99	0.89
T	North European + Siberian	–	0.90
P	North European + Siberian	1.00	0.99
P	North European	0.94	1.03
P	Siberian	0.76	0.70
P	North European + Siberian	–	0.89
C	North European + Siberian	–	0.83

Table 7.2 Values of the Earth's paleoradius estimated from paleomagnetic data plotted on the reconstructions of ancient megacontinents. (McElhinny et al. 1978b)

Period	Megacontinent	Number of paleomagnetic data	R_a/R	K_{max}
K	Eurasia	28	1.06	46
T-J	Gondwana + N. America	56	1.03	62
T	Eurasia	58	1.03	55
P-T	Gondwana	23	1.03	44
P	Euramerica	77	0.97	90
C-P	Gondwana	16	1.07	101
C	Euramerica	63	0.99	63
D	Euramerica	30	1.05	56

However, the problem is awaiting its final solution. All the calculations given above proceed from a supposition that, during expansion, the continental parts of the plates remained rigid and were not pulled apart but their shapes may have adjusted to changes in the skewness of the Earth's surface. If this "pulling apart" had occurred and the Earth had expanded (or compressed) uniformly, and not only within the oceanic limits, then as the paleomagnetic observations deal only with the angular values, these cannot provide evidence on the changes of the Earth's dimensions.

The problem of expansion (or compression) of the Earth, whether of irreversible or cyclic (pulsation) nature, is worth further examination by paleomagnetic methods. The scale of this phenomenon does not seem to be as great, as suggested earlier, however, the precision of the paleomagnetic determinations in this case should be higher.

Furthermore, all of the estimations discussed here refer only to the Phanerozoic. The problem of the Earth's expansion in the Precambrian has just been raised. It was pointed out earlier (Sect. 4.5.3) that paleomagnetic data on the Precambrian can give satisfactory agreement with the model of an abrupt expansion of the Earth in the 1600–1000 m.y. interval (Schmidt and Embleton 1981).

7.3 Paleomagnetism of the Precambrian

From the very start of paleomagnetic studies, Precambrian paleomagnetism was one of the most important aspects. This is mainly explained by the fact that the distant geologic past of the geomagnetic field might provide clues to the principal problems of the physical theory of geomagnetism; important progress can be expected only if the process of the appearance of the geomagnetic field and the first stages of its evolution are known. Actually, the Phanerozoic period represents only one-eighth of the entire geologic history and knowledge of the essential evolution of geomagnetic parameters during this time is hardly possible. During the Phanerozoic, the characteristics of the geomagnetic field, e.g. the frequency of reversals, the mean value of the field, amplitude and frequence of secular variations, seem to be subjected to complex rhythmic changes rather than to any directed changes, and their evolutional trend for these 600 m.y. should not be considered as established.

Another important aspect of Precambrian paleomagnetism is the concept of plate tectonics. Certain data indicate a peculiar situation with reference to plate tectonics during the Precambrian which seems to have been in an "embryonic" state at that time and thus differed from the Phanerozoic plate tectonics. Therefore, studies of Precambrian paleomagnetism will most likely allow the elucidation of this major geotectonic problem.

Finally, the paleomagnetic method in geochronology and stratigraphy of the Precambrian is much more important than in the Phanerozoic period. This can be primarily explained by the increasing errors in dating by paleontological and absolute age methods. Errors due to the first method are due to the low rate of biological evolution and its relatively sparcity; errors of the second method are due to the "exhaustion" of relatively quickly decaying isotopes. In contrast, we have no data in favour of slower geomagnetic evolution during the Precambrian in comparison with that during the Phanerozoic. The paleomagnetic method is capable of solving Precambrian problems, i.e. the inverse tasks of paleomagnetology. However, the factors stated above make the realization of the direct task of paleomagnetology, i.e. study of the history of the geomagnetic field, difficult. Thus, complex and ambiguous interpretations of the reconstructions and correlations of the chronological series of paleomagnetic determinations occur.

The very long Precambrian history of the Earth (more than 3500 m.y.) does not as yet have a reliable stratigraphy. New methods of division of the Precambrian are necessary in order to understand the Precambrian periods and the chronology of these events. The biostratigraphic method, i.e. the principal one in the Phanerozoic, is only satisfactorily used for the uppermost Precambrian; the faunistic and phytologic characteristics are now known for the Vendian and Riphean sequences.

Another complex problem is related to the peculiarities of Precambrian rocks as objects of paleomagnetic studies. The main peculiarity of the paleomagnetism of Precambrian rocks lies in the much higher probability of remagnetization of rocks by later fields due to the extremely long evolution of these rocks. All of the problems of remagnetization of middle-early Paleozoic rocks examined in Sect. 7.1 remain valid for Precambrian rocks, but the recognition of their primary magnetization is even more difficult. The Southern Urals sequences exemplify the regional remagnetization of the Vendian and Riphean rocks. The rocks of this region are characterized by a stable location of the paleomagnetic poles starting from the lower Riphean through the Vendian; the general displacement of the pole amounts to a few degrees, whereby the complete amplitude of oscillations is about 20° (Khramov 1971, 1973, 1975, 1979; Komissarova and Mikhailov 1981).

This stable situation in the Urals shows a striking difference in comparison with the world paleomagnetic data on the Precambrian. All of the cratons (Baltic, Aldan, Canadian, African, Indian, Australian), to say nothing of the fold systems, demonstrate complicated and significant displacements of poles (Pesonen and

Neuvonen 1981; Piper 1982a). One of the following suppositions could probably explain this contrast: (1) all of the cratons were moving quickly relative to the pole and relative to each other during the late Precambrian, while the Urals region was stable during this entire period; (2) the time of formation of all older nonfossiliferous strata of the Urals embraces a short interval within the limits of the early Paleozoic and Vendian, which was about 100 m.y., i.e. 15 times shorter than it is now considered; (3) the time interval of formation of the magnetization recognized is much shorter than the age range of the strata examined, which means that this magnetization is metachronous.

The first supposition is most unlikely from the view point of geotectonics and consequently, all of the paleomagnetic data on the Precambrian in the world are doubtful, except for the Urals region. The second supposition returns us to the past when the nonfossiliferous, older strata of the Urals were commonly assigned to the Paleozoic of which many were considered as analogues of each other, which means that almost all of the absolute age determinations for the Southern Urals are refuted. Therefore, the third supposition on the metachroneity of the magnetization recognized seems to be the most realistic (Danukalov et al. 1977).

Paleomagnetic data on the sedimentary and volcanogenic rocks of the Proterozoic are quite numerous. Their interpretation is problematic not only because of the wide distribution of superimposed processes of regional remagnetization, but also because of the extremely complicated tectonics. We now have sufficient data on paleomagnetism from the Upper Precambrian, i.e. Vendian and Riphean. The basis for reconstructions is the seris of paleomagnetic poles for different blocks of the Earth's crust. The APWP is plotted according to these series, but in the entire time range it is not traced with sufficient reliability. In the Russian and Siberian platforms the polarity must be selected for the early Paleozoic and even older periods, as data on the middle Paleozoic interval are not available. The older the rocks, the less is the consistency of the paleomagnetic data. Several groups of determinations exist for the same age, which differ greatly in the average positions of their paleomagnetic poles. Thus, the paleomagnetic poles of the middle Proterozoic of the Baltic Shield form two groups (Fig. 7.3) (Pesonen and Neuvonen 1981). However, an even younger remagnetization may causes this separation.

A series of paleomagnetic poles from the Precambrian of the Siberian platform revealed the existence of four separate blocks: the Aldan, Anabaran, West Baikalian and North Baikalian, which were situated in the lower latitudes of the Southern Hemisphere (Fig. 7.4). At the beginning of the early Paleozoic these blocks were brought together and in the middle of Cambrian they were welded under the effect of the Baikalian orogeny and thus formed the Siberian plate which became a single block since the end of Cambrian and then commenced its northward movement (Khramov et al. 1985).

A compilation of the general paleomagnetic sequence of the Vendian and Riphean strata of the Siberian platform exemplifies the possibility of the correlation of Precambrian sections using the paleomagnetic method. This work is the result of a paleomagnetic study of the Siberian Precambrian, which has been conducted by the paleomagnetic laboratory of VNIGRI since the 1960's (Sidorova 1965). The correlation of the sequence was done according to the paleomagnetic

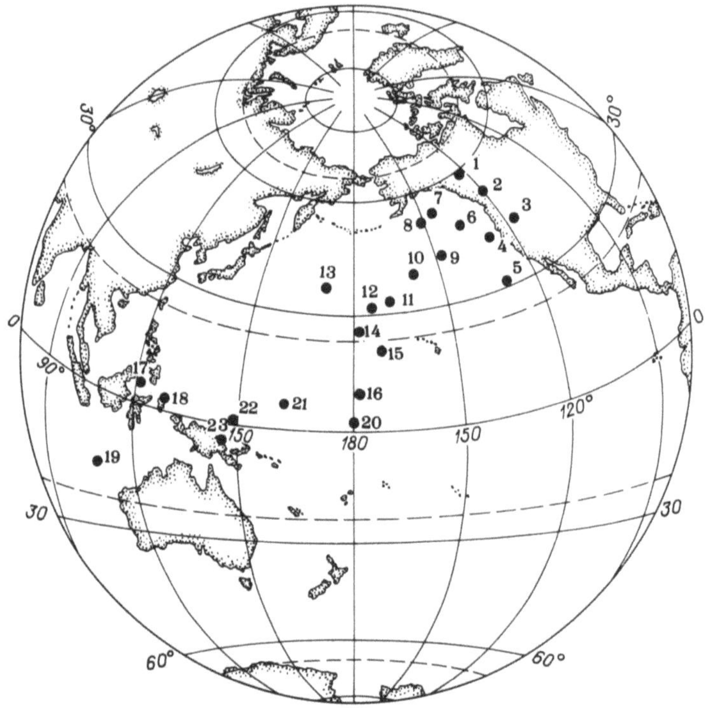

Fig. 7.3. The middle Proterozoic paleomagnetic poles (1900–1560 m.y.) from the eastern Baltic Shield. Poles *1, 2, 4, 5, 7, 8, 9, 10, 11, 17, 18, 19* (from Khramov 1971, 1973, 1975, 1979). Poles *3, 6, 12, 13-16, 20-23* (Pesonen and Neuvonen 1981). Poles *20-23* are ~1300 m.y. old

zones, data on the existing stratigraphic schemes and on the absolute age determinations were also used. As the geomagnetic field reversals are ubiquitously synchronous, the boundaries of the paleomagnetic zones are coeval and thus serve as the bench marks in the correlation. The boundary between the Middle and Upper Riphean was the level referred to for all of the sequences of the Siberian platform (Fig. 7.5). This level was preferable because of the reliable radiologic dating available and because of the geomagnetic reversal at the middle-late Riphean boundary. This boundary for the paleomagnetic zones was established in the sequences of the western and northern Baikal regions and in the Uchur-Maya region. It can be seen (Fig. 7.5) that a reversed magnetization is dominant in the Vendian and Riphean rocks and thus the zones of normal polarity serve as the bench marks in the correlation. The final result of the correlation is a generalized paleomagnetic section. Its control by the absolute age scale permits the disclosure of the regularities in the changes of the geomagnetic field with time.

During correlation, it is necessary to take into the possibility of a partial remagnetization of rocks and the breaks in the sections. The proposed generalized section, is thus, to be regarded as a working scheme that might be used as a basis for correlation, but later it should be modified and refined.

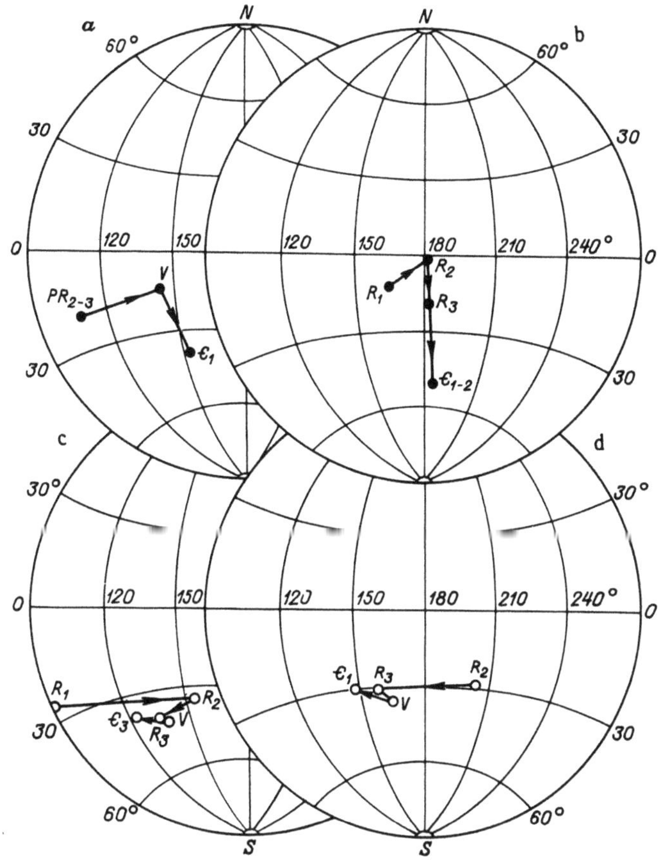

Fig. 7.4a-d. The Vendian-Riphean paleomagnetic APWP's for different blocks of the Siberian platform. **a** Anabaran; **b** Aldanian; **c** North Baikalian; **d** West Baikalian blocks

It has been seen in this generalized paleomagnetic section that the geomagnetic field during the Riphean and Vendian was mostly of reversed polarity, whereas normal polarity was dominant only in the early Riphean. The entire time interval since the early Riphean through the Vendian can be divided into several stages according to the pattern of geomagnetic reversals. The beginning of the early Riphean is the first stage when a normal polarity field was dominant for a long time, about 130–150 m.y. In the second half of the early Riphean, approximately of the same duration, the field was of reversed polarity. At the early-middle Riphean boundary a stage of subsequent field reversals started, which lasted for about 200 m.y. In the second half of the middle Riphean the geomagnetic field was of reversed polarity. At the middle-late Riphean boundary, during ~50 m.y., normal polarity characterized the geomagnetic field and the corresponding zone serves as a correlation bench mark. In the first half of the late Riphean the field reversed several times; this stage lasted about 200 m.y. The second half of the late Riphean is typified

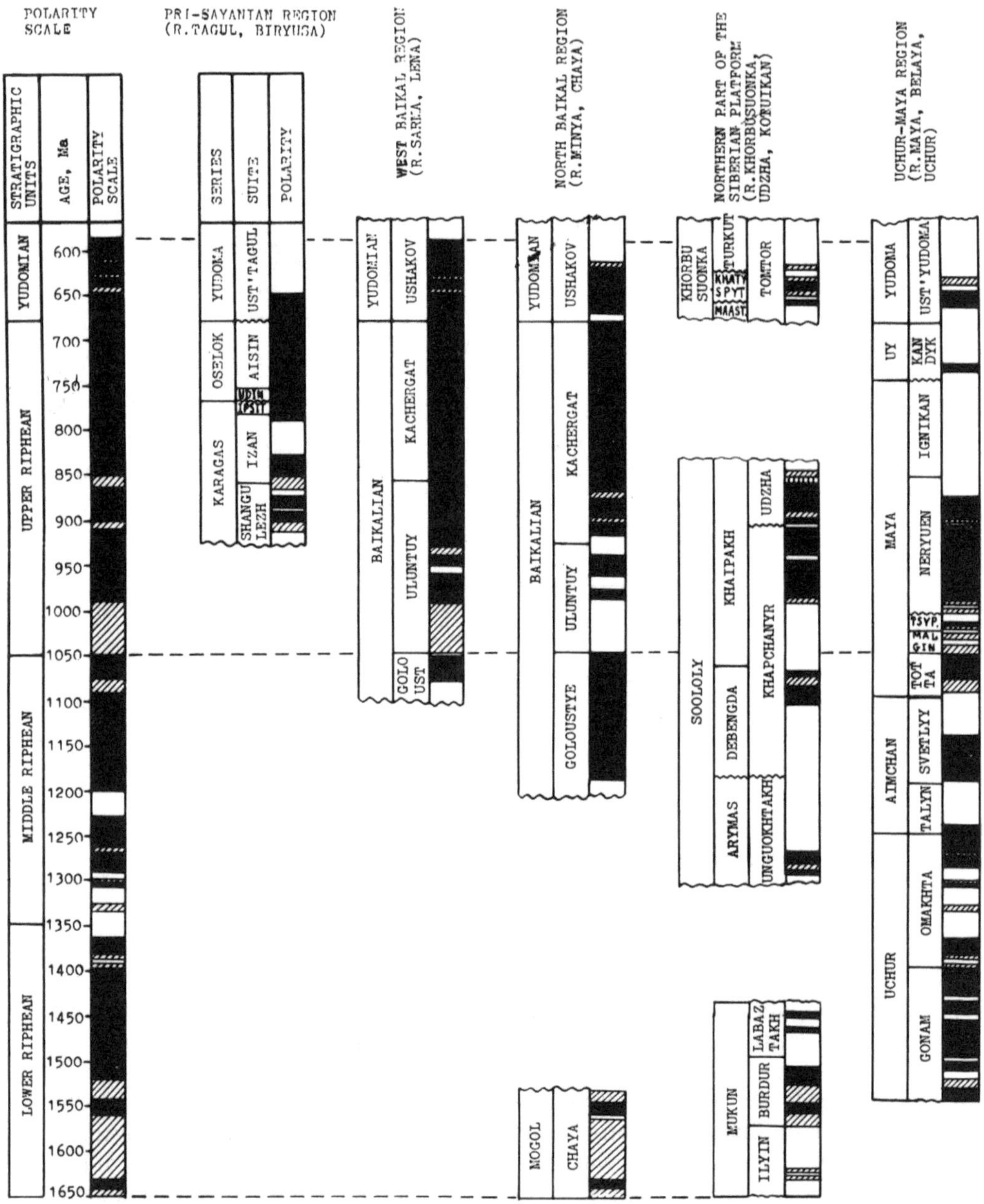

Fig. 7.5. The Precambrian magnetic polarity time scale for the Siberian platform and the scheme of section correlation

by a long interval of reversed geomagnetic polarity (200 m.y.); this interval embraces the beginning of Yudomian. During the last 500 m.y., near the Yudomian-Phanerozoic boundary, the geomagnetic field changed its polarity six times, but since the beginning of Phanerozoic a field of reversed polarity has existed for a long time. Overall, in the Precambrian section studied (Fig. 7.5), a geomagnetic field of reversed polarity was dominant. Some paleomagnetic R-zones have an age range up to 200 m.y., while the N-polarity zones do not exceed 100 m.y. duration.

Thus, the problems of Precambrian paleomagnetism, e.g. regional remagnetization of rocks of certain time intervals, implication of regional and global tectonics and the basis of exact datings of objects, have no common solution at this stage of knowledge. However, the solution of these problems appears more or less satisfactory when statistically determined as in the particular cases given in every section and region.

Appendix. Mean Paleomagnetic Poles for Plates and Blocks

Period epoch	Pole position		Statistics			Index numbers and references[a]
	Φ,N	Λ,E	K	A_{95}	N	
North European Plate						
₽	76°	162°	76	4°	19	II000I, II00I2 + I4 + 506 + 572,II00I7, II00I8 + I9 + 21 + 23 + I99 + + 535,II0145, 156, 110200 + 224 + 662,110220 237, 110325 + 326 + 695 + 703, 110580,656,675,693; 14/128, 14/143(N = 2) + 14/153 + 16/59 + (Dagley and Musset 1978), 14/163,15/70,16/34
K	77°	180°	37	8°	10	4—28,36,38–40,45; 100001,51,88; 12/79
J	77°	162°	35	10°	7	5—01,4; 090001; (Kruczyk et al. 1980a,1980b N = 2), (Hijab and Tarling 1982; N = 2)
T_{2-3}	55°	137°	29	13°	8	6—01-3; 080006,7,116,123; (Bylund 1973)
T_1	51°	153°	40	5°	26	6—05-9,11,13,14,16,17,30,31,43-46,65,66; 08[illegible]001 + 2,080008,080070 + 124; (Nairn 1960; N = 4), (Turner and Ixer 1977)
P_2	45°	162°	59	3°	38	7—01,2,6,9,11,13-18,23-29,31-39,47-51; (Krs 1966; N = 3), (Patrascu et al. 1964); 14/307 + 14/308, 14/325
P_1	40°	167°	58	3°	33	7—19-21,55-58; 060089,070008,9,12,13,0700[illegible]4 + 87 + 88 + 89, 070015 + 168, 070016 + 18,070069,103,113,070122 + 123,070[illegible]26 + 127,129,136,137,153,154,155; 1/82, 1/83, 7/35, 14/324; (Krs 1966), (Berho[illegible] et al. 1975; N = 2)
C_3	41°	169°	63	4°	20	7—29,8—01-3,32,33,55,56; 060001,2,060018 + 19 + 30 + 31,060094,99,100,101,104,310,311; 14/341; (Bylund and Patchett 1977)
C_2	32°	163°	59	5°	17	8—05-11,24,35,51,52,57,58; 060096-98; 14/3[illegible]3
C_1I	46°	158°	82	7°	6	8—12-15,61; 060303
DI	31°	161°	71	6°	10	9—01-5,16; 050001,53,75; (Jones et al. 1979)
(S-D)II	5°	134°	111	9°	4	10—01,5; 050080,142
SI	38°	155°	22	16°	5	10—08-10; 040017,30
O	30°	143°	47	9°	5	11—09,10; (Buchá et al. 1976; N = 3)
Є	−13°	135°	–	–	2	11—01; (Buchá et al. 1976)

Appendix *(continued)*

Period epoch	Pole position		Statistics			Index numbers and references[a]
	Φ,N	Λ,E	K	A_{95}	N	
Scotland						
C_1I	41°	150°	30	14°	5	060305, 312–315
DI	26°	152°	44	12°	5	050005,6,146,169,170
DII	1°	144°	32	9°	9	050004,61,62,82,83,180; 14/377, 14/379; (Turner and Archer 1975)
O_3	13°	166°	14	21°	5	(Carmichael and Storetvedt 1981), (Piper 1978b), (Watts and Briden 1984; N=3)
O_{1-2}	16°	197°	21	15°	6	030034,100; (Carmichael and Storetvedt 1981), (Piper 1979), (Watts and Briden 1984; N=2)
England, Wales and Ireland						
C_1I	43°	133°	–	–	2	060014,355
C_1II	25°	155°	14	13°	11	060015–17,21,22,24,25,27,28,33,34
DI	26°	154°	23	14°	6	050067,173;16/162; (Piper et al. 1978; Piper 1979; N=3)
DII	9°	128°	16	24°	4	050003,050168; 15/224; (Piper 1979)
S	−7°	121°	20	16°	4	040074,75,81; (Morris et al. 1973)+(Morris 1976)
O_3	9°	186°	57	12°	4	030099,102; 14/390; (Piper 1979)+(Piper and Stearn 1975; N=2)
O_{1-2}	11°	187°	22	14°	6	030098,107; 16/180; (Piper 1978b; N=3)
$Є_1$	−26°	63°	21	11°	8	(Piper 1982b; N=2), (Piper 1976; N=2), (Seguin 1983; N=4)
Armorica						
O_3	−28°	175°	10	25°	5	(Duff 1978,1979; N=2), (Perroud and Bonhommet 1981), (Perroud et al. 1983a, 1983b; N=2)
ЄI	−26°	154°	10	21°	6	(Duff 1979, 1980; N=2), (Hagstrum et al. 1980; N=2), (Russel et al. 1983), (Van der Voo 1982)
ЄII	−32°	63°	11	28°	4	(Briden and Duff 1980), (Hagstrum et al. 1980; N=2), (Seguin 1983)
Bohemian Massif						
O_3	10°	66°	48	18°	3	030011,12; (Buchá 1965)
O_1	−44°	86°	137	6°	5	(Buchá et al. 1976; N=3), (Krs et al. 1983; N=2)
Є	−38°	85°	7	24°	7	020032; (Buchá et al. 1976; N=2), (Buchá 1965; N=4)

Iberian Plate						
Ᵽ	64°	201°	17	30°	3	110313,318; 14/141
K_2-$Ᵽ_3$	74°	188°	200	24°	2	100059+60+83, 100064
J	71°	236°	36	7°	12	(Schott et al. 1981; N=12)
T	48°	220°	16	15°	7	080073,102,103; (Van der Voo 1967), (Mäkel et al. 1984; N=3)
P	45°	219°	43	14°	4	070021+55+56+79, 070105+106+107, 070 09,112
Lesser Caucasus, Turkey and Iranian Plate						
Ᵽ	65°	293°	37	9°	7	3–06,22–24;110271,272; (Soffel et al. 1975)
K	65°	286°	18	12°	9	3–24,4–41,42,46,49,51;100053,54,55
J	64°	254°	23	10°	8	5–13,14,26+27,32,35,39; 090078; (Lauer 198)
P	25°	296°	13	19°	4	6–78,7–75; 070070; (Soffel et al. 1975)
D_3-C_1	0°	212°	–	–	I	16/156
D_1	51°	196°	–	–	I	(Wensink 1983)
Transdanubian Block						
K	87°	322°	22	13°	5	100056,250; 16/66, 16/67, 16/68
J_{1-3}	35°	294°	15	21°	3	(Marton and Marton 1983; N=3)
T_{1-3}	43°	283°	–	–	2	(Marton and Marton 1983; N=2)
Adriatic and Sardinian Plates						
$Ᵽ_2$	62°	251°	38	8°	8	110289,368,668; 14/116+14/119, 14/117+(Edei and Lörtscher 1977; N=2), 14/118+(DeJong et al. 1973)+(Edei and Lörscher 1977; N=2), 16/40; (Lowrie and Alvarez 1975)+(Vandenberg et al. 1978)
K	58°	254°	32	7°	15	100057,236,100238+239,100242,247,249;15/40+15/41+15/42+15/43, 16/64, 16/65, 16/79, 16/81; (Alvarez and Lowrie 1978), (Vandenberg et a 1978; N=3)
J	45°	287°	31	11°	5	(Horner and Lowrie 1981), (Vandenberg et al 1978), (Vandenberg 1979; N=3)
T	58°	249°	16	14°	6	080087,88,99,109; 14/323; (Horner and Lowr 1981)
P	45°	235°	37	6°	15	060103,070019+20+102+175,070068,75,070077+78+81+128,070080,82,85,133,176; 14/300, 14/327,15/94; (Westphal 1976; N=2

Appendix *(continued)*

Period epoch	Pole position Φ,N	Pole position Λ,E	Statistics K	Statistics A_{95}	Statistics N	Index numbers and references[a]	
Sicilian Block							
Ꝑ$_2$	71°	218°	–	–	1	(Besse et al. 1984)	
K_2	65°	225°	193	7°	4	(Grasso et al. 1983; N = 4)	
Levant Block							
K	47°	271°	35	21°	3	I00037; 12/67, 12/71	
Scyphian – Turan Plate							
Ꝑ	77°	210°	45	5°	15	2–79,80, 3–01-5, 7-9, 13,18,26,27; (Nozharov and Petkov 1976) + (Velchev et al. 1976) + 16/36	
K	75°	176°	22	7°	19	4–01,2,6-8,13-15,17,19,27,29-31,34,35,50,53,56	
J	72°	163°	24	10°	12	4–37 + 5–21 + 5–23, 5–02,5,11,17,24,30,33,34,36-38	
P_2-T_1	48°	149°	40	13°	3	6–17, 7–09,47	
P_1	31°	146°	23	10°	6	7–54, 7–73/1,2,3,4, 7–74/3	
D_{2-3}	29°	149°	12	20°	4	9–37, 50-52	
S-D_1	16°	136°	–	–	2	9–55,10–15	
O-S	6°	128°	–	–	1	10–05	
Urals							
C_1	19°	155°	13	20°	6	8–18,28,29,39,41,42	Group A
D_{2-3}	24°	160°	28	9°	10	9–17,18,21,24,27,29,38,39,48,49	Group A
S-D	15°	151°	13	26°	4	9–15,57, 10–12,14	Group A
C_3	62°	202°	–	–	1	8–23	
D_3-C_1	46°	179°	84	7°	6	8–17,31,40,44,63,65	Group B
D_{2-3}	39°	175°	24	19°	4	9–07,8, 9–09 + 10 + 11 + 12, 9–30	Group B
S	13°	169°	13	20°	6	10–02,3, 16-19	Group B

Kazakhstan Block						
P_2	54°	169°	–	–	1	13/43
C_2-P_1	54°	167°	46	10°	4	7–22, 8–04; 179*, 180*
C_1	69°	201°	1250	4°	3	13/60-13/62
D_3	56°	193°	–	–	1	202*
Altai-Sayan Block						
C_3-P	37°	141°	28	13°	4	7–41,42,45, 8–34
C_1	16°	128°	37	12°	4	8–46,47,49,50
D_3	–1°	138°	9	23°	4	8–48, 9–26,40,44
S-D_1	–21°	134°	46	9°	5	9–41, 45-47, 10–11
S_1	–3°	141°	–	–	2	10–11/1,2
Є$_1$-O_1	–19°	110°	–	–	1	12–02
Є$_{1-2}$	–33°	62°	24	14°	4	12–31/1,2, 12–45; (Rodionov 1983)
Siberian Plate						
K	74°	180°	41	10°	7	3–17, 4–09-12,16,20
J	65°	148°	10	22°	4	4–54, 5–03,18,40
T_{2-3}	59°	137°	37	8°	11	5–12, 6–04, 32-39, 47
T_1	49°	149°	46	4°	24	6–19-29, 48-56, 68-71
P	45°	141°	19	10°	13	6–57-62, 7-10,22,30,40-42,45
C_{1-2}	30°	150°	24	14°	4	8–36-38, 45
D_3	20°	140°	34	12°	4	9–19,43; 208*, 213*
D_1	15°	116°	11	24°	3	9–32,34,36
S_1	–10°	104°	–	–	2	10–07,20
O_3	–20°	131°	130	6°	6	11–02,3, 11–12/1,2,3, 11–14
O_2	–23°	134°	113	4°	11	11–04-08, 11–13/1,2, 11–15,23,24,26
O_1	–40°	136°	57	6°	9	11–16-19, 11–20/1,2,3,4, 11–30
Є$_3$	–36°	128°	118	3°	19	12–03-5, 12–06/1,2,3, 12–07,08, 17–19, 21, 26, 12–27/2,3,4, 12–41-43
Є$_2$	–40°	164°	20	15°	6	12–09-11,13,29,30
Є$_1$	–34°	146°	20	15°	6	12–32,33; (Rodionov 1984; N = 2), (Osipova 1983; N = 2)

Appendix *(continued)*

Period epoch	Pole position		Statistics			Index numbers and references[a]
	Φ,N	Λ,E	K	A_{95}	N	
North American Plate						
P	86°	223°	38	6°	13	110167,212,213,238,454,573,579,597; 10/46+(Hoblitt and Larson 1975), 11/30+11/31, 14/121, 14/169, 16/48
K_2	70°	206°	25	11°	8	100048,49,50,71,188,216,222; 15/52
K_1	70°	184°	184	4°	8	100011,12,13,17,35,72,73,90
$J\text{-}K_1$	66°	187°	12	20°	6	100029,30,67,79,85,86
J	68°	140°	60	5°	16	090072,91,137,154,167,176,177; 15/78+15/79; (De Boer 1967; N=8)
$T_{2\text{-}3}$	64°	91°	38	4°	28	080032,33,080035+36+38+39, 080037,42,43,44,62,63,65,68,71,111,128,132, 080168+169+170+171; 14/276, 14/280, 15/86; (Bowker 1960; N=6), (Grubbs and Van der Voo 1976; N=3)
T_1	53°	107°	50	6°	13	080066+67, 72,89,90,120; 14/295, 16/133; (Collinson and Runcorn 1960; N=5), (Grubbs and Van der Voo 1976)
P_2	52°	114°	100	4°	12	070148; 3/45, 3/47, 3/50, 3/51+3/52,3/53,3/54+3/46,9/53+9/54; (Larochelle 1967), (Peterson and Nairn 1971; N=3)
P_1	44°	118°	40	5°	20	070044,45,47,50,51,52,73,93,94,95,96,97,147,149; 13/52; (Graham 1955), (Peterson and Nairn 1971; N=3), (Runcorn 1956)
C_3	37°	124°	42	9°	6	060064,65,87,116,140,172
$C_{1\text{-}2}$	34°	128°	56	7°	10	060063, 060088+173, 060135–138, 060171,174,240,309
D	44°	122°	517	4°	4	050191+192; 16/157; (Kent 1979), (Van der Voo et al. 1979)
S	29°	123°	24	19°	4	040011,12,17; 16/163
$O_{1\text{-}3}$	31°	138°	18	13°	8	030006, 104, 109, 120; 16/173; (Hower 1979), (Seguin 1977), (Watts and Van der Voo 1979)
O_2	−19°	106°	—	—	2	030007,08
$Є_3$	−6°	170°	14	15°	6	020045,020148; 1/123, 13/69, 13/70, 15/142
$Є_{2\text{-}3}$	38°	136°	21	9°	11	020020,24,54; 16/186; (Patterson et al. 1978), (Watts et al. 1980a; N=4; 1980b; N=2)
$Є_1$	9°	154°	19	14°	7	020065,66; 15/143, 16/189, 16/196+16/197; (Stevan et al. 1977), (Van der Voo and Watts 1978)

Avalon Plate						
S-D (group A)	22°	121°	16	13°	9	050074,115; 15/128; (Kent and Opdyke 1980), (Lapointe 1979), (Murthy 1983a, 1983b; N=2), (Roy et al. 1979; N=2)
S-D (group B)	−42°	99°	11	24°	5	(Lapointe 1979; N=2), (Murthy 1983a, 1983b; N=2), (Roy et al. 1979)
Greenland Block						
P	64°	184°	78	7°	6	110222,643,659; 14/173; (Faller 1975), (Schwarz et al. 1979)
T_3	61°	151°	23	20°	4	080045,257,090163; (Bidgood and Harland 1971)
P-T	37°	174°	–	17°	2	070053, 080256
Caribbean Plate						
K	43°	203°	11	21°	6	100042,43,189,190; 16/90; (Vincenz and Dasgupta 1978)
South American Plate						
K	82°	228°	61	5°	15	100019,108,110,182,193; 14/227, 16/96, 16/15[illegible]; (Ernesto et al. 1979), (Pacca and Hioto 1976; N=2), (Schult and Guerriero 1979), (Valencio et al. 1977a; N=2), (Vilas 1976)
J	83°	254°	70	6°	8	080051,88,92,105,163, 090084; 14/231; (Creer 1962)
T_{2-3}	77°	86°	76	7°	7	080080+81, 080082; 13/42, 14/287, 14/291; (Creer et al. 1970), (Vilas and Valencio 1970)
P_2-T_1	74°	45°	92	5°	12	070114,115,116,080049,080078+161,080079,080104+164, 080107, 162; (Hargraves 1978; N=3)
P_2	75°	135°	87	8°	5	070060,195; 16/136; (Creer et al. 1970), (Valencio et al. 1975)
P_1	62°	170°	108	7°	5	070072; 14/286; (Embleton 1970; N=3)
C_3	53°	164°	53	10°	6	060079,230,231,233,236; (Sinito et al. 1979)
O-D	−1°	127°	15	18°	4	030054, 050109,110; (Creer 1970a)
$Є_3$-O	−13°	154°	34	2°	4	030051–53; (Creer 1962, 1965, 1970b)
$Є_1$	−29°	214°	17	18°	4	14/417–14/419, 14/421
PЄ-Є	−26°	139°	15	20°	3	010090,91, 020038

Appendix *(continued)*

Period epoch	Pole position		Statistics			Index numbers and references[a]
	Φ,N	Λ,E	K	A_{95}	N	
African Plate						
Ᵽ$_3$-N$_1$	82°	111°	45	12°	5	110160,459,509,670; 14/127
Ᵽ$_{1-2}$	77°	177°	57	7°	8	110159,186,526,677,694; (Hussain et al. 1979; N = 2), (Schult et al. 1978)
K$_2$	71°	212°	40	9°	8	100032,97,210,211; 16/72, 16/73, 16/93; (Ito et al. 1977)
K$_1$	60°	253°	32	9°	10	100015,16,80,81,252,254,257; 13/32; (Bardon et al. 1978), (Martin et al. 1978)
J	67°	258°	60	5°	16	090024,25,26,27,50,58,59,66,67,111,112,157,158,175; 14/248; (McElhinny et al. 1976)
T	67°	265°	18	9°	16	080026,27,52,53,55,60,159, 080259 + 260; 13/40, 14/303, 15/87, 15/88, 15/89, 15/90; (Martin et al. 1978), (Valencio et al. 1978)
P$_1$	38°	245°	24	13°	7	070061,91,92; 15/114, 15/115, 16/142; (Creer 1965)
C$_{1-2}$	21°	211°	21	17°	5	060056 + 57; (Martin et al. 1978), (McElhinny and Opdyke 1968; N = 3)
D$_3$	0°	205°	–	–	1	14/361
Є$_3$-S	–48°	219°	48	10°	4	(Bucur 1971), (Daly et al. 1977), (Morris and Carmichael 1978; N = 2)
Є$_1$-Є$_3$	–20°	160°	–	–	2	010127, 030030
Madagascar Plate						
K$_2$	68°	222°	35	8°	10	100008,27,69,70;15/44,15/45,15/46,15/47,15/48,15/49
P$_2$-T$_1$	68°	291°	34	4°	4	14/269, 15/95 + 15/96, 15/97 + 15/107; (McElhinny and Embleton 1976)
P$_1$	56°	262°	–	–	1	070071 + 15/113
C$_3$	48°	264°	–	–	1	15/120
Indian Plate						
Ᵽ$_3$-N$_1$	72	249°	–	–	1	110565
Ᵽ$_{1-2}$	45°	283°	98	8°	4	110698, 701; 14/181, 14/185
Ᵽ$_1$	37°	278°	76	9°	5	110204, 370, 663; 14/178, 14/180

K_2-Ᵽ$_1$	33°	281°	42	14°	4	110612, 685; 14/182, 14/184
K_2	33°	280°	98	5°	11	100022,23,24,44,45,61,112,113; 12/52, 13/21, 13/22
K_1	19°	298°	30	17°	4	100044 + 47, 100046; 12/72, 12/73
J	14°	295°	100	8°	3	090044, 71, 100046
T_3	20°	308°	–	–	2	080086, 96
P_2-T_1	–5°	307°	24	13°	7	070118,119,121,297, 080085,243; 14/302
Є$_{1-2}$	25°	212°	–	–	2	020044,020144
PЄ-Є	44°	209°	36	13°	5	(McElhinny et al. 1978a; N = 5)
Antarctic Plate						
Ᵽ	81°	237°	115	8°	3	110126; (Kellog and Reynolds 1978; N = 2)
K	80°	279°	11	21°	6	100038; 15/57, 15/58, 15/219, 15/220, 16/10[illegible]
J	54°	36°	69	5°	10	080093, 090037,38,40,41,42,56,138; 9/63, 15/[illegible]21 + 15/222 + 15/223
T	50°	29°	50	14°	3	080093; 9/63; (Bull et al. 1961)
O	–13°	22°	24	17°	3	10/140, 14/408; (Manzoni and Nanni 1977)
Australian Plate						
Ᵽ$_3$-N_1	74°	276°	214	4°	6	110152,234,674; 14/111, 14/126; (Idnurm and Senior 1978)
Ᵽ$_{1-2}$	66°	312°	402	4°	5	110093,94,124,235; 14/149
K	54°	327°	36	15°	4	100009, 10; 15/35; (Idnurm and Senior 1978)
J	48°	343°	20	9°	11	090029,30,31,32,33; 15/62, 15/67, 15/71, 15/[illegible]2; (Boesen et al. 1961), (Schmidt 1976)
T	49°	330°	65	15°	3	080028,29; (Boesen et al. 1961)
P_2	42°	307°	133	7°	3	070042, 43, 64
C_3	49°	319°	100	7°	4	060058; (Irving 1966; N = 3)
C_2	82°	306°	–	14°	2	060059, 60
DI	54°	234°	19	21°	4	050033, 143, 181; 16/153
(S_2-D_3)II	70°	162°	61	10°	5	050032,36,78,144,145
S	44°	82°	–	24°	2	040010, 65
Є$_{2-3}$-O	12°	210°	20	12°	7	020018 + (Luck 1972), 020158, 030048; 14/39[illegible], 14/405, 14/415, 15/226
Є$_1$	17°	172°	7	19°	8	020017; 14/416,16/190; (McElhinny and Luck 1970; N = 5)

Appendix *(continued)*

Period epoch	Pole position		Statistics			Index numbers and references[a]
	Φ,N	Λ,E	K	A_{95}	N	
Indo-China Plate						
₽	72°	235°	–	–	1	110600
J	69°	177°	–	–	1	(Haile and Tarling 1975)
C-T	61°	166°	58	9°	4	060328, 070300, 301; 14/294
China Plate						
K-₽	58°	176°	26	14°	4	9/32, 10/51 + 10/52 + 10/57 + 10/59 + 10/60, 10/58, 10/61
J	55°	151°	– .	–	2	090077; (Lee et al. 1963)
P	41°	220°	6	33°	3	070058,59; (McElhinny et al. 1981)
D	61°	311°	–	–	1	050047
Amur Plate						
₽-N	87°	233°	63	9°	4	2–28, 29, 64, 81
K_2-₽	78°	142°	9	17°	7	3–14, 16, 25, 4–32, 33, 43; 100076
K	75°	164°	33	14°	3	4–11, 44; 100031
P-T	32°	205°	17	17°	4	080046; (Gurary et al. 1966), (Vlasov and Popova 1963; N = 2)
$Є_{1-2}$	–11°	137°	–	–	1	020058 + 59
Japan, South-West Block						
₽-N	57°	213°	14	13°	8	100021, 110230; (Kawai et al. 1971; N = 3), (Otofuji and Matsuda 1982; N = 3)
K	51°	205°	14	11°	12	100007, 116, 117, 100194 + 195 + 196 + 197 + 198 + 199 + 200 + 201, 100202 + 203 + 204 + 205 + 206, 10070 + 208; (Domen 1982; N = 5), (Otofuji and Matsuda 1982)
T-J	72°	232°	23	13°	5	(Kawai et al. 1961; N = 5)

P-T	67°	254°	5	21°	9	070150; (Kawai et al. 1961; N=4), (Kawai et al. 1971), (Fujiwara 1968), (Sasajima 1981; N=2)
C	11°	243°	3	45°	3	(Domen 1966), (Fujiwara 1967), (Sasajima 1981)
Japan, Central Block						
₽$_{1-2}$	68°	261°	–	–	1	(Kawai et al. 1971)
K	78°	33°	5	33°	4	100074, 100150+151+152+153, 100159; (Tosha 1982)
T-J	68°	73°	–	–	1	080091
C_1-T	59°	270°	6	30°	4	060083, 190, 070086; (Sasajima 1981)
D	39°	200°	–	–	1	050052
Japan, Hokkaido						
K	86°	153°	7	24°	5	100089, 253; (Fujiwara 1978), (Sasajima 1981; N=2)
Omolon Massif						
K	58°	184°	–	–	2	4—22, 24
J	63°	126°	14	22°	3	5—15,19,20
T	61°	252°	27	12°	5	6—40,41,42,64; (Lozhkina 1982)
P_2	59°	225°	6	33°	3	7—46/1,2,3
Prikolyma Massif						
D_{2-3}	47°	215°	70	8°	5	(Gagiev et al. 1982; N=5)
Omulev Uplift						
D_{2-3}	−19°	108°	13	23°	3	(Pisarevsky, Pogarskaya, Slautsitais, new data; N=3)
S_2	−22°	77°	–	–	2	(Pisarevsky, Pogarskaya, Slautsitais, new data; N=2)
O_2	−22°	140°	–	–	2	(Pisarevsky, Pogarskaya, Slautsitais, new data; N=2)

Appendix *(continued)*

Period epoch	Pole position		Statistics			Index numbers and references[a]
	Φ,N	Λ,E	K	A_{95}	N	
Chukotka Block						
₽	68°	177°	–	–	1	3–15
K	61°	152°	130	6°	4	4–21,23,25,26
J_{2-3}	67°	124°	–	–	1	5–16
D_3-P	42°	156°	471	4°	3	7–44; (Pechersky 1970; N=2)
North American Cordilleras Terranes						
1. Stikinia Terrane						
K_1	76°	327°	–	–	1	(Irving et al. 1980)
J_1	70°	57°	–	–	1	(Irving et al. 1980)
T_3	31°	140°	–	16°	2	(Irving et al. 1980; N=2)
2. Wrangellia Terrane						
T_3	52°	23°	–	–	1	(Irving et al. 1980)

[a]Explanation:
1) 4–28,36 . . . Index numbers (from Khramov 1971, 1973, 1975, 1979, 1982).
2) 15/70, 16/34 . . . Index numbers (from Irving 1964; McElhinny 1972, 1973; McElhinny and Cowley 1977, 1978, 1980).
3) 080070, 090072 . . . Index numbers (from Irving and Hastie 1975; Irving et al. 1976a, 1976b, 1976c).
4) 208* . . . Index numbers (from Khramov and Sholpo 1967).
5) (Bylund 1973; N=3) . . . Determinations from publications; N = number of determinations.
6) 110325+326+695; 15/97+15/107; 14/118+(DeJong et al. 1973), (Piper 1979)+(Piper and Stearn 1975; N=2) . . . Unit weight is given for every resultant determination.
7) 3-01-5,11012–15, 13/60–13/62 denotes that all successive determinations are included.
8) 12–31/1,2,3 . . . Unit weight is given for each determination 12/31-1, 12/31-2 and 12/31-3.

References

Adam NV, Ben'kova NP, Khramov AN, Cherevko TN (1975) Spherical harmonic analysis of the geomagnetic field of the Brunhes epoch. Stud Geophys Geod 19:141–149

Agabekov VA, Metallova VV (1972) Incomplete self-reversal of a partial thermoremanent magnetization in natural titanomagnetites. Izv Acad Sci USSR Ser Fisika Zemli 3:44–49

Agadzhanyan AK, Virina EI, Faustov SS (1981) Paleomagnetism of the late Cenazoic deposits of the Middle Don River. In: II Vsesoyuzn s'yezd Postoyan geomagn pole, magn gornykh porod i paleomagn II. Izd Tbilisi Univ, Tbilisi, p 106

Ahmad Z, Abbas SG (1979) The Muslim Bagh ophiolites. In: Farah A, De Yong KA (eds) Geodynamics of Pakistan, Quetta. Geol Surv Pakistan, pp 243–249

Akhmatov PG (1966) Paleomagnetic studies of the Upper Paleozoic igneous rocks in the Chatkal-Kurama region (Middle Tien-Shan). In: Tezisy dokl VII Vsesoyuzn konf po postoyan geomagn polyu i paloeomagn. Izd Acad Sci USSR, Moscow, pp 70–72

Aleksandrov SM, Zanyukov VN (1969) On the meridional and sublatitudinal faults of the Lyutoch depression (Southern Sakhalin). Dokl Acad Sci USSR 189(3):589–591

Alekseyeva VA, Zubov AG, Kochegura VV (1979) Natural remanent magnetization of tephra with reference to the paleosecular variations from its study. In: Problemy izuch paleovek variats magn polya Zemli. Del'nevost Nauch Tsentr Acad Sci USSR Vladivostok, pp 36–51

Ali-Zade AA, Aleskerov JA, Pevzner MA (1973) Paleomagnetic studies of the Pliocene deposits of the Apsheron Peninsula. In: Paleomagn analiz pri izuchen chetvert otlozh i vulkanitov. Nauka, Moscow, pp 10–14

Alldredge LD, Hurwitz L (1964) Radial dipoles as sources of the Earth's main magnetic field. J Geophys Res 69:2631–2640

Alvarez W, Lowrie W (1978) Upper Cretaceous palaeomagnetic stratigraphy at Moria (Umbrian Apennines, Italy) a verification of the Gubbio section. Geophys J R Astron Soc 55(1):1–17

Alvarez W, Arthur MA, Fischer AG, Lowrie W, Napoleone G, Premoli-Silva I, Roggenthen WM (1977) Upper Cretaceous-Paleocene magnetic stragigraphy at Gubbio, Italy. V. Type section for the late Cretaceous-Paleocene geomagnetic time scale. Geol Soc Am Bull 88(3):383–389

Ananiashvili GD, Pevzner MA (1977) Paleomagnetic studies of Miocene deposits of the western Georgia. Izv Acad Sci USSR Ser Geol 9:43–51

Aparin VP (1983) Correlation of periodical alternation of continental plate relative velocity, sedimentary tempo and oil pools time of formation. Docl Acad Sci USSR 270(1):161–165

Aparin VP, Vlasov AY (1965) Certain problems in the geomagnetic field evolution at the end of the Middle Paleozoic and in the Upper Paleozoic. In: Nastoyashcheye i proshloye magn polya Zemli. Nauka, Moscow, pp 213–220

Aparin VP, Kirillov VM, Kuznetsov AA (1970) Paleomagnetic section of the coal-bearing strata in the Kuzbass according to examination of cores. In: Materialy VIII konf po postoyan geomagn polyu i paleomagn II. Naukova Dumka, Kiev, pp 25–30

As JA (1967) The A C demagnetization technique. In: Collinson DW, Creer KM (eds) Methods in palaeomagnetism. Elsevier, Amsterdam New York, pp 221–223

Asanidze BZ, Pechersky DM (1979) Paleomagnetic studies of Jurassic rocks in Georgia and Northern Caucasus. Izv Acad Sci USSR Ser Fisika Zemli 10:77–92

Asrarullah, Ahmad Z, Abbas SG (1979) Ophiolites in Pakistan: an introduction. In: Farah A, De Yong KA (eds) Geodynamics of Pakistan. Geol Surv Pakistan, Quetta, pp 181–192

Baag C, Helsley CE (1974a) Evidence for penecontemporaneous magnetization of the Moenkopi formation. J Geophys Res 79(23):3308–3320

Baag C, Helsley CE (1974b) Geomagnetic secular variation model E. J Geophys Res 79(32):4918–4922

Bagina OL (1970) Determination of the NRM type in sedimentary rocks by the redeposition coefficient. In: Sb tezisov dokl VIII konf po vopr postoyan geomagn polya, magn gornykh porod i paleomagn II, Izd Inst Fisiki Zemli Acad Sci USSR, Moscow, pp 27–31

Bagina OL, Minasyan DO, Petrova GN (1973) A possibility to determine the ancient field intensity by analysis of anhysteretic magnetization (H_e). In: Materialy IX konf po vopr postoyan geomagn polya, magn gornykh porod i paleomagn II. Izd Acad Sci Azerb SSR, Baku, pp 28–29

Bardon C, Bosset A, Manzen R, Westphal M (1978) Paléomagnetisme des formations volcaniques du Crétacé inferieur dans l'Atlas de Beni-Mellal (Maroc). Notes Mem Serv Geol Maroc 39(272): 7–26

Barton CE, McElhinny MW, Edwards DJ (1980) Laboratory studies of depositional DRM. Geophys J R Astron Soc 61(2):355–377

Bazhenov ML (1979) Studies of local tectonic deformations using paleomagnetic method. Izv Acad Sci USSR Ser Fisika Zemli 11:53–59

Bazhenov LM (1980) Methods of paleomagnetic studies of arc-island fold structures (Pamir, Kopetdag, Northern Carpathians). Synopsis of thesis. Izd Inst Fisiki Zemli Acad Sci USSR, Moscow, 21 pp

Bazhenov ML, Burtman VS, Gurary GZ (1978) Studies of the Pamir arc curvature in Paleogene using paleomagnetic method. Dokl Acad Sci USSR 242(5):1137–1139

Bazhenov ML, Ryabushkin PK (1978) Application of statistical criteria of consistency in paleomagnetic studies. Izv Acad Sci USSR Ser Fisika Zemli 7:100–104

Belokon' VI, Kochegura VV, Sholpo LE (1973) Methods of paleomagnetic studies of rocks. Nedra, Leningrad, 248 pp

Ben'kova NP, Khramov AN, Cherevko TN, Adam NV (1973) Spherical harmonic analyses of the paleomagnetic field. Earth Planet Sci Lett 18(2):141–147

Berberian M (1984) Structural evolution and tectonics of the Iran Plateau: a plate tectonic approach. In: 27th Int Geol Congr Abstr, vol 3. Nauka, Moscow, p 140

Berhold G, Nairn AEM, Negendank IFW (1975) A paleomagnetic investigation of some of the igneous rocks of the Saar-Nahe basin. Neues Jahrb Geol Plalaeontol Monatsh 3:134–150

Besse Y, Pozzi YP, Mascle G, Feinberg H (1984) Paleomagnetic study of Sicily: consequences for the deformation of Italian and African margins over last 100 million years. Earth Planet Sci Lett 67(3):377–390

Bidgood DET, Harland WB (1961) Palaeomagnetism in some east Greenland sedimentary rocks. Nature 189(4765):633–634

Blow RA, Hamilton N (1978) Effect of compaction on the acquisition of a detrital remanent magnetization in fine-grained sediments. Geophys J R Astron Soc 52(1):13–23

Boesen R, Irving E, Robertson WA (1961) The palaeomagnetism of some igneous rock bodies in New South Wales. J Proc R Soc New South Wales 94(6):227–232

Bogdanov AA (1969) Tectonic epochs (on the tectonic division of the Earth's evolution). Byull Mosk Ova Ispyt Prir Otd Geol 5:5–17

Bol'shakov AS, Shcherbakova VV (1979) Thermomagnetic criterium for definition of the domain structure of ferromagnetic materials. Izv Acad Sci USSR Ser Fisika Zemli 2:38–47

Bol'shakov AS, Solodovnikov GM (1969) Determination of intensity of the ancient geomagnetic field by magnetization of baked rocks. In: Magnetism gornykh porod i paleomagn. Izd Inst Fisiki Zemli Acad Sci USSR, Moscow, pp 129–131

Bol'shakov AS, Solodovnikov GM (1975) Paleointensity of the Earth's magnetic field in the late Triassic. Dokl Acad Sci USSR 221(4):828–831

Bol'shakov AS, Solodovnikov GM (1980a) Paleomagnetic data on the intensity of the Earth's magnetic field. Izv Acad Sci USSR Ser Fisika Zemli 8:87–100

Bol'shakov AS, Solodovnikov GM (1980b) On the geomagnetic field intensity in the late Jurassic. Izv Acad Sci USSR Ser Fisika Zemli 11:74–86

Bol'shakov AS, Solodovnikov GM (1981a) Intensity of the geomagnetic field for the last 400 million years. Dokl Acad Sci USSR 260(6):1340–1343

Bol'shakov AS, Solodovnikov GM (1981b) On the geomagnetic field intensity in the late Cretaceous. Izv Acad Sci USSR Ser Fisika Zemli 10:58–68

Bol'shakov AS, Mel'nikov BN, Dashevskaya DM (1973) Partial self-reversal of magnetization in the Kovdor magnetites. Izv Acad Sci USSR Ser Fisika Zemli 6:103–109

Bol'shakov AS, Solodovnikov GM, Vechfinsky VS (1978) Determination of paleointensity of the geomagnetic field in the late Cretaceous by magnetization of baked rocks. Izv Acad Sci USSR Ser Fisika Zemli 12:83–92

Bol'shakov AS, Gapeev AK, Nguen Thi Kim Thoa, Solodovnikov GM (1981) Determination of the geomagnetic field intensity of the late Cretaceous time by magnetization of baked rocks. Izv Acad Sci USSR Ser Fisika Zemli 4:86–90

Bondarchuk VG (1970) Motion and structure of tectonosphere (Theoretical principals of the Earth's crust structure). Naukova Dumka, Kiev, 190 pp

Bonhommet N, Zähringer J (1969) Paleomagnetism and potassium-argon age determinations of the Laschamp geomagnetic polarity event. Earth Planet Sci Lett 6(1):43–46

Boronin VP (1973) Paleomagnetism of the Upper Permian and Lower Triassic stratotypical sections in the Volga-Urals area. In: Materialy IX konf po vopr postoayn geomagn polya magn gornykh porod i paleomagn III. Izd Acad Sci Azerb SSR, pp 16–17

Boronin VP (ed) (1979) Methods of paleomagnetic studies of red beds. Izd Kazan Univ, Kazan, 132 pp

Bowker DD (1960) Remanent magnetization of eastern United States Triassic rocks. Thesis Massachucetts Inst Technol, Cambridge

Boyko GE (1982) Forecasting of oil-and-gas bearing deposits according to genetic aspects. Naukova Dumka, Kiev, 252 pp

Braginsky SI (1964) Kinematic models of hydromagnetic dynamo of the Earth. Geomagn Aeronom 4(4):732–747

Braginsky SI (1978) Geomagnetic dynamo. Izv Acad Sci USSR Ser Fisika Zemli 9:74–90

Bretshtein YS, Klimova AV, Karetnikov AS (1982) Paleomagnetism of the Meso-Cenozoic volcanogenic-sedimentary and magmatic complexes of the Southern Kuril Islands. In: Paleomagn stratigr mesokainoz otlozh Naukova Dumka, Kiev, pp 41–50

Briden JC, Duff BA (1980) Pre-carboniferous paleomagnetism of Europe north of the Alpine orogenic belt. In: McElhinny MW, Valencio DA (eds) Paleoreconstruction of the continents. Geodynamic series 2. Am Geophys Union, Washington DC, Boulder, pp 137–149

Briden JC, Smith AG, Sallomy JT (1970) The geomagnetic field in Permo-Triassic time. Geophys J R Astron Soc 23(1):101–117

Briden JC, Drewry DE, Smith AG (1974) Phanerozoic equal-area world maps. J Geol 2(5):555–574

Brock A (1981) Paleomagnetism of Africa and Madagascar. In: Paleoreconstruction of the continents. Geodynamic series, vol 2. Am Geophys Union, Washington DC, Boulder, pp 65–76

Brunhes B (1906) Récherches sur la direction d'aimantation des roches volcaniques. J Phys (Paris) 5:705–724

Brunhes B, David P (1903) Sur la direction de l'aimantation permanente dans diverses roches volcaniques. CR Acad Sci (Paris) 137(2):975

Bubnoff S von (1930) Geologie von Europa. Borntraeger, Berlin, 690 pp

Bubnoff S von (1954) Grundprobleme der Geologie, 3. Aufl, Academie, Berlin, 234 pp

Buchá V (1965) Results of the palaeomagnetic research on rocks of Pre-Cambrian and Lower Palaeozoic age in Czechoslovakia. J Geomagn Geoelectr 17(3/4):435–444

Buchá V (1973) The continuous pattern of variations of the geomagnetic field in the Quaternary and their causes. Stud Geophys Geod 17(3):218–231

Buchá V (1975) Paleomagnetism and plate tectonics since the Lower Paleozoic. Stud Geophys Geod 19:42–60

Buchá V (ed) (1983) Magnetic field and the processes in the Earth's interior. Academia, Prague, 514 pp

Buchá V, Malkovsky Z, Petrova GN, Rodionov VP, Rother K, Khramov AN (1976) Problem of continental drift in Eurasia according to paleomagnetic data on the Lower Paleozoic. In: Dreif Kontinentov. Nauka, Moscow, pp 86–202

Bucur I (1971) Etude paléomagnétique d'une formation sèdimentaire du Sahara algèrient, d'age Cambro-Ordovicien. Ann Géophys 27(2):255–261

Bugayev MS, Zvegintsev AG, Belobrova IA (1972) Self-reversal of magnetization of magnesian titanomagnetites. Izv Acad Sci USSR Ser Fisika Zemli 5:53–59

Bull C, Irving E, Wills I (1961) Further palaeomagnetic results from south Victoria Land, Antarctica. Geophys J R Astron Soc 6(3):322–336

Bullard EC (1949) Electromagnetic induction in rotating sphere. Proc R Soc Ser A 199(1059): 413–443

Bullard EC, Gellman H (1954) Homogenous dynamos and terrestrial magnetism. Philos Trans R Soc Ser A 247(928):213–278

Burakov KS, Ryabushkin PK (1975) About magnetization of coarse-grained sediments. Izv Acad Sci USSR Ser Fisika Zemli 7:127–128

Burakov KS, Didenko AN, Pechersky DM (1984) Evaluation of the geomagnetic field in the middle Devonian by the baked chert and gabbro in Southern Mugodzhary. Izv Acad Sci USSR Ser Fisika Zemli 8:45–59

Burakov KS, Gurary GZ, Khramov AN, Petrova GN, Rassanova GV, Rodionov VP (1976) Some peculiarities of the virtual pole positions during reversals. J Geomagn Geoelectr 28(4):295–307

Burek PJ (1971) An advanced device for chemical demagnetization of red beds. J Geophys 37(3): 493–498

Burlatskaya SP (1965) Archeomagnetism. Studies of the Earth magnetic field in the past epochs. Nauka, Moscow, 127 pp

Burlatskaya SP (1983) On the dipole character of the ancient magnetic field. Izv Acad Sci USSR Ser Fisika Zemli 5:76–82

Burov BV, Boronin VP (1973) On the usefulness of factor Q in relative evaluation of intensity of the ancient geomagnetic field. In: Materialy IX konf po vopr postoyan geomagn polya, magn gornykh porod i paleomagn III. Izd Acad Sci Azerb SSR, Baku, pp 24–25

Burov BV, Yasonov PG (1979) Introduction into differential thermomagnetic analysis of rocks. Izd Kazan Univ, Kazan', 160 pp

Burov BV, Komissarova RA, Slautsitais IP, Khramov AN (1977) The geomagnetic inversions at the Paleozoic-Mesozoic boundary. In: Paleomagnetism i voprosy tektoniki plit. Trudy VNIGRI, N 394, Leningrad, pp 13–26

Burtman VS (1976) Structural evolution of the Paleozoic fold systems. Nauka, Moscow, 164 pp

Butler RF, Lindsay EH (1981) Magnetic polarity stratigraphy of latest Cretaceous to early Eocene continental deposits, San Juan Basin and Clark's Fork Basin. In: Programme and abstracts. 4th IAGA Sci Assembly August 3–15, 1981, Edinburgh. Paris, p 212

Bylund G (1973) Paleomagnetic study of Scanian dolerites and basalts. Geologiska Inst Lunds Univer, Lund, p 53

Bylund G, Patchett PJ (1977) Palaeomagnetic and Rb-Sr isotopic evidence for the age of the Särna alkaline complex, Western Central Sweden. Lithos 10(1):73–79

Carey SW (1976) The expanding Earth. Elsevier, Amsterdam, 488 pp

Carmichael CM, Storetvedt KM (1981) Multicomponent magnetization of the Ordovician age Younger Gabbros, Aberdeenshire, Scotland. Geophys J R Astron Soc 67(1):199–215

Chamalaun FH, Creer KM (1964) Thermal demagnetization studies on the Old Red sandstone of the Anglo-Welsh cuvette. J Geophys Res 69(8):1607–1616

Channell JET, Medizza F (1981) Upper Cretaceous and Paleogene magnetic stratigraphy and biostratigraphy from Venetian (Southern) Alps. Earth Planet Sci Lett 55(3):419–432

Chebanenko II (1963) Basic regularities of the Earth's crust fault tectonics. Izd Acad Sci Ukraine SSR, Kiev, 156 pp

Collinson DW (1967) Chemical demagnetization. In: Collinson DW, Creer KM, Runcorn SK (eds) Methods in paleomagnetism. Elsevier, Amsterdam, pp 306–310

Collinson DW (1983) Methods in rock magnetism and palaeomagnetism. Techniques and instrumentation. Chapman and Hall, London, 503 pp

Collinson DW, Runcorn SK (1960) Polar wandering and continental drift. Geol Soc Am Bull 71(7): 915–958

Collinson DW, Creer KM, Runcorn SK (eds) (1967) Methods in palaeomagnetism. Elsevier, Amsterdam, 609 pp
Coupland DH, Van der Voo R (1980) Long-term non-dipole components in the geomagnetic field during the last 130 m.y. J Geophys Res 85:3529–3548
Cowling TG (1934) The magnetic field of sunspots. Monthly Not R Astron Soc 94(1):1–39
Cox A (1962) Analysis of present geomagnetic field for comparison with paleomagnetic results. J Geomagn Geoelectr 13(2):101–112
Cox A (1968) Lengths of geomagnetic polarity intervals. J Geophys Res 73(10):3241–3260
Cox A (1969) Geomagnetic reversals. Science 163(3864):237–245
Cox A (1970) Latitude dependence of the angular dispersion of the geomagnetic field. Geophys J R Astron Soc 20(3):253–269
Cox A, Doell RR (1960) Review of paleomagnetism. Geol Soc Am Bull 71(6):645–768
Cox A, Doell RR, Dalrymple GB (1963) Geomagnetic polarity epochs and Pleistocene geochronometry. Nature 198(4885):1049–1051
Creer KM (1962) Palaeomagnetic data from South America. J Geomagn Geoelectr 13(1/2):154–165
Creer KM (1964) Palaeolatitudes of the continents in Devonian as reveals by the NRM of rocks. In: Problems in Palaeoclimatology. Wiley, London, pp 260–284
Creer KM (1965) Palaeomagnetic data from the Gondwanic continents. Philos Trans R Soc Lond Ser A 258(1088):27–40
Creer KM (1968) Paleozoic palaeomagnetism. Nature 219(5151):246–250
Creer KM (1970a) A review of paleomagnetism. Earth Sci Rev 6(6):369–466
Creer KM (1970b) Palaeomagnetic survey of South American rocks I–V. Philos Trans R Soc Lond Ser A 267(1183):457–558
Creer KM (1977) Geomagnetic secular variations during the last 25,000 years: an interpretation of data obtained from rapidly deposited sediments. Geophys J R Astron Soc 48(1):91–109
Creer KM, Embleton BJJ (1967) Devonian palaeomagnetic pole for Europe and North America. Nature 214(5083):42–43
Creer KM, Ispir Y (1970) An interpretation of behavior of the geomagnetic field during polarity transitions. Phys Earth Planet Interior 12(4):283–293
Creer KM, Kopper JS (1976) Secular oscillations of the geomagnetic field recorded by sediments deposited in caves in the Mediterranean region. Geophys J R Astron Soc 45(1):35–58
Creer KM, Readman PW (1980) Palaeomagnetic and palaeontological dating of a section at Yioia Tauro, Italy: identification of the Blake event. Earth Planet Sci Lett 50(1):289
Creer KM, Irving E, Runcorn SK (1957) Geophysical interpretation of paleomagnetic directions from Great Britain. Philos Trans R Soc Lond Ser A 250(974):144–156
Creer KM, Irving E, Nairn AEM (1959) Palaeomagnetism of the Great Whin Sill. Geophys J R Astron Soc 2(3):306–323
Creer KM, Embleton BJ, Valencio DA (1970) Triassic and Permo-Triassic data for South America. Earth Planet Sci Lett 8(2):173–178
Creer KM, Mackereth FM, Molyneux L, Thompson R (1972) Geomagnetic secular variations recorded in the stable magnetic remanence of Recent sediments. Earth Planet Sci Lett 14(3):115–127
Creer KM, Gross DL, Lineback JA (1976) Origin of regional geomagnetic variations recorded by Wisconsinian and Holocene sediments from Lake Michigan, USA and Lake Windermere, England. Geol Soc Am Bull 87(4):531–540
Creer KM, Tucholka P, Barton CE (1983) Geomagnetism of baked clays and recent sediments. Elsevier, Amsterdam, 324 pp
Dagley P, Musset AE (1978) Palaeomagnetism of the Fishnish dykes Mull, Scotland. Geophys J R Astron Soc 55(3):553–558
Daly L, Leblance M, Pozzi JP (1977) Mise en evidence de plusieurs phases d'aimantations et d'un noveau pôle pléomagnétique cambrien dans les formations de Bou Azzer-ElGraara (Anti-Atlas, Maroc). CR Acad Sci Paris D284(4):271–274
Dankers P (1982) Implications of early Devonian poles from the Canadian Arctic Archipelago for the North American apparent polar wander path. Can J Earth Sci 19(9):1802–1809

Danukalov NF (1975) Paleomagnetism and magnetism of intrusive formations in the Southern Urals. Nauka, Moscow, 152 pp

Danukalov NF, Komissarova RA, Khramov AN (1977) Metachronous magnetization. In: Geomagnitn issled N 19. Sov Radio, Moscow, pp 12–17

Danukkalov NF, Kondruchina LS, Chernikov AP (1983) Paleomagnetism of Paleozoic in the Southern and Central Urals. Izd Bashkir Filial Acad Sci USSR, Ufa, 120 pp

Darwin GH (1911) The tides and kindred phenomena in the solar system, 3rd edn. Murray, London, 438 pp

Davidzon RM, Rzhevky YS (1979) Reconstruction of sedimentary basin in the Paleogene in the area of the Tajik depression. Dokl Acad Sci Tajik SSR 22(11):83–85

Davydov VI, Komissarova RA, Khramov AN, Chediya IO (1982) On the paleomagnetic characteristics of the Upper Permian strata in the south-eastern Pamir. Dokl Acad Sci USSR 267(5): 1177–1181

Day R (1977) TRM and it's variations with the grain's size. J Geomagn Geoelectr 29(4):233–265

De Boer J (1967) Palaeomagnetic tectonic study of Mesozoic dike swarm in the Appalachians. J Geophys Res 72(8):2237–2250

De Jong KA, Manzoni M, Stavenga T, Van Dijk F, Vander Voo R, Zijderveld JDA (1973) Palaeomagnetic evidence for rotation of Sardinia during the early Miocene. Nature 243(5405):281–283

Demenitskaya RM, Gorodnitsky AM, Kaminsky VD, Litvinov EM (1978) Seamounts (problems of their geophysical studies). Nedra, Leningrad, 164 pp

Denham CR, Anderson RF, Bacon MP (1977) Paleomagnetism and radiochemical age estimates for late Brunhes polarity episodes. Earth Planet Sci Lett 35(3):384–397

Dewey JF, Bird J (1970) Plate tectonics and geosynclines. Tectonophysics 10(5/6):625–638

Dianov-Klokov VI (1960) Device for measurement of weak remanent magnetization of rocks. Izd Acad Sci USSR Ser Geophys 1:142–147

Dolginov SS (1977) Magnetism of planets. Geomagn Aeronom 17(4):569–595

Domen H (1962) Piezo-remanent magnetization in rocks and its field evidence. J Geomagn Geoelectr 13(3):66–72

Domen H (1966) A memoir on the natural remanent magnetization of Kawajirimisaki basalts, Yamaguchi, West Japan. Yamaguchi Daigaku Kyoikugakubu Kenkyu Nonso Dai 2bu 15:13–16

Domen H (1977) A single heating method of paleomagnetic field intensity determination applied to old roof tiles and rocks. Phys Earth Planet Interior 13(4):315–318

Domen H (1982) A couple of paleomagnetic data of the Cretaceous rocks from Yamaguchi and Shimane prefecture, West Japan. Rocks Magn Paleogeophys 4:61–65

Dortman NB (ed) (1976) Physical properties of rocks and useful minerals (petrophycis). Reference book for geophysicists. Nedra, Moscow, 528 pp

Duff BA (1978) Palaeomagnetic and rock magnetic studies of Lower Palaeozoic rocks in Jersey and the adjacent regions of the Armorican Massif. Ph D thesis, Univ Leeds, 338 pp

Duff BA (1979) The palaeomagnetism of Cambro-Ordovician red beds, the Erquy spilite series and Tregastel-Ploumanach granite complex, Armorican Massif (France and the Channel Islands). Geophys J R Astron Soc 59(2):345–365

Duff BA (1980) The palaeomagnetism of Jersey volcanics and dykes and the Lower Palaeozoic apparent polar wander path for Europe. Geophys J R Astron Soc 60(3):355–375

Dunlop DJ (1969) Interaction in rocks and the reliability of paleointensity data. Earth Planet Sci Lett 7(2):178–182

Dunlop DJ (1979) On the use of Zijderveld vector diagrams in multicomponent paleomagnetic studies. Phys Earth Planet Interiors 20(1):12–24

Edel JB, Lörtscher A (1977) Paleomagnetisme du volcanisme tertiare de Sardaigne. Nouveaux resultats et synthese. Bull Soc Geol France 19(4):815–824

Edwards S (1980) Comparisons between the generation and properties of rotational remanent magnetization and anhysteretic remanent magnetization. Geophys J R Astron Soc 62(2): 379–392

Egyed L (1961) Palaeomagnetism and the ancient radii of the Earth. Nature 190(4781):1097–1098

Einarsson T (1957a) Der Paläomagnetismus der isländischen Basalte und seine stratigraphische Bedeutung. Neues Jahrb Geol Palaeontol Monatsheft 4:159–175

Einarsson T (1957b) Magnetogeological mapping in Iceland with the use of a compass. Adv Phys 6(22):232–239

Embleton BJJ (1970) Palaeomagnetic results for the Permian of South America and a comparison with the African and Australian data. Geophys J R Astron Soc 21(2):105–118

Ernesto M, Hioto FY, Pacca JG (1979) Estudo paleomagnetico de seqüência de Derrames bassalticos du formacão Serra Jeral em Santa Catarina, Anais da Academia Brasileria de Ciencias 5(2):327–332

Faller AM (1975) Palaeomagnetism of the oldest Tertiary basalts in the Kangerdlugsuag area of East Greenland. Bull Geol Soc Denmark 24(3/4):173–178

Fischer AG, Judson S (eds) (1978) Oil-and-gas-bearing deposits and global tectonics. Nedra, Moscow, 237 pp

Fisher RA (1953) Dispersion on a sphere. Proc R Soc Lond Ser A 217:295–305

Forsh NN (1981) On the Proterozoic drift of the crustal plates in the USSR region. In: Magnitostr i geodinamika. Radio i svyaz', Moscow, pp 109–113

Foster JH (1966) A paleomagnetic spinner magnetometer using a fluxgate gradiometer. Earth Planet Sci Lett 1(1):163–166

Fujita K, Newberry JT (1983) Accretionary terranes and tectonic evolution of northeast Siberia. In: Accretion tectonics Circum-Pacific Reg Proc Oji Int Semin Tomakomai 10–16 Sept 1981. Dordrecht, Tokyo, pp 43–57

Fujiwara Y (1967) Paleomagnetism of Upper Carboniferous rocks in Akiyoshi Province, S.W. Honshu, Japan. J Fac Sci Hokkaido Univ Ser 4 13(4):395–399

Fujiwara Y (1968) Changing of the palaeolatitude in the Japanese islands through the Palaeozoic and Mesozoic. Geol Mineral 14(2):159–168

Fujiwara Y (1978) Paleomagnetic properties on the alkaline dolerite complex in the Nemuro Peninsula, Hokkaido, Northeast Japan. Geol Fac Sci Hokkaido Univ Ser 4 18(1/2):237–242

Fuller M, Williams I, Hoffman KA (1979) Paleomagnetic records of geomagnetic field reversals and morphology of the transitional fields. Rev Geophys Space Phys 17(2):179–203

Gagiev MK, Kolesov EV, Ivanov YY (1982) Preliminary results of magnetobiostratigraphic researches of Devonian sediments of the Prikolyma uplift. In: Metody paleomagnetisma v reshenii geol zadach (na primere Dal'nego Vostoka). Izd DV Nauchn Tsentra Acad Sci USSR, Vladivostok, pp 5–12

Galushkin YI, Ushakov SA (1979) Geometry of plate motion on a sphere. In: Okeanologiya. Geofisika okeana, vol 2. Geodinamika. Nauka, Moscow, pp 162–173

Gamov LN, Pen'kov AV (1970) Generalized paleomagnetic section for Mesozoic of the southern Tajikistan. In: Materialy VIII konf po postoyan geomagn polyu i paleomagn II. Naukova Dumka, Kiev, pp 35–38

Gamov LN, Pen'kov AV (1976) On the geomagnetic episodes in the Brunhes epoch. In: Glavnoye geomagn pole i problemy paleomagn III. Izd Acad Sci USSR, Moscow, p 28

Gansser A (1964) Geology of the Himalayas. Wiley, London, 189 pp

Gansser A (1979) Reconnaissance visit to the ophiolites in Baluchistan and the Himalaya. In: Farah A, De Jong KA (eds) Geodynamics of Pakistan. Geol Surv Pakistan, Quetta, pp 193–213

Gapeyev AK, Pechersky DM (1983) Changes of titanomagnetites and their effects on the magnetic properties. In: Sovremennoye sostoyan issled v oblasti geomagnetisma. Nauka, Moscow, pp 182–191

Gasanova KL, Gasanov AZ, Isayeva M, Ismail-Zade TA, Ragimova ZA (1973) Paleomagnetic correlation of the Jurassic formations of the NKAO. In: Materialy IX konf po vopr postoyan geomagn polya, magn gornykh porod i paleomagn III. Izd Acad Sci Azerb SSR, pp 41–42

Gavrish VK (1978) Role of abyssal faults in migration and accumulation of oil and gas. Naukova Dumka, Kiev, 172 pp

Girdler RW (1963) Sur l'application de pressions hydrostatiques a des aimantations thermoremanentes. Ann Géophys 19(2):118–121

Glevasskaya AM (1983) Magnetic minerals and magnetism of volcanics. Naukova Dumka, Kiev, 208 pp

Goncharov GI (1970) On the critical paleoparallels. In: Materialy VIII konf po postoyan geomagn polyu i paleomagn II. Naukova Dumka, Kiev, pp 43–47

Goncharov GI (1973) On the possibility of oil potential forecasting using paleolatitudinal data. In: Postoyan magn pole Zemli, paleomagn i magnet gornykh porod. Naukova Dumka, Kiev, pp 87–88

Goncharov GI (1974) Paleolatitudinal zonation of certain geotectonic phenomena and distribution of mineral deposits. In: Khramov AN (ed) Paleomagnetism paleozoya. Nedra, Leningrad, pp 158–168

Goncharov GI (1976) Certain geologic applications of paleomagnetism. In: XXVIII Gertsenovskiye chteniya. Geologiya. Izd LGPI, Leningrad, pp 9–17

Goncharov GI (1977) Evolution of the structural elements of lithosphere and periodicity of transgressions according to paleolatitudinal data. In: Paleomagn i voprosy tect plit. Trudy VNIGRI, Issue 394, Leningrad, pp 141–153

Goncharov GI (1981) Paleomagnetic-tectonic criteria of oil-and-gas bearing deposits. In: II Vsesoyuzn s'yezd Postoyan geomagn pole, magn gornykh porod i paleomagn II. Izd Tbilisi Univ, Tbilisi, p 161

Gonnov VV, Molostovsky EA, Kitovani TG (1981) Paleomagnetic section of the Upper Pliocene and the ancient Euxine of Guria. In: II Vsesoyuzn s'yezd Postoyan geomagn pole, magn gorykh porod i paleomagn II. Izd Tbilisi Univ, Tbilisi, p 115

Gorodnitsky AM, Zonenshain LP (1979) Paleogeodynamic reconstructions of Phanerozoic. In: Geofisika okeana, vol 2. Geodynamika. Nauka, Moscow, pp 338–369

Graham JW (1949) The stability and significance of magnetism in sedimentary rocks. J Geophys Res 54(1):131–167

Graham JW (1955) Evidence of polar shift since Triassic time. J Geophys Res 60(3):329–347

Graham KWT, Helsley CE, Hales AL (1964) Determination of relative positions of continents from paleomagnetic data. J Geophys Res 69(18):3895–3900

Grishanov AN, Molostovsky EA (1981) Paleomagnetic characteristics of the Mesozoic deposits of the European part of the USSR. In: II Vsesoyuzn s'yezd Postoyan geomagn pole, magn gornykh porod i paleomagn II. Izd Tbilisi Univ, Tbilisi, p 117

Grasso M, Lentini F, Nairn AEM, Vigliotti L (1983) A geological and paleomagnetic study of the Eblean volcanic rocks. Sicily. Tectonophysics 98(3/4):271–295

Grubbs KL, Van der Voo R (1976) Structural deformation of the Odaho-Wyoming overthrust belt (USA) as determined by Triassic paleomagnetism. Tectonophysics 33(3/4):321–336

Gurary GZ (1969) Data on the geomagnetic field on the Siberian platform during the early Paleozoic. Izv Acad Sci USSR Ser Fisika Zemli 6:105–113

Gurary GZ (1977a) Some reversals of the geomagnetic field in the late Cenozoic with reference to Western Turkmenia. Izv Acad Sci USSR Ser Fisika Zemli 7:81–90

Gurary GZ (1977b) Studies of transitional regimes. Geomagn issled N 19. Sov radio, Moscow, pp 5–11

Gurary GZ (1981) The Matuyama-Jaramillo geomagnetic reversal in Western Turkmenia. Izv Acad Sci USSR Ser Fisika Zemli 3:73–82

Gurary GZ, Trubikhin VM (1981) Paleomagnetic characteristics of the late Pliocene in the eastern Azerbaijan. In: II Vsesoyuzn s'yezd Postoyan geomagn pole, magn gornykh porod i paleomagn II. Izd Tbilisi Univ, Tbilisi, p 118

Gurary GZ, Kropotkin PN, Pevzner MA, Ro Su Von, Trubikhin VM (1966) Laboratory estimation of paleomagnetic fitness of North Korean sedimentary rocks. Izv Acad Sci USSR Ser Fisika Zemli 11:128–135

Gusev BV (1962) Paleomagnetic correlation of volcanic members of northern Siberian platform. In: Inform Sbornik Inst Geol Arktiki, N 29, pp 30–34

Gusev BV (1970) Field methods of Quaternary pebblestones age determination using viscous magnetization. In: Materialy VIII konf po postoyan geomagn polyu i paleomagn II. Naukova Dumka, Kiev, pp 50–54

Gusev BV (1973a) Paleomagnetic reconstruction of a relative situation of fold structures in the Russian and Siberian Platforms during the Precambrian. In: Materialy IX konf po vopr postoyan geomagn polya, magn gornykh porod i paleomagn III. Izd Acad Sci Azerb SSR, Baku, pp 55–56

Gusev BV (1973b) Reconstructions of the Precambrian anomalous magnetic field in the Russian and Siberian platforms. In: Geofiz metody razvedki v Arktike, N 8. Izd Inst Geol Ark, Leningrad, pp 49–51

Gusev BV (1974) Horizontal movements of the Earth's crust and the evolution of the Mesozoic depressions and adjoining structures in the central part of the Soviet Arctic. In: Geotekton predposylki k poiskam polezn iskop na shel'fe Severn Ledov okeana. Izd Inst Geol Arkt, Leningrad, pp 68–81

Hagstrum JT, Van der Voo R, Auvray B, Bonhommet N (1980) Eocambrian-Cambrian palaeomagnetism of the Armorican massif, France. Geophys J R Astron Soc 61(4):489–517

Haigh G (1958) The process of magnetization by chemical change. Philes Mag 3(27):267–286

Haile NS, Tarling DH (1975) Note on reconnaissance paleomagnetic measurements of Jurassic redbeds from Thailand. Pacif Geol 10:101–103

Halls HC (1978) The use of converging remagnetization circles in paleomagnetism. Phys Earth Planet Interiors 16(1):1–11

Hamilton N, King RF (1964) Comparison of the bedding errors of artificially and naturally deposited sediments with those predicted from a simple model. Geophys J R Astron Soc 8(4):370–374

Hargraves RB (1978) Problems in paleomagnetic synthesis illustrated by results from Permo-Triassic dolerites in Guyana. Phys Earth Planet Interior 16(3):277–284

Harrison CGA (1966) The paleomagnetism of deep sea sediments. J Geophys Res 71(12):3033–3043

Harrison CGA, Jarrard RD, Vacquier V, Larson RL (1975) Palaeomagnetism of Cretaceous Pacific Seamounts. Geophys J R Astron Soc 42(3):859–882

Harrison CGA, McDougall I, Watkins ND (1979) A geomagnetic field reversal time scale back to 13.0 million years before present. Earth Planet Sci Lett 42(2):143–152

Hedberg HD (ed) (1976) International stratigraphic guide. Wiley, New York, 200 pp

Heirtzler JR (1981) Marine magnetic anomaly studies during the 1970's; In: McElhinny MW, Valencio DA (ed) Geodynamics series, vol 2. Am Geophys Union Washington DC, Geol Soc Am Boulder, Colorado, pp 50–54

Heirtzler JR, Dickson GO, Herron EM, Pitman WC, Le Pichon X (1968) Marine magnetic anomalies, geomagnetic field reversals and motions of the ocean floor and continents. J Geophys Res 73(6):2119–2136

Heller F (1977) Palaeomagnetism of Upper Jurassic limestones from Southern Germany. Geophys J R Astron Soc 42(5):475–488

Helsley CE (1969) Magnetic reversal stratrigraphy of the Lower Triassic Moenkopi formation of Western Colorado. Geol Soc Am Bull 80:2431–2460

Herrero-Bervera E, Helsley CE (1983) Paleomagnetism of a polarity transition in the Lower(?) Triassic Chugwater Formation, Wyoming. J Geophys Res D88(4):3506–3522

Hijab BR, Tarling DH (1982) Lower Jurassic palaeomagnetic results from Yorkshire, England, and their implications. Earth Planet Sci Lett 60(1):147–154

Hillhouse JWA (1977) A method for the removal of rotational remanent magnetization acquired during alternating field demagnetization. Geophys J R Astron Soc 50(1):29–34

Hills ES (1953) Outlines of structural geology, 3rd edn revis. Methuen, London, 182 pp

Hoblitt R, Larson RL (1975) Paleomagnetic and geochronologic data bearing on the structural evolution of the northeastern margin of the Front Range, Colorado. Geol Soc Am Bull 86(2): 237–242

Hoffman KA (1983) Transitional behaviour of the geomagnetic field. J Geomagn Geoelec 37(1): 139–146

Horner F, Lowrie W (1981) Paleomagnetic evidence from Mesozoic carbonate rocks for the rotation of Sardinia. J Geophys 49(1):11–19

Hospers J (1953) Reversals of the main geomagnetic field. Proc Kon Nederl Akad Wet B 56: 467–476

Hospers J (1954) Magnetic correlation in volcanic districts. Geol Mag 91(5):352–360

Hospers J, Van Andel SI (1970) Paleomagnetism and tectonics, a review. Earth Sci Rev 6(5):5–44

Hower SC (1979) A Lower Ordovician geomagnetic pole from Pennsylvania. Earth Planet Sci Lett 44(1):65–72

Hussain A, Shult A, Soffel H (1979) Palaeomagnetism of the basalts of Wadi Abu Tereifiya Mandisha and dioritic dykes of Wadi Abu Shihat, Egypt. Geophys J R Astron Soc 56(1):55–61

Idnurm M, Senior BR (1978) Palaeomagnetic ages of Late Cretaceous and Tertiary weathered profiles in the Eromanga Basin, Queensland. Palaeogeogr Palaeoclim Palaeoecol 24(4):263–277

Imnadze ZA, Kitovani TG, Dzhashi OV, Grishanov AN, Yeremin VN, Molostovsky EA (1982) Stratigraphy and paleomagnetism of the reference sections of the Kimmerian and Pontian stages in the western Georgia. In: Paleomagn Stratigr mesokainoz otlozh Mater Vsesoyuzn seminara. Naukova Dumka, Kiev, pp 37–41

Irving E (1956) Palaeomagnetic and palaeoclimatological aspects of polar wandering. Geofis Pura Appl 33:23–41

Irving E (1957) The origin of the palaeomagnetism of the Torridonian sandstones of North-West Scotland. Phil Trans R Soc Lond Ser A 250:100–110

Irving E (1964) Palaeomagnetism and its application to geological and geophysical problems. Wiley, New York, 399 pp

Irving E (1966) Paleomagnetism of some Carboniferous rocks from New South Wales and its relation to geological events. J Geophys Res 71(24):6025–6051

Irving E (1967) Evidence for paleomagnetic inclination error in sediment. Nature 213(5075): 483–484

Irving E, Gaskell TF (1962) The palaeogeographic latitude of oil fields. Geophys J R Astron Soc 7(1):54–63

Irving E, Hastie J (1975) Catalogue of paleomagnetic directions and poles, 2nd issue. Precambrian results 1957–1974. Geomagn Serv Can Geomagn Ser 3:42 pp

Irving E, Major E (1964) Post-depositional detrital remanent magnetization in a synthetic sediment. Sedimentology 3(2):135–143

Irving E, Ward MA (1964) A statistical model of the geomagnetic field. Geofis Pura Appl 57:25–30

Irving E, North FK, Couilard R (1974) Oil, climate and tectonics. Can J Earth Sci 11(1):1–17

Irving E, Tanczyk E, Hastie J (1976a) Catalogue of paleomagnetic directions and poles, 3rd issue. Paleozoic results 1949–1975. Geomagn Serv Can Geomagn Ser 5:99 pp

Irving E, Tanczyk E, Hastie J (1976b) Catalogue of paleomagnetic directions and poles, 4th issue. Mesozoic results 1954–1975 and results from seamounts. Geomagn Serv Can Geomagn Ser 6: 70 pp

Irving E, Tanczyk E, Hastie J (1976c) Catalogue of paleomagnetic directions and poles, 5th issue. Cenozoic results 1927–1975. Geomagn Serv Can Geomagn Ser 10:87 pp

Irving E, Monger JWH, Yole RW (1980) New paleomagnetic evidence for displaced terranes in British Columbia. In: The continental crust and its mineral deposits. Geol Assoc Canada, Spec Pap N 20, pp 441–456

Ismail-Zade TA, Gasanov AZ (1973) Paleomagnetic studies of Paleogene deposits of Talysh. In: Postoyan magn pole Zemli, paleomagn i magn gornykh porod. Naukova Dumka, Kiev, pp 69–71

Ito H, Tokieda K, Suwa K, Kume S (1977) Palaeomagnetism of South African kimberlites. Prelim Rept Afr Stud Ngoya Univ N 2, pp 194–198

Ivanov LB (1975) Geotectonic grid of the Earth and the prerequisites for creation of a new general metallogenic theory. In: Informats sbornik UGN Acad Sci Kazakh SSR. Nauka, Alma-Ata, pp 65–75

Ivanov NA (1968) Application of paleomagnetic methods for difinition of certain tectonic elements in the rocks of the Urals. In: Glubinnoye stroyeniye Urala. Nauka, Moscow, pp 52–54

Ivkin SV, Aparin VP (1980) Application of the magnetic anisotropy characteristics for determination of the geomagnetic field paleointensity. Izv Acad Sci USSR Ser Fisika Zemli 6:101–106

Jacobs JA (1975) The Earth Core. Academic Press, London, 307 pp

Johnson EA, Murphy T, Torreson OW (1948) Pre-history of the Earth magnetic field. Terr Magn Atmos Elect 53(4):349–372

Jones M, Van der Voo R, Bonhommet N (1979) Late Devonian to early Carboniferous palaeomagnetic poles from the Armorican Massif, France. Geophys J R Astron Soc 58(2):287–308

Kalashnikov AG (1961) History of the geomagnetic field (according to paleomagnetic data). Izv Acad Sci USSR Ser Geofiz 9:1244–1279

Karakhanyan AK (1981) Results of paleomagnetic studies of Paleogene in Armenia. In: II Vsesoyuzn s'yezd Postoyan geomagn pole, magn gornykh porod i paleomagn II. Izd Tbilisi Univ, Tbilisi, p 130

Karasik AM (1978) Anomalous magnetic field of oceans. In: Tectonosfera Zemli. Nauka, Moscow, pp 302–314

Katterfeld GN (1962) The Earth's image and its origin. Geografgiz, Moscow, 152 pp

Kawai N, Ito H, Kume S (1961) Deformation of the Japanese islands as inferred from rock magnetism. Geophys J R Astron Soc 6(1):124–130

Kawai N, Nakajima T, Hirooka K (1971) The evolution of the Island arc of Japan and the formation of granites in the Circum-Pacific belt. J Geomagn Geoelectr 23(3):267–293

Kellogg KS, Reynolds RL (1978) Palaeomagnetic results from Lassiter Coast, Antarctica. J Geophys Res 83(B5):2293–2299

Kent DV (1979) Paleomagnetism of the Devonian Onondaga limestone revisited. J Geophys Res 84(B7):3576–3578

Kent DV, Opdyke ND (1980) Paleomagnetism of Siluro-Devonian rocks from eastern Maine. Can J Earth Sci 17(12):1653–1665

Khaburzaniya IA, Shubladze RL, Edilashvili VYA, Sholpo LE (1979) The paleomagnetic section of volcanogenic rocks in Southern Georgia. Izv Acad Sci USSR Ser Fisika Zemli 4:71–78

Khain VE (1957) Recent tectonic hypotheses. Bull Moscow Ova Ispyt Prir Otd Geol 30(2):137–138

Khain VE (1971) Regional geotectonics. Nedra, Moscow, 548 pp

Khain VE (1978) Tectonics of lineaments and neomobilism. Geol Balcan 3:3–6

Khramov AN (1958) Paleomagnetic correlation of sedimentary strata. Gostoptekhizdat, Leningrad, 218 pp

Khramov AN (1963) Paleomagnetic studies of the Upper Permian and Lower Triassic in the north and east of the Russian platform. In: Paleomagn strat issled. Trudy VNIGRI, N 204, Gostoptekhizdat, Leningrad, pp 118–143

Khramov AN (1965) Compensation of the secondary magnetization of rocks. In: Nastoyashcheye i proshl magn polya Zemli. Nauka, Moscow, pp 197–203

Khramov AN (1968) Detrital magnetization of fine dispersed sediments. Izv Acad Sci USSR Ser Fizika Zemli 1:115–118

Khramov AN (1970) Horizontal movements in the Earth's crust and palaeomagnetism. Phys Earth Planet Interiors 2(5):367–372

Khramov AN (ed) (1971) Paleomagnetic directions and paleomagnetic poles (reference book for the USSR). Trudy VNIGRI, Leningrad, 124 pp

Khramov AN (ed) (1973) Paleomagnetic directions and paleomagnetic poles. Reference book for the USSR, Issue 2. Materials of the WDC-B, Soviet Geophys Committee, Moscow, 88 pp

Khramov AN (ed) (1975) Paleomagnetic directions and paleomagnetic poles. Reference book for the USSR, Issue 3. Materials of the WDC-B, Soviet Geophys Committee, Moscow, 43 pp

Khramov AN (1978) Paleomagnetism and geotectonic problems. In: Tectonosfera Zemli. Nauka, Moscow, pp 280–302

Khramov AN (ed) (1979) Paleomagnetic directions and paleomagnetic poles. Reference book for the USSR, Issue 4. Materials of the WDC-B, Soviet Geophys Committee, Moscow, 51 pp

Khramov AN (ed) (1982) Paleomagnetic directions and paleomagnetic poles. Reference book for the USSR, Issue 5. Materials of the WDC-B, Soviet Geophys Committee, Moscow, 47 pp

Khramov AN (1983) Global reconstructions of the ancient craton positions in the Late Precambrian. In: Paleomagn Verkhnego dokembr USSR. Izd VNIGRI, Leningrad, pp 127–137

Khramov AN (1984) The paleomagnetic method. In: Nikitin IF, Zhamoyda AI (eds) Praktich stratigraphiya. Nedra, Leningrad, pp 139–163
Khramov AN, Kravchinsky AY (1984) Geomagnetic and geotectonic cyclicity. In: 27th Int Geol Congr Repts, vol 8. Geophysics. Nauka, Moscow, pp 161–169
Khramov AN, Pechersky DM (1982) Paleomagnetism and tectonics. In: Sovremen sostoyan issled v oblasti geomagn. Nauka, Moscow, pp 128–143
Khramov AN, Rodionov VP (1977) The problem of Laurasia in the Early and Middle Paleozoic in the light of paleomagnetic data. In: Paleomagnetism i voprosy tektoniki plit. Trudy VNIGRI, N 394, Leningrad, pp 108–140
Khramov AN, Rodionov VP (1980a) Paleomagnetism and reconstruction of palaeogeographic positions of the Siberian and Russian plates during the Later Proterozoic and Palaeozoic. J Geomagn Geoelectr 32 (Suppl 3):23–27
Khramov AN, Rodionov VP (1980b) The geomagnetic field during Palaeozoic time. J Geomagn Geoelectr 32 (Suppl 3):99–115
Khramov AN, Sholpo LE (1967) Paleomagnetism. Principles, methods and geological applications of paleomagnetology. Nedra, Leningrad, 252 pp
Khramov AN, Rodionov VP, Komissarova RA (1965) New data on the history of the Earth's magnetic field in Paleozoic in the USSR. In: Nastoyashcheye i proshloye magn polya Zemli. Nauka, Moscow, pp 206–213
Khramov AN, Goncharov GI, Komissarova RA, Osipova EP, Pogarskaya IA, Rodionov VP, Slautsitais IP, Smirnov LS, Forsh NN (1974) Paleomagnetism of Paleozoic. Nedra, Leningrad, 238 pp
Khramov AN, Pechersky DM, Molostovsky EA, Rodionov VP (1981) Paleomagnetic scale of Paleozoic and Mesozoic of the USSR. In: Magnitostr i geodynamika. Radio i svyaz', Moscow, pp 38–51
Khramov AN, Gurevich EL, Komissarova RA, Osipova EP, Pisarevsky SA, Rodionov VP, Slautsitais IP (1985) Paleomagnetism, microplates and Siberian plate consolidation. J Geodyn 2(2/3):127–139
Khristov EV, Mikolaychuk AV (1983) On the pre-geosynclinal crustal basement in the Fergana-Koshan Hercynides. Geotektonika 3:76–86
King RF (1955) Remanent magnetism of artificially deposited sediments. Mon Notic R Astron Soc Geophys Suppl 7(1):115–134
Kittel N (1956) Introduction to solid state physics, 2nd edn. New York, 591 pp
Klemme HD (1975) Giant oil fields related to their geologic setting a possible guide to exploration. Bull Can Petr Geol 23:30–60
Klootwijk CT (1979a) A review of paleomagnetic data from the Indo-Pakistan fragment of Gondwanaland. In: Farah A, De Jong KA (eds) Geodynamics of Pakistan. Geol Surv Pakistan, Quetta, pp 41–80
Klootwijk CT (1979b) A summary of paleomagnetic data from Extrapeninsular Indo-Pakistan and South Central Asia: Implications for collision tectonics. In: Saklani PS (ed) Structural geology of the Himalaya. Today and Tomorrow, Delhi, pp 307–360
Klootwijk CT, Bingham DK (1980) The extent of Greater India III. Palaeomagnetic data from the Tibetan sedimentary series, Thakkhola region, Nepal Himalaya. Earth Planet Sci Lett 51:381–405
Kochegura VV, Balesta ST (ed) (1979) Problems of studies of paleosecular variations of the Earth's magnetic field. Izd DV Nauch. Tsentra Acad Sci USSR, Vladivostik, 128 pp
Kođi A (1973) Paleomagnetism of the Quaternary sediments of the locality Cervený Kopec (Red Hill). Stud Geophys Geodaet 17(3):232–239
Kođi A (1974) Paleomagnetic investigation of sediments. In: Quaternary glaciations in northern hemisphere. Report No 1. Prague, pp 110–120
Koenigsberger JG (1930) Größenverhältnis von Remanenten zu induzierten Magnetismus in Gesteinen, Größe und Richtung des remanenten Magnetismus. Z Geophys 6(4–7):190–207
Komissarova RA (1969) On the age and origin of magnetization of the Katav Formation limestones in the Southern Urals. In: Magnet gornykh porod i paleomagnet. Izd Inst Fisiki Zemli, Moscow, pp 149–152
Komissarova RA, Mikhaylov PN (1981) Paleomagnetism of the Riphean and Vendian sedimentary strata in the western slope of the Southern Urals. In: Paleomagn i voprosy paleogeogr. Trudy VNIGRI, Leningrad, pp 35–45

Kono M (1972) Mathematical models of the Earth's magnetic field. Phys Earth Planet Interiors 5:140–150

Kono M (1973) Uniqueness of the spherical harmonic analysis of the geomagnetic field based on the inclination and declination data. Rock Magnetism Paleogeophys 1:118–123

Kono M, Ueno N (1977) Paleointensity determination by a modified Thellier method. Phys Earth Planet Interiors 13(4):305–314

Kopystyansky RS (1978) Rock fractures and its role in oil and gas geology. Naukova Dumka, Kiev, 216 pp

Kravchenko KN (1979) Tectonic evolution of the Tien Shan, Pamir and Karakorum. In: Farah A, De Jong KA (eds) Geodynamics of Pakistan. Geol Surv Pakistan, Quetta, pp 25–40

Kravchinsky AY (1970) Characteristics of the late Precambrian and early Ordovician geomagnetic reversals in the Siberian platform. In: Trudy Irkutsk Plitekhn. Inst i Irkutsk geol upr, N 51, Irkutsk, pp 135–138

Kravchinsky AY (1977) Paleomagnetic and paleogeographic resettings in the Precambrian platforms. Nedra, Moscow, 102 pp

Kravchinsky AY, Dovydov VF (1972) Construction of a detailed paleomagnetic section of the late Cambrian strate in the Irkutsk amphitheatre. In: Neftegazonosnost' yuga Vost Sibiri. Nedra, Moscow, pp 19–25

Krasovsky FN (1953) Review and results of recent latitude measurements. In: Izbrannye sochinaniya vol 1. Izd Geodezich Liter, Moscow, pp 87–183

Krs M (1966) The palaeomagnetism and palaeomagnetic stability of rocks of some formations of the Bohemian massif (Devonian, Carboniferous, Triassic, Tertiary, Quaternary). Geofis Sbor 251:339–358

Krs M (1969) Paleomagnetism. Akademia, Praha, 207 pp

Krs M, Krsova M, Pruner P, Havlíček V (1983) Palaeomagnetism, palaeogeography and multicomponent analysis of magnetization of Ordovician rocks from the Barrandian area of the Bohemian Massif. Preprint Ustav Uzité geofysiky, Praha, 24 pp

Kruczyk J, Jelenska M, Kadzialko-Hofmokl M (1980a) Paleomagnetism of Jurassic sediments from the western Tatra Mts. Acta Geophys Polonica 28(4):297–307

Kruczyk J, Kadzialko-Hofmokl M, Jelenska M (1980b) Paleomagnetism of the Jurassic sediments from the Cracov upland. Acta Geophys Plonica 28(4):309–322

Krupička S (1973) Physik der Ferrite und der verwandten magnetischen Oxide. Academia, Praga, B2, 501 pp

Krutikhovskaya ZA, Zavoysky BN, Podolyanko SM, Savenko TY (1964) Rock magnetization of iron-formations in the area of Bolshoy Krivoy Rog and Kursk Magnetic Anomaly. Naukova Dumka, Kiev, 179 pp

Kudryavtseva GP, Garanin VK, Zhilyaeva VA, Trukhin VI (1982) Magnetism and mineralogy of natural ferrimagnetic substances. Izd Moscow Univ, Moscow, 294 pp

Kulikova LS (1980) Definition of magnetochronologic reference levels according to paleomagnetic studies of the late Pleistocene strata of the Molodova V an Korman' IV sections. In: Geokhronologiya chetvertichn perioda. Nauka, Moscow, pp 116–130

Kulikova LS, Pospelova GA (1976) Secular variations of the geomagnetic field in the late Pleistocene from studies of the alluvial deposits in the Ob' River area. In: Paleomagn Mesoz i Kaynos Sibiri i Dal'n Vost. Izd Sib Otd Acad Sci USSR, Novosibirsk, pp 96–112

Kume S (1962) Sur des changements d'aimantation rémanent de corps ferromagnétiques soumis a des pressions hydrostatiques. Am Géophys 18(1):18–22

Kumpan AS, Rusinov BS, Sholpo LE (1968) Results of paleomagnetic studies in Central Kazakhstan. Izv Acad Sci USSR Ser Fisika Zemli 11:96–103

LeBreque JL, Kent DV, Cande SC (1977) Revised magnetic polarity time scale for Late Cretaceous and Cenozoic time. Geology 5:330–335

Lapointe PL (1979) Paleomagnetism and orogenic history of the Botwood Group and Mount Peyton Batholith, Central Mobile Belt, Newfoundland. Can J Earth Sci 16(4):866–876

Larochelle A (1967) Paleomagnetic directions of a basic sill in Prince Edward Island. Geol Surv Can N 1–6, paper 39

Larson RL, Helsley CE (1975) Mesozoic reversal sequence. Rev Geophys Space Phys 13(3):174–176

Larson RL, Hilde TWC (1975) A revised times scale of magnetic reversals for the early Cretaceous and late Jurassic. J Geophys Res 80(17):2586–2594
Larson RL, Pitman WC (1972) World-wide correlation of Mesozoic magnetic anomalies and its implications. Geol Soc Am Bull 83:3645–3668
Lauer JP (1981) Origine méridionale deş Pontides d'après de nouveaux résultats paléomagnétiques obtenus en Turquie. Bull Soc Géol France Ser 7, 23(6):619–624
Lawley EA (1970) The intensity of the geomagnetic field in Iceland during Neogene transitions and systematic deviations. Earth Planet Sci Lett 10(2):145–149
Lazarev PP (1924) On the method of deformation of the Earth's magnetic field intensity in geologic epochs. Dokl Rossiyskoy Akademii Nauk Ser A 10–11:183
Lee C, Lee H, Lio H, Lio G, Yeh S (1963) Preliminary study of paleomagnetism of some Mesozoic and Cenozoic redbeds of South China. Acta Geol Sin 43:261–264
Le Pichon X (1968) Sea-floor spreading and continental drift. J Geophys Res 73(12):3661–3697
Le Pichon X, Francheteau J, Bonnin J (1973) Plate Tectonics. Elsevier, Amsterdam, 300 pp
Lerbekmo JF, Evans ME, Baadsgaard H (1979) Magnetostratigraphy, biostratigraphy and geochronology of Cretaceous-Tertiary boundary sediments, Red Deer Valley. Nature 279)5708): 27–30
Leybenzon LS (1955) Deformation of an elastic sphere with reference to the Earth's structure. In: Sobraniye trudov, vol 4 (Gidroaerodynam, Geofiz). Izd Acad Sci USSR, pp 186–266
Lichkov BL (1965) On the principles of the Earth's recent theory. Leningr Univ, Leningrad, 120 pp
Lin'kova TI (1976) Stratigraphy of deep-sea bottom sediments in the Pacific Ocean based on paleomagnetic data. In: Geol-geofis issled v yugo-vost chasti Tikhogo okeana. Nauka, Moscow, pp 234–240
Lin'kova TI (1984) Paleomagnetism of the Upper Cenozoic sediments of the World Ocean. Nauka, Moscow, 132 pp
Lordkipanidze MB, Adamiya ShA, Asanidze BZ (1984) Evolution of the active margins of the Tethys Ocean (with reference to the Caucasus). In: 27 IGC Col 03 Paleookeanologiya Dokl, vol 3. Nauka, Moscow, pp 72–83
Lowrie W, Alvarez W (1975) Paleomagnetic evidence for rotation of the Italian peninsula. J Geophys Res 80(11):1579–1592
Lowrie W, Alvarez W (1977) Upper Cretaceous-Paleocene magnetic stratigraphy at Gubbio, Italy III. Upper Cretaceous magnetic stratigraphy. Geol Soc Am Bull 88:374–377
Lozhkina NV (1982) Magnetic properties of the early Mesozoic sediments and the paleomagnetic section in the north-eastern part of the USSR. In: Metody paleomagn v reshen geol zadach na primere Dal'nego Vostoka. Izd DV Nauch Tsentra Acad Sci USSR, Vladivostok, pp 13–29
Luck GK (1972) Palaeomagnetic results from Palaeozoic sediments of Northern Australia. Geophys J R Astron Soc 28(5):475–487
Maack R (1957) Über Vereisungsperioden und Vereisungsspuren in Brasilien. Geol Rundsch 45(3): 547–595
Magnetostratigraphic polarity units-a supplementary chapter of the ISSC International Stratigraphic Guide (1979) Geology 7(12):578–583
Mäkel GH, Rondell HE, Vandenberg J (1984) Triassic paleomagnetic data from the Subbetic and Malaguide complex of the Betic Cordilleras (Southeast Spain). Tectonophysics 101(1/2):131–141
Malkus WVR (1963) Precessional torques as the cause of geomagnetism. J Geophys Res 68(10): 2871–2886
Manzoni M, Nanni T (1977) Paleomagnetism of Ordovician lamprophyres from Taylor Valley Victoria Land, Antarctica. Pure Appl Geophys 115(4):961–977
Martin DL, Nairn AEM, Noltimier HC, Petty MH, Schmitt TJ (1978) Paleozoic and Mesozoic paleomagnetic results from Morocco. Tectonophysics 44(1/4):91–114
Márton E, Márton P (1983) A refined apparent polar wander curve for the Transdanubian Central Mountains and its bearing on the Mediterranean tectonic history. Tectonophysics 98(1/2):43–57
Matveyev AK (ed) (1972) Map of coal deposits of the world 1:7.500.000. 21 maps. Moscow Univ, Moscow

McDougall I, Watkins ND, Walker PL, Kristjansson L (1976) Potassium-Argon and paleomagnetic analysis of Icelandic lava flows: limits on the age of anomaly 5. J Geophys Res 81:1505–1512
McElhinny MW (1971) Geomagnetic reversals during the Phanerozoic. Science 172(3979):157–159
McElhinny MW (1972) Palaeomagnetic directions and pole positions XIII. Pole numbers 13/1 to 13/94. Geophys J R Astron Soc 30(3):281–293
McElhinny MW (1973) Palaeomagnetism and plate tectonics. Cambridge Univ Press, London, 358 pp
McElhinny MW (1981) Plate tectonics and continental drift in Precambrian. In: Magnitostratigraphiya i geodinamika, Radio i svyaz', Moscow, pp 106–108
McElhinny MW, Cowley JA (1977) Palaeomagnetic directions and pole positions XIV. Pole numbers 14/1 to 14/574. Geophys J R Astron Soc 49(2):313–356
McElhinny MW, Cowley JA (1978) Palaeomagnetic directions and pole positions XV. Pole numbers 15/1 to 15/232. Geophys J R Astron Soc 52(2):259–276
McElhinny MW, Cowley JA (1980) Palaeomagnetic directions and pole position XVI. Pole numbers 16/1 to 16/296. Geophys J R Astron Soc 61(3):549–571
McElhinny MW, Embleton BJJ (1976) The palaeoposition of Madagascar: remanence and magnetic properties of the late Palaeozoic sediments. Earth Planet Sci Lett 31(2):101–112
McElhinny MW, Luck GR (1970) The palaeomagnetism of the Antrim Plateau volcanics of northern Australia. Geophys J R Astron Soc 20(2):191–203
McElhinny MW, Merrill RT (1975) Geomagnetic secular variation over the past 5 m.y. Rev Geophys Space Phys 13(5):687–708
McElhinny MW, Opdyke ND (1968) The paleomagnetism of some Carboniferous glacial varves from Central Africa. J Geophys Res 73(2):689–696
McElhinny MW, Opdyke ND (1973) Remagnetization hypothesis discounted: a palaeomagnetic study of the Trenton limestone, New York state. Geol Soc Am Bull 84(11):3697–3708
McElhinny MW, Valencio DA (eds) (1981) Paleoreceonstruction of the continents. Geodynamic series. vol 2. Am Geophys Union, Washington, Boulder, 194 pp
McElhinny MW, Luck GR, Edwards D (1971) A large volume magnetic field free space for thermal demagnetization and other experiments in paleomagnetism. Pure Appl Geophys 90(7):126–130
McElhinny MW, Embleton BJJ, Daly L, Pozzi JP (1976) Paleomagnetic evidence for the location of Madagascar in Gondwanaland. Geology 4(8):455–457
McElhinny MW, Cowley JA, Edwards DJ (1978a) Palaeomagnetism of some rocks from peninsular India and Kashmir. Tectonophysics 50(1):41–54
McElhinny MW, Taylor SR, Stevenson DJ (1978b) Limits to the expansion of Earth, Moon, Mars and Mercury and to changes in the gravitational constant. Nature 271(5643):316–321
McElhinny MW, Embleton BJJ, Ma XH, Zhang ZK (1981) Fragmentation of Asia in the Permian. Nature 293(5829):212–216
McMahon BE, Strangway DW (1968) Investigation of Kiaman magnetic division in Colorado redbeds. Geophys J R Astron Soc 15(3):265–285
McNish AG, Johnson EA (1938) Magnetization of unmetamorphosed varves and marine sediments. Terr Magnet Atmos Electr 43(4):401–407
Melloni M, Arago M (1853) Sur l'aimantation des roches volcaniques. CR Acad Sci Paris 37:229–231
Mel'nikov TN, Khisina NR (1976) Spinodal decay and the related partial self-reversal of magnetization in titanomagnetites of the African rift zones. Izv Acad Sci USSR Ser Fisika Zemli 10: 84–91
Mercanton PL (1926) Aimantation de roches volcaniques australiennes. CR Acad Sci Paris 182: 1231–1232
Mercanton PL (1932) Inversion de l'inclinasion magnêtique aux âges geologiques Nouvelles observationes. CR Acad Sci Paris 194(16):1371
Metallova VV, Shashkanov VA (1978) Results of determination of the geomagnetic field paleointensity at the baked contact by the stepwise remagnetization method. Izv Acad Sci USSR Ser Fisika Zemli 2:110–112

Metallova VV, Pisarevsky SA, Rodionov VP, Khramov AN, Lipatov SN (1980) Change of the geomagnetic field intensity during the late Cambrian reversals. Izv Acad Sci USSR Ser Fisika Zemli 10:65–70

Metallova VV, Iosifidi AG, Mostryukov AO, Rodionov VP, Khramov AN (1984a) Changes in the geomagnetic field intensity during the reversals in the early Ordovician. Izv Acad Sci USSR Ser Fisika Zemli 10:66–70

Metallova VV, Mostrukov AO, Iositidi AG, Rodionov VP, Khramov AN (1984b) Secular variation of geomagnetic field intensity during polarity transitions. Geophys J R Astron Soc 76(2):289–298

Mikhailova NP, Tsykora VN, Glevasskaya AM (1970) Thermomagnetic studies of N- and R-magnetized effusives from the Trans-Carpathian Neogene depression. Geofiz Sbornik 36:73–84

Mikhailova NP, Glevasskaya AM, Tsykora VN (1974) Paleomagnetism of volcanogenic rocks and reconstruction of the geomagnetic field during Neogene. Naukova Dumka, Kiev, 252 pp

Minasyan DO (1973) About the geomagnetic field intensity in the late Pliocene and middle Quaternary time in Armenia. In: Materialy IX konf po vopr postoyan geomagn polya, magn gornykh porod i paleomagn III. Izd Acad Sci Azerb SSR, Baku, pp 106–107

Minster JB, Jordan TH, Molnar P, Haines E (1974) Numerical modelling of instantaneous plate tectonics. Geophys J R Astron Soc 34(3):541–576

Molostovsky EA (1970) Paleomagnetic data on the Upper Permian and Triassic stratigraphy in the eastern Russian platform and in the Fore-Urals depression. In: Materialy VIII konf po postoyan geomagn polyu i paleomagn II. Naukova Dumka, Kiev, pp 98–101

Molostovsky EA (1983a) New data on the paleomagnetic scale of the USSR and some general problems of magnetostratigraphy. In: Sovremen sost issled v oblasti geomagn. Nauka, Moscow, pp 143–162

Molostovsky EA (1983b) Paleomagnetic stratigraphy of the Upper Permian and Triassic in the eastern European part of the USSR. Izd Saratov Univ, Saratov, 167 pp

Molostovsky EA, Khramov AN (1984) Paleomagnetic scale for Phanerozoic and the problems of magnetostratigraphy. In: 27th Intern Geol Congr Rep, vol 1. Nauka, Moscow, pp 16–23

Molostovsky EA, Pevzner MA, Pechersky DM, Rodionov VP, Khramov AN (1976) Magnetostratigraphic scale of Phanerozoic and regime of reversals of the geomagnetic field. Geomagn issled 17 Sov Radio, Moscow, pp 45–52

Moody JD, Hill MJ (1956) Wrench-fault tectonics. Geol Soc Am Bull 67(9):1207–1246

Morel P, Irving E (1978) Tentative palaeocontinental maps for the Early Phanerozoic and Proterozoic. J Geol 86(5):535–561

Mörner NA (1974) Magnetic changes within the late Brunhes epoch. In: Late Cenozoic Magnetostratigraphy. Workshop Conf Nat Sci Mus, Tokyo and Otsu, pp 29–31

Mörner NA (1981) Paleomagnetism of the Maastrichtian/Danian stratotype. In: Programs and abstracts 4th IAGA Sci Assembly, Edinburgh August 3–15. Paris, p 212

Morris WA (1976) Palaeomagnetic results from Lower Paleozoic of Ireland. Can J Earth Sci 13(2):294–304

Morris WA, Briden JC, Piper JDA, Sallomy JT (1973) Palaeomagnetic studies in the British Caledonides V. Miscellaneous new data. Geophys J R Astron Soc 34(1):69–105

Morris WA, Carmichael CM (1978) Paleomagnetism of some late Precambrian and lower Paleozoic sediments from L'Adrar de Mauritanie, West Africa. Can J Earth Sci 15(2):253–262

Muratov MV (ed) (1977) Faults in the Earth's crust. Nauka, Moscow, 191 pp

Murthy GS (1983a) Paleomagnetism of the Deadman's Bay diabase dikes from northeastern central Newfoundland. Can J Earth Sci 20(2):195–205

Murthy GS (1983b) Paleomagnetism of diabase dikes from the Bonavista Bay area of northeastern central Newfoundland. Can J Earth Sci 20(2):206–216

Nagata T (1953) Self-reversal of thermo-remanent magnetization of igneous rocks. Nature 172 (4384):850–852

Nagata T (1961) Rock magnetism. Maruzen company LTD, Tokyo, 350 pp

Nagata T (1965) Low temperature characteristics of rock magnetism. J Geomagn Geoelectr 17 (3–4):315–324

Nairn AEM (1960) Paleomagnetic results from Europe. J Geol 68(3):285–306

Nalivkin VD (1962) On the cyclicity of geologic history. In: Geographich sbornik. Izd Acad Sci USSR Moscow-Leningrad, vol 15, pp 188–197
Nalivkin VD, Pozner VM (1975) Paleogeography of the USSR, vol 2. Devonian, Carboniferous and Permian periods. Nedra, Moscow, 179 pp
Nazarov K (1973) Paleomagnetic horizons in the Jurassic and Cretaceous strata of Turkmenia. In: Materialy IX konf po vopr postoyan geomagn polya, magn gornykh porod i paleomagn III. Izd Acad Sci Azerb SSR, Baku, pp 116–118
Nazarov K, Davtyan MT (1970) Paleomagnetic division and correlation of Paleogene. Cretaceous and Jurassic strata of Turkmenistan. In: Materialy VIII konf po postoyan geomagn polyu i paleomagn II. Naukova Dumka, Kiev, pp 101–103
Néel L (1951) L'inversion de l'aimantation permanente des roches. Ann Géophys 7(2):90–103
Néel L (1955) Some theoretical aspects of rock magnetism. Adv Phys 4(14):191–243
Ninkovich D, Opdyke ND, Heezen BC, Foster JH (1966) Paleomagnetic stratigraphy, rates of deposition and tephrachronology in North Pacific deep-sea sediments. Earth Planet Sci Lett 1(6): 476–492
Nozharov P, Petkov N (1976) Paleomagnetism and paleomagnetic stability of certain magmatic rocks in the Majarovo ore field. Bulg Geophys J 2(4):90–103
Nurgaliyev DK (1983) Methods of studies of the natural remanent magnetization directions in sedimentary rocks. Synopsis of thesis. Izd Inst Geophys Acad Sci Ukraine SSR, Kiev, 16 pp
Oil-and-gas-bearing deposits and global tectonics (1978) Nedra, Moscow, 237 pp
Opdyke ND, Henry KW (1969) A test of the dipole hypothesis. Earth Planet Sci Lett 6:138–151
Osipova EP (1983) Problem of paleomagnetic correlation of Upper Precambrian of Siberian and North American platforms. In: Paleomagnetizm verkhnego dokembriya SSRR. Trudy VNIGRI, Leningtad, pp 89–98
Otofuji Y, Matsuda T (1982) Paleomagnetism of igneous rocks of Late Cretaceous to Miocene age from to Go River area – clockwise rotation of Southwest Japan. In: Rock magnetism and paleogeophysics, vol 9. Tokyo, pp 41–48
Owens WH (1978) Towards an explanation of rotational remanence. Geophys J R Astron Soc 53(1):155
Ozima M, Ozima M, Nagata T (1964) Low temperature treatment as an effective means of "magnetic cleaning" of natural remanent magnetization. J Geomagn Geoelectr 16(1):37–40
Pacca JG, Hioto EY (1976) Paleomagnetic analysis of Mesozoic Serra Geral basaltic lawa flows in Southern Brazil. Ann Accad Brasil Ciênc 48 (Suppl):207–214
Panasenko OM (1981) Old rift zones in the south-western margin of the Turan plate. In: Zakonomern tecton struct. Sredney Asii, Donish, Dushanbe, pp 117–120
Patrascu S, Costa-Foru A, Apostol E (1964) Récherches paléomagnetiques sur les roches d'age Paléozoique en Romanie. Rev Roumanie Geol Geophys Geogr Ser Geophys 8:103–109
Patterson PE, Rosenbaum JG, Larson EE (1978) Refinement of a Cambro-Ordovician paleomagnetic pole from plutonic rocks in southwestern Colorado. EOS Trans AGU 59(43):1060
Pechersky DM (1969) Paleomagnetism of Mesozoic rocks of the north-eastern Asia. In: Magnetism porod i paleomagnetism. Inst Fisiki Zemli Acad Sci USSR, Moscow, pp 142–144
Pechersky DM (1970) Paleomagnetic investigations of Mesozoic rocks of the North-East of the USSR. Izv Acad Sci USSR Ser Fisika Zemli 6:69–83
Pechersky DM, Bagin VI, Brodskaya SY, Sharonova ZV (1975) Magnetism and environments of igneous rock formation. Nauka, Moscow, 288 pp
Peive AB (1956) General characteristics, classification and location of deep faults (fractures). Izv Acad Sci USSR Geol Ser 1:90–106
Pen'kov AV (1970) Some of the principles of paleomagnetic correlation applied in studies of the Neogenic strata of the southern Tajikistan. In: Materialy VIII konf po postoyan geomagn polyu i paleomagn II. Naukova Dumka, Kiev, pp 109–112
Pen'kov AV, Gamov LN (1980) Paleomagnetic reference levels in the Pliocene-Quaternary strata of the Southern Tajikistan. In: Granitsa neogena i chetvert systemy. Nauka, Moscow, pp 184–194
Perfil'yev AS (1979) Formation of the Earth's crust in the region of the Urals eugeosyncline (Trudy Geol Inst Acad Sci USSR N 328). Nauka, Moscow, 188 p

Pergament MA, Pechersky DM, Khramov AN (1971) On the paleomagnetic scale of Mesozoic. Izv Acad Sci USSR Ser Geol 10:3–11
Permyakov SN (1949) Tectonic jointing of the Russian platform In: Mater k poznan geologich stroyeniya USSR. Izd Mosc Ova Ispyt Prir, nov ser Issue 12(16), 215 pp
Perroud H, Beonhommet N (1981) Paleomagnetism of the Ibero-Armorican arc and the Hercynian orogeny in Western Europe. Nature 292(5822):445–448
Perroud H, Bonhommet N, Van der Voo R (1983a) Paleomagnetism of the Ordovician dolerites of the Crozon peninsula (France). Geophys J R Astron Soc 72(2):307–319
Perroud H, Van der Voo R, Bonhommet N (1983b) Timing of the Paleozoic drift of the Armorica plate on the basis of new paleomagnetic results. EOS Trans AGU 64(5):218
Pesonen LJ, Neuvonen KJ (1981) Palaeomagnetism of the Baltic Shield-Implications for Precambrian tectonics. In: Precambrian plate tectonics. Elsevier, Amsterdam, vol 25, pp 623–648
Petersen N, Heller F, Lowrie W (1981) Magnetostratigraphy of the Cretaceous-Tertiary boundary. In: Programs and abstracts 4th IAGA Sci Assembly, Edinburgh, August 3–15 Paris, p 211
Peterson DN, Nairn AEM (1971) Paleomagnetism of Permian redbeds from southwestern United States. Geophys J R Astron Soc 23(2):191–205
Petrova GN (1977) Laboratory methods in paleomagnetic studies. In: Geomagn issled N 19. Sov Radio, Moscow, pp 40–49
Petrova GN, Bagina OL (1976) On the non-heating method for determination of intensity of the ancient geomagnetic field. Izv Acad Sci USSR Ser Fisika Zemli 4:54–62
Petrova GN, Bagina OL (1984) Paleointensity of the geomagnetic field in the late Cretaceous by method with and without heating. Izv Acad Sci USSR Ser Fisika Zemli 1:87–99
Petrova GN, Burlatskaya SP (1979) Actual concepts on secular variations. In: Problemy izuch paleovekov variats magnits polya Zemli. DV Nauch Tsentr Acad Sci USSR, Vladivostok, pp 5–15
Petrova GN, Khramov AN (1970) Paleomagnetism and the structure of the Earth's interior. Izv Acad Sci USSR Ser Fisika Zemli 4:65–83
Petrova GN, Bagina OL, Solodovnikov GM (1979) Determination of intensity of the Pliocene-Quaternary geomagnetic field in Armenia by the Thellier and H_e methods. Izv Acad Sci USSR Ser Fisika Zemli 1:101–106
Petrova GN, Bucha VV, Gamov LN, Gurary GZ, Davydov VF, Ismail-Zade TA, Kalinin YD, Kravchinsky AYa, Pospelova GA, Rodionov VP (1972) Typical aspects of transitional regimes of the geomagnetic field. Izv Acad Sci USSR Ser Fisika Zemli 6:53–75
Pevzner MA (1970) Paleomagnetic studies of the Pliocene-Quaternary terraces of the Dniester River. In: Materialy VIII konf po postoyan geomagn polyu i paleomagn II. Naukova Dumka, Kiev, pp 103–109
Pevzner MA, Chikovani AA (1978) Paleomagnetic studies of the Upper Miocene and Lower Pliocene marine sediments of the Taman' Peninsula. Izv Acad Sci USSR Ser Geol 8:61–66
Pevzner MA, Semenko VN (1979) Correlation of the Upper Miocene and Pliocene of the Pontian-Caspian region according to biostratigraphic and paleomagnetic data. Izv Acad Sci USSR Ser Geol 1:3–8
Piper JDA (1976) Magnetic properties of Precambrian pillow lavas of the Mona complex and related dyke swarm, Anglesey, Wales. Geol J 11(2):189–201
Piper JDA (1978a) Palaeomagnetic survey of the (Palaeozoic) Shelve Inlier and Berwyn Hills, Welsh Borderlands. Geophys J R Astron Soc 53(2):355–371
Piper JDA (1978b) Paleomagnetism and paleogeography of the Southern Uplands block in Ordovician times. Scot J Geol 14(2):93–107
Piper JDA (1979) Aspects of Caledonian palaeomagnetism and their tectonic implications. Earth Planet Sci Lett 44(2):176–192
Piper JDA (1982a) The Precambrian palaeomagnetic record: the case for the Proterozoic supercontinent. Earth Planet Sci Lett 59(1):61–89
Piper JDA (1982b) A palaeomagnetic investigation of the Malvernian and Old Radnor Precambrian, Welsh Borderlands. Geol J 17(2):69–88
Piper JDA, Stearn JEF (1975) Paleomagnetism of the Breiden Hill (Palaeozoic) inlier, Welsh Borderlands. Geophys J R Astron Soc 43(3):1013–1016

Piper JDA, McCook AS, Watkins KP, Brown GC, Morris WA (1978) Palaeomagnetism and chronology of Caledonian igneous episodes in the Cross Fell Inlier and Northern Lake District. Geol J 13(1):73–92

Pisarevsky SA, Iosifidi AG (1981) New paleomagnetic data on the paleogeographic situation and deformation of the north-eastern areas surrounding the Siberian platform in the late Paleozoic and Mesozoic. In: Paleomagnet i voprosy paleogeogr. Trudy VNIGRI, Leningrad, pp 56–66

Pogarskaya IA, Goncharov GI, Nazarov K (1976) Geomagnetic transition and paleosecular variations in the late Cretaceous. In: Glavnoye geomagn pole i problemy paleomagn III. Izd Inst Fisiki Zemli i Inst Zemn Magnet Acad Sci USSR, Moscow, p 76

Pogarskaya IA, Komissarova RA (1977) Paleomagnetism of the middle Paleozoic of the Russian and Siberian Platforms. In: Paleomagn i vopr tectoniki plit. Trudy VNIGRI, N 394, Leningrad, pp 63–107

Poore RZ, Tauxe L, Percival SF, La Breque JL (1982) Late Eocene-Oligocene magnetostratigraphy and biostratigraphy at South Atlantic DSDP Site 522. Geology 10(10):508–511

Poorter RPE (1981) Precambrian palaeomagnetism of Europe and position of the Balto-Russian plate relative to Laurentia. In: Kroner A (ed) Precambrian plate tectonics, chapter 24. Elsevier, Amsterdam, pp 599–622

Popov VI (1947) Paleomagnetic studies with reference to application of the magnetometric method. Sov Geol 25:21–28

Pospelova GA (1971) Anormalous behavior of the geomagnetic field in Pliocene-Pleistocene with reference to data of Paleomagnetic studies of Ob' plateau area. Geol i Geofiz 6:117–122

Pospelova GA (1976a) Paleomagnetic scale of the Jurassic and early Cretaceous time. In: Paleomagn mezozoya i kaynozoya Sibiri i Dal'nego Vostoka. Izd Sibir Otd Acad Sci USSR, Novosibirsk, pp 27–46

Pospelova GA (1976b) Fine time structure of the geomagnetic field in the late Cenozoic. In: Paleomagn mezozoya i kaynozoya Sibiri i Dal'nego Vost. Izd Sibirsk Otd Acad Sci USSR, Novosibirsk, pp 129–142

Pospelova GA, Gnibidenko ZN (1977) Paleomagnetism of the Upper Cenozoic strata in the south of the Western and Eastern Siberia. In: Geofiz metody v pozn zemnoy kory v Sibiri. Trudy Sib nauch-issled inst geol, geofiz i miner syr'ya, 249. Novosibirsk, pp 137–144

Pospelova GA, Kazansky AY, Larionova GY (1982) Reconnaissance paleomagnetic studies of the Miocene strata in the Zaysan depression. In: Paleomagn strat mezokaynoz otlozh Materialy Vsesoyuzn seminara. Naukova Dumka, Kiev, pp 56–58

Pospelova GA, Larionova GA, Anuchin AV (1967) Paleomagnetic studies of the Jurassic and Lower Cretaceous sediments of Siberia. Geol Geofiz 9:3–15

Remizovsky RI (1982a) Paleomagnetic stratigraphy of the Neogenic strata of Kamchatka. In: Metody paleomagn v reshenii geol zadach na prim Del'nego Vost Mat I Dal'nevost semin po paleomagn i magn por. Izd DV Nauch Tsentra Acad Sci USSR, Vladivostok, pp 30–42

Remizovsky VI (1982b) Paleomagnetism of the Dagin Formation strata (northern Sakhalin). In: Metody paleomagn v reshenii geol zadach na prim Dal'nego Vost Mater I Dal'nevost semin po paleomagn i magn por. Izd DV Nauch Tsentra Acad Sci USSR, Vladivostok, pp 43–49

Rikitake T (1966) Electromagnetism and the Earth Interior. Elsevier, Amsterdam, 331 pp

Rochester MG, Jacobs JA, Smylie DE, Chong KF (1975) Can precession power the geomagnetic dynamo? Geophys J R Astron Soc 43(3):661–678

Rodionov VP (1966) On the dipole character of the Earth magnetic field in the late Cambrian and Ordovician in the southern part of the Siberian platform. Geol i Geofis 1:94–101

Rodionov VP (1969) Natural remanent magnetization of sedimentary rocks in the regions of trappean activity. In: Magnetism gornykh porod i paleomagnetism. Izd Inst Fisiki Zemly Acad Sci USSR, Moscow, pp 152–156

Rodionov VP (183) Paleomagnetic data on the interrelation of Siberian and Russian platforms during the late Precambrian and Paleozoic. In: Paleomagnetizm verkhnego dokembriya SSSR. Trudy VNIGRI, Leningrad, pp 98–107

Rodionov VP (1984) Paleomagnetism of Upper Precambrian and Lower Paleozoic of Udzha River region. In: Paleomagnitn metody v stratigraphii. Trudy VNIGRI, Leningrad, pp 18–29

Rodionov VP, Osipova EP (1973) Paleomagnetic stratigraphy of the Lower Paleozoic in the Siberian platform. In: Materialy IX konf po vopr postoyan geomagn polya, magn gornykh porod i paleomagn III. Izd Acad Sci Azerb SSR, Baku, pp 133–135

Rodionov VP, Sidorova EP (1963) Paleomagnetic studies of the Upper Cambrian, Ordovician and Lower Silurian sections in the south of the Siberian platform. In: Paleomagn strat issled. Trudy VNIGRI, N 204. Gostoptekhizdat, Leningrad, pp 50–68

Rodionov VP, Osipova EP, Slautsitais IP (1977) Paleomagnetic data on the horizontal tectonic movement in the west of the Werkhoyansk fold area. In: Paleomagnetism i voprosy tectoniki plit. Trudy VNIGRI N 394, Leningrad, pp 44–62

Roy JL, Morris WA (1983) A review of palaeomagnetic results from the Carboniferous of North America: the concept of Carboniferous geomagnetic field markers. Earth Planet Sci Lett 65(1):167–181

Roy JL, Park JK (1974) The magnetization process of certain redbeds: vector analysis of chemical and thermal results. Can J Earth Sci 11(3):437–471

Roy JL, Reynolds J, Sanders E (1973) An alternating field demagnetizer for rock magnetism studies. Publ Earth Phys Branch Dept Energy Mines Resourc 44(3):37–45

Roy JL, Anderson P, Lapointe PL (1979) Paleomagnetic results from three rock units of New Brunswick and their bearing on the Lower Paleozoic tectonics of North America. Can J Earth Sci 16(6):1210–1227

Runcorn SK (1956) Paleomagnetic survey in Arizona and Utah: preliminary results. Geol Soc Am Bull 67(3):301–316

Rusakov OM (1969) The Earth magnetic field in mesozoic. Naukova Dumka, Kiev, 142 pp

Russel P, Van der Voo R, Bernard A, Bonhommet N (1983) Palaeomagnetism of late Precambrian volcanics and intrusives from the Armorican Massif, France. Geophys J R Astron Soc 75(1): 235–260

Ruzhentsev SV, Sokolov SD (1983) Tectonics of the North-Nuratin suture. Geotectonica 3:63–75

Ryabushkin PK, Pevzner MA (1978) Statistics of widely dispersed directions in the NRM studies of moraines. Izv Acad Sci Ser Fisika Zemli 12:103–106

Rzhevsky YS (1966) Results of using paleomagnetic method for identification of shear movements in the Tajik depression. Izv Acad Sci USSR ser Fisika Zemli 8:101–108

Rzhevsky YS (1973) Results of paleomagnetic investigation of horizontal deformation of Nuranch anticline. Izv Acad Sci USSR Ser Fisika Zemli 5:111–113

Rzhevsky YS (1977) Palinspastic scheme of the Tajik depression according to paleomagnetic data. Dokl Acad Sci Tajik SSR 20(2):42–44

Rzhevsky YS, Khramov AN (1985) Palinspastic restorations for the South Tien Shan mountains and Tajik Depression based on paleomagnetic data. J Geodyn 2(2/3):119–126

Rzhevsky YS, Slautsitias IP (1976) Implication of the fold axis tilting data to correct determination of primary rock magnetization. Izv Acad Sci USSR Ser Fisika Zemli 4:92–95

Sager WW (1983) A Late Eocene paleomagnetic pole for the Pacific plate. Earth Planet Sci Lett 63(3):408–422

Sager WW, Keating BH (1984) Paleomagnetism of line Islands seamounts: evidence for Late Cretaceous and Early Tertiary volcanism. J Geophys Res B89(13):11135–11151

Sarwar G, De Jong KA (1979) Arcs, Oroclines, Syntaxes: the curvatures of mountain belts in Pakistan. In: Farah A, De Jong KA (eds) Geodynamics of Pakistan. Geol Surv Pakistan, Quetta, pp 341–349

Sasajima S (1981) Pre-Neogene paleomagnetism of Japanese islands (and vicinities). In: McElhinny MW, Valencio DA (eds) Paleoreconstruction of the continents. Geodynamics series, vol 2. Am Geophys Union, Washington DC, Boulder, pp 115–128

Schmidt A (1899) Mathematische Entwicklungen zur allgemeinen Theorie des Erdmagnetismus. Arch Dtsch Seewarte 12(3):1–29

Schmidt PW (1976) The non-uniqueness of the Australian Mesozoic palaeomagnetic pole position. Geophys J R Astron Soc 47(2):285–300

Schmidt PW, Embleton BJJ (1981) A geotectonic paradox: has the Earth expanded? Geophys J R Astron Soc 49(1):20–25

Schott JJ, Montigny R, Thuizat R (1981) Paleomagnetism and potassium-argon age of the Messejana Dike (Portugal and Spain): angular limitation to the rotation of the Iberian Peninsula since the Middle Jurassic. Earth Planet Sci Lett 53(3):457–470

Schult A, Guerriero DC (1979) Paleomagnetism of Mesozoic igneous rocks from the Maranhao Basin, Brazil and the time of oponing of the South Atlantic. Earth Planet Sci Lett 42(3): 427–436

Schult A, Soffel HC, Hussain G (1978) Paleomagnetism of Cretaceous Nubian sandstone, Egypt. J Geophys 44(4):333–340

Schwarz EJ, Symons DTA (1969) Geomagnetic intensity between 100 million and 2500 million years ago. Phys Earth Planet Interiors 2(1):11–18

Schwarz EJ, Coleman LC, Cattrol HM (1979) Paleomagnetic results from the Skaergard intrusion, East Greenland. Earth Planet Sci Lett 42(3):437–443

Sclater JC, Cox A (1970) Paleolatitudes from JOIDES deep-sea sediment cores. Nature 226(5242): 934–935

Scotese CR, Bambach RK, Barton C, Van der Voo R, Ziegler AM (1979) Paleozoic base maps. J Geol 87(3):217–277

Seguin MK (1977) Palaeomagnetism of Middle Ordovician volcanic rocks from Quebec. Phys Earth Planet Interior 15(4):363–373

Seguin MK (1983) Palaeomagnetism of some lithological units from the Lower Palaeozoic of the Vendèe, France. Tectonophysics 96(3/4):257–279

Shablinskaya NV (1977) New data on the global fault pattern in the platforms. Dokl Acad Sci USSR 257(5):1159–1162

Shashkanov VA, Metallova VV (1982) Non-heating methods of determination of the ancient geomagnetic field intensity. Izd Leningr Univ, Leningrad, 142 pp

Shashkanov VA, Metallova VV, Smirnov VV (1975) Determination of the ancient geomagnetic field intensity by method of stepwise remagnetization. Izv Acad Sci Ser Fisika Zemli 11:79–90

Shatsky NS (1955) About an origin of the Pachelma depression. Bull Mosk Ova Ispyt Prir Otd Geol 30(5):5–27

Shcherbakov VP, Shcherbakova VV (1977) On the calculation of the thermoremanent and anhysteretic magnetization of an ensemble of interacting single-domain grains. Izv Acad Sci USSR Ser Fisika Zemli 6:69–83

Sheinmann YM (1975) Some inductive regularities deriving from paleomagnetic measurements. Izv Acad Sci USSR Ser Fisika Zemli 1:66–75

Shevlyagin EV (1982) Geologic interpretation of paleomagnetic vectors of rocks. Izd Rostov Univ, Rostov-Don, 152 pp

Shmeleva AN (1963) On the methods of studies of the NRM aquisition in sediments. In: Paleomagn stratigr issled. Trudy VNIGRI N 204, Gostoptekhizdat, Leningrad, pp 30–37

Sholpo LE (1977) Application of rock magnetism for solving geologic tasks. Nedra, Leningrad, 272 pp

Sholpo LE, Mamedov SA (1969) Determination of direction of the rock primary magnetization. Izv Acad Sci USSR Ser Fisika Zemli 1:95–102

Shul'ts SS (1964) Studies of global jointing. In: Dokl sovet geol (IGC 22nd session) Problema 4. Nauka, Moscow, pp 147–153

Shul'ts SS (ed) (1973) Planetary fracture. LGU, Leningrad, 176 pp

Shult'ts SS (1979) Tectonics of the Earth's crust (with reference to analysis of the recent movements). Nedra, Leningrad, 272 pp

Sidorenkov ND (1982) The Earth's rotation irregularity and the pole motions. Priroda 4:82–91

Sidorova EP (1965) Paleomagnetic studies of the Sinian and Cambrian strata in the area of the Maya River. In: Nastoyashch i proshloye magn polya Zemli. Nauka, Moscow, pp 304–309

Sigurgeirsson T (1959) Direction of magnetization in Icelandic basalts. Adv Phys 6(22):240–246

Sinito AM, Valencio DA, Vilas JF (1979) Palaeomagnetism of a sequence of Upper Palaeozoic-Lower Mesozoic redbeds from Argentina. Geophys J R Astron Soc 58(2):237–247

Sinitsyn VM (1967) Introduction into paleoclimatology. Nedra, Leningrad, 232 pp

Slautsitais IP (1969) Method of redeposition in estimation of paleointensity of the geomagnetic field. In: Magnet gornykh porod i paleomagn. Inst Fisiki Zemli Acad Sci USSR, Moscow, pp 134–136

Slautsitais IP (1973) Some data on the nature of the R-N transition in the early Traissic strata of Mangyshlak. In: Postoyan magn pole Zemli, paleomagn i magn gornykh porod. Naukova Dumka, Kiev, pp 39–41

Slautsitais IP (1975) Stratigraphy of the Permian and Triassic strata of Mangyshlak based on paleomagnetic data. In: Kontinent krasnotsvetn otlozh permi i triasa. Izd Saratov Univ, Saratov, pp 39–41

Smith JD, Foster JH (1969) Geomagnetic reversal in Brunhes normal polariti epoch. Science 163 (3867):565–567

Smith G, Merril RT (1980) The origin of rotational remanent magnetization. Geophys J R Astron Soc 61(2):329–336

Smith PJ (1970) The intensity of the ancient geomagnetic field. A summary of conclusions. In: Palaeogeophysics, chapter 10. Academic Press, London, pp 79–90

Soffel H, Forster H, Becker H (1975) Preliminary polar wander path of Central Iran. J Geophys 41(5):541–543

Sorokhtin OG (1979a) Nature of global marine transgressions into continents. In: Okeanologiya. Geofis okeana, vol 2. Geodinamika. Nauka, Moscow, pp 262–265

Sorokhtin OG (1979b) Plate tectonics and forecasting of mineral deposits. In: Okeanologiya. Geofix okeana, vol 2. Geodinamika. Nauka, Moscow, pp 371–383

Stacey FD (1963) The physical theory of rock magnetism. Adv Phys 12:46–133

Stacey FD (1972) On the role of Brownian motion in the control of detrital remanent magnetization of sediments. Pure Appl Geophys 98(1):139–145

Stacey FD, Banerjee SK (1974) The physical principles of rock magnetism. Elsevier, London, 195 pp

Steiner MB, Helsley CE (1975) Late Jurassic magnetic polarity sequence. Earth Planet Sci Lett 27(1):108–112

Stephenson RW (1976) A study of rotational remanent magnetization. Geophys J R Astron Soc 47(2):363–373

Stevan HG, Mauk FJ, Van der Voo R (1977) Paleomagnetism of the upper Keweenawan sediments: the Nonesuch shale and Freda Sanstone. Can J Earth Sci 14(5):1128–1138

Stott PM, Stacey FD (1960) Magnetostriction and paleomagnetism of igneous rocks. J Geophys Res 65(8):2419–2424

Stovas MV (1955) Critical parallels of the Earth's ellipsoid and geotectonics. Izv VGO 87(6): 561–563

Stovas MV (1975) Selected papers, vol 1. Naukova Dumka, Kiev, 155 pp

Strakhov NM (1949) On periodicity and irreversibility of sedimentation processes in the Earth's evolution. Izv Acad Sci USSR Ser Geol 6:70–111

Strakhov NM (1962) Theoretical principles of lithogenesis, vol 2. Izd Acad Sci USSR, Moscow, 574 pp

Strakhov NM (1963) Types of lithogenesis and their evolution in the Earth history. Gosgeoltekhizdat, Moscow, 535 pp

Tahirkheli RAK, Mattauer M, Proust F, Tapponnier P (1979) The india-Eurasia suture zone in Northern Pakistan: synthesis and interpretation of recent data at plate scale. In: Farah A, De Jong KA (eds) Geodynamics of Pakistan. Geol Surv Pakistan, Quetta, pp 125–130

Tarling DH (1971) Principles and applications of palaeomagnetism. Chapman and Hall, London, 164 pp

Tarling DH, Mitchell JG (1976) Revised Cenozoic polarity time scale. Geology 4(3):133–136

Tarling DH (1983) Palaeomagnetism: principles and applications in geology, geophysics and archaeology. N.Y., Methuen Inc, 379 pp

Thellier E (1941) Sur les propriétés d l'aimantation thermoremanente des terres cuites. CR Acad Sci Paris 213(26):1019–1022

Thellier E, Thellier O (1959) Sur l'intensité du champ magnetique terrestre dans le passé historique et géologique. Ann Géophys 15(3):285–377

Theyer F, Hammond SR (1974) Cenozoic magnetic time scale in deep-sea cores: completion of the Neogene. Geology 2(10):487–492

Thompson R (1978) European paleomagnetic secular variation 13 000 B.P. Pol Arch Hydrobiol 25 (1–2):413–418

Thompson R, Mitchell JG (1972) Palaeomagnetic and radiometric evidence for the age of the lower boundary of the Kiaman magnetic interval in the South America. Geophys J R Astron Soc 27(2):207–214

Tillo AA (1887) Distribution of continental centres on the Earth's surface. Izv RGO 23(6):750–753

Tosha T (1982) Paleomagnetic measurement of the mijako Group in the Kitakami massif, Northeast Japan. Rock Magn Paleogeophys 9:56–60

Tretyak AN (1973) Results of paleomagnetic studies of the Neogene sedimentary complexes in the Alpine geosynclinal area of the Ukraine. Geofiz Sbornik 51:66–68

Tretyak AN (1983) Natural remanent magnetization and the question of stratification of sedimentary strata. Naukova Dumka, Kiev, 256 pp

Tretyak AN, Molostovsky EA, Veselov AA, Volok ZE (1980) Preliminary results of paleomagnetic studies of the Maykop Group in the Crimea-Caucasus area. Geofiz Zhurn 2(3):106–109

Trubikhin VM (1977) Paleomagnetism and stratigraphy of the Akchagyl strata of the western Turkmenia. Trudy Geol Inst Acad Sci USSR, N 301, 79 pp

Trukhin VI, Zhilyaeva VA, Katerenchuk AV, Savrasov DI, Safroshkin VYu (1984a) Magnetism of rocks from the kimberlite pipes of Yakutia. Izv Acad Sci USSR Ser Fisika Zemli 9:57–70

Trukhin VI, Zhilyaeva VA, Savrasov DI, Safroshkin VY, Bubnov AV (1984b) Self-reversal of thermoremanent magnetization of rocks in the kimberlite pipes of Yakutia. Izv Acad Sci USSR Ser Fisika Zemli 11:78–89

Tsaregradsky VA (1963) On deformation of the Earth's crust. In: Problemy planet geol. Gosgeoltekhizdat, Moscow, pp 149–221

Tsirel' VS, Burtsev YA, Rose EN, Ysonov PG (1983) Status in the actual equipment. In: Sovremen sostoyan issled v oblasti geomagnet. Nauka, Moscow, pp 207–224

Tsykora VN, Mikhailova NP, Glevasskaya AM (1973) Determination of the ancient geomagnetic field intensity according to effusive rock studies. In: Materialy IX konf po vopr psotoyan geomgn polya, magn gornykh porod i paleomagn III. Izd Acad Sci Azerb SSR, Baku, pp 151–152

Turner P, Archer R (1975) Magnetization history of lower Old Red Sandstones from Gamrie Outlier, Scotland. Earth Planet Sci Lett 27(2):240–250

Turner P, Ixer RA (1977) Diagenetic development of unstable and stable magnetization in the St. Bees Sandstone (Triassic) of northern England. Earth Planet Sci Lett 34(1):113–124

Tursunov BN (1976) The first paleomagnetic scale of Carboniferous and Permian for the Southern Tien Shan. In: Glavnoye geomagn pole i problemy paleomagn III. Izd Inst Fisiki Zemli i Inst Zemn Magnet Acad Sci USSR, Moscow, p 94

Uspenskaya NY, Tauson NN (1972) Oil and gas bearing aboard provinces and areas. Nedra, Moscow, 296 pp

Vadkovsky VN, Gurary GZ, Mamikonyan MR (1980) Analysis of a change in the geomagnetic field polarity. Izv Acad Sci USSR Ser Fisika Zemli 7:55–69

Vagabova RI, Vorob'yeva GI, Ismail-Zade TA, Mamedzade RN, Ragimova ZA (1973) Paleomagnetic correlation of Cretaceous deposits in Azerbaijan. In: Materialy IX konf po vopr postoyan geomagn polya, magn gornykh porod i paleomagn III. Izd Acad Sci Azerb SSR, Baku, pp 28–31

Valdiya KS (1984) Tectonics of the folded fringe of the Indian platform. In: 27th Int Geol Congr, vol 5. Tectonics of Asia, Nauka, Moscow, pp 89–112

Valencio DA, Mendia JE, Vilas JFA (1975) Paleomagnetism and K-Ar ages of Triassic igneous rocks from the Ischigualasto-Ischichuca basin and Puesto Viejo formation, Argentina. Earth Planet Sci Lett 26(3):319–330

Valencio DA, Mendia GE, Giutici A, Cascon YO (1977a) Paleomagnetism of the Cretaceous Pirgua Subgua subgroup (Argentina) and age of the opening of the South Atlantic. Geophys J R Astron Soc 51(1):47–58

Valencio DA, Vilas JFA. Mendia JE (1977b) Palaeomagnetism of a sequence of bedbeds of the Middle and Upper sections of Paganzo group (Argentina) and the correlation of Upper Palaeozoic-Lower Mesozoic rocks. Geophys J R Astron Soc 51(1):59–74

Valencio DA, Rocha-Campos AG, Pacca IG (1978) Paleomagnetism of the Cassanji series ("Karoo system"), Angola. Ann Acad Brasil Cienc 50(3):353–364

Vandenberg J (1979) Palaeomagnetic data from the Western Mediterranean: a review. Geol Mijnbouw 58(2):161–174

Vandenberg J, Klootwijk CT, Wonders AAN (1978) Late Mesozoic and Cenozoic movements of Italian peninsula: new paleomagnetic data from Umbria. Geol Soc Am Bull 89(1):133–150

Van der Voo R (1967) The rotation of Spain: paleomagnetic evidence from the Spanish Meseta. Palaeogeogr Palaeoclim Palaeoecol 3(4):393–416

Van der Voo R (1982) Pre-Mesozoic paleomagnetism and plate tectonics. Ann Rev Earth Planet Sci, vol 10. Palo Alto, Valif, pp 191–220

Van der Voo R (1983) Paleomagnetic constraints on the assembly of the Old Red Continent. Tectonophysics 91(3–4):271–283

Van der Voo R, Channell JET (1980) Palaeomagnetism in orogenic belts. Rev Geophys Space Phys 18:455–481

Van der Voo R, French AN, French RB (1979) A palaeomagnetic pole position from the folded Upper Devonian Catskill redbeds and its tectonic implications. Geology 7(7):345–348

Van der Voo R, Johnson RJ, Perroud H (1984) A Caledonian to Hercynian history of collisions in the North American Appalachians of the basis of paleomagnetic data. In: 27th Int Geol Congr, vol 8. Geophysics. Najka, Moscow, pp 175–184

Van der Voo R, Watts DR (1978) Paleomagnetic results from igneous and sedimentary rocks from the Michigan basin borehole. J Geophys Res B83(12):5844–5848

Van Hilten D (1963) Palaeomagnetic indications of an increase in the Earth's radius. Nature 200(4913):1277–1279

Van Zijl ISV, Graham KWT, Hales A (1962) The paleomagnetism of the Stormberg lavas of South Africa. Geophys J R Astron Soc 7(2):169–182

Velchev Z, Mavrudchiev B, Kamenoev B (1976) Deep structure and late Alpine magmatic activity in southeast Bulgaria. Godishnic Sophia Univ Geol-geogr Facult 1974/1975, t 61, Sofia, pp 249–294

Vereshchagin VA, Nalivkin VD (ed) (1972) Atlas of lithologic-paleogeographic maps of the USSR. A concise explanatory note. Nedra, Leningrad, 64 pp

Verhoogen J (1956) Ionic ordering and self-reversal of magnetization in impure magnetites. J Geophys Res 61(2):201–210

Verhoogen J (1962) Oxidation of iron-titanium oxides in igneous rocks. J Geol 70(2):168–181

Veronnet A (1912) Rotation de l'ellipsoide heterogene et figure exacte de la Terre. J Matem Pures Appl 8:22–35

Verosub KL (1977) Depositional and post-depositional processes in the magnetization of sediments. Rev Geophys Space Phys 15(2):129–143

Verosub KL, Banerjee SK (1977) Geomagnetic excursions and their palaeomagnetic record. Rev Geophys Space Phys 15(2):145–155

Vilas JFA (1976) Palaeomagnetism of the Lower Cretaceous Sierra de Los Condores group, Cordoba Province, Argentina. Geophys J R Astron Soc 46(2):295–305

Vilas JFA, Valencio DA (1970) The recurrent Mesozoic drift of South America and Africa. Earth Planet Sci Lett 7(5):441–444

Vincenz SA, Dasgupta SN (1978) Paleomagnetic study of some cretaceous and tertiary rocks on Hispaniola. Pure Appl Geophys 116(6):1200–1210

Vine FJ, Mattews DH (1963) Magnetic anomalies over oceanic ridges. Nature 199(4897):947–949

Visotsky IV (ed) (1976) References on the oil and gas fields in the world. Nedra, Moscow, vol 1, 600 pp; vol 2, 584 pp

Vlasov AY, Popova AV (1963) Pole position in Permian, Triassic and Cretaceous according to paleomagnetic data from sedimentary rocks of Maritime area (Far East). In: Magnetism gornykh porod i paleomagnetism. Sib Otd Acad Sci USSR, Krasnoyarsk, pp 333–339

Vlasov AY, Kovalenko GV, Tropin YD (1961) Effects of compaction of sediments experimentally deposited on the remanent magnetization. Izv Acad Sci USSR Ser Geofiz 8:1189–1192
Vonsovsky SV (1971) Magnetism. Nauka, Moscow, 1032 pp
Vorob'yeva GP, Mirzakhanov RG, Khalafov AA, Gasan-Zade FG, Malumyan LM (1981) Paleomagnetism of Cretaceous deposits in Azerbaijan. In: II Vsesoyuz s'yezd Postoyan geomagn pole, magn gornykh porod i paleomagn II. Izd Tibilisi Univ, Tbilisi, p 112
Voronov PS (1968) Outlines of the regularities in morphometry of the global-scale relief of the Earth. Nauka, Leningrad, 124 pp
Voronov PS, Gakkel YY, Sytinsky AD (1962) On a possible effect of rotational forces of the Earth on morphostructures of Arctic and Antarctic. In: Problemy Arktiki i Antarktiki. Izd Morskoy Transport, Leningrad, issue 10, pp 17–26
Ward MA (1963) On detecting changes in the Earth's radius. Geophys J R Astron Soc 8(2):217–225
Watkins ND (1972) Review of the development of the geomagnetic polarity time scale and discussion of prospects for its finer definition. Geol Soc Am Bull 83:551–574
Watson GS (1956) Analysis of dispersion on a sphere. Mon Notic R Astron Soc Geophys Suppl 7(1):153–159
Watson GS, Irving E (1957) Statistical methods in rock magnetism. Monthly Not R Astron Soc Geophys Suppl 7(6):289–299
Watts DR, Briden JC (1984) Palaeomagnetic signature of slow post-orogenic cooling of the north-east Highlands of Scotland recorded in the Newer Gabbros of Aberdeenshire. Geophys J R Astron Soc 77(3):775–788
Watts DR, Van der Voo R (1979) Paleomagnetic results from the Ordovician Moccasin Bays and Chapman Ridge formations of the Valley and Ridge province, eastern Tennessee. J Geophys Res B84(2):645–655
Watts DR, Van der Voo R, Reeve SC (1980a) Cambrian paleomagnetism of the Llano uplift, Texas. J Geophys Res B85(10):5316–5330
Watts DR, Van der Voo R, French RB (1980b) Paleomagnetic investigation of the Cambrian Waynesboro and Rome formations of the Valley and Ridge Province of the Appalachian Mountains. J Geophys Res B85(10):5331–5343
Weeks LG, Lewis CR (1952) Factors of sedimentary basin development that control oil occurrence. Am Assoc Petrol Geol Bull 36(11):23–61
Wegener A (1926) Palaogeographische Darstellung der Theorie der Kontinentalverschiebungen. In: Enzyklopädie der Erdkunde, Bd 8, Leipzig, pp 174–189
Wegener A (1929) Die Entstehung der Kontinente und Ozeane, 4. Aufl, Bd 66. Sammlung Wiss, Braunschweig, 231 pp
Wensink H (1982) Tectonic inferences of paleomagnetic data from some Mesozoic formations in Central Iran. Geophys J R Astron Soc 51(1):12–23
Wensink H (1983) Paleomagnetism of red beds of Early Devonian age from Central Iran. Earth Planet Sci Lett 63(2):325–334
Wensink H, Warekamp JC (1980) Paleomagnetism of basalts from Alborz: Iran part of Asia in the Cretaceous. Tectonophysics 68:113–129
Westphal M (1976) Paléomagnetisme des formations permiennes de Corse. Comparison avec La Sardaigne et l'Esterel. Bull Soc Géol France 18(5):1209–1215
Wilson RL (1962) The palaeomagnetism of baked contact rocks and reversals of the Earth's magnetic field. Geophys J R Astron Soc 7(2):194–202
Wilson RL, Lomax R (1972) Magnetic remanence related to slow rotation of ferromagnetic material in alternating magnetic fields. Geophys J R Astron Soc 30(3):295–303
Wilson RL, McElhinny MW (1974) Investigation of the largescale nature of the palaeomagnetic field over the past 25 million years. Geophys J R Astron Soc 39:570–586
Wilson RL, Watkins ND (1967) Correlation of petrology and natural magnetic polarity in Columbia Plateau basalts. Geophys J R Astron Soc 12(4):405–424
Yanovsky BM (1978) The Earth's magnetism. Izd Leningr Univ, Leningrad, 592 pp
Yanshin AL (1963) Basic features of tectonic structure and evolution of Eurasia. In: Tezisy dokl soveshchan po probl tekton v 1963. Izd Sci USSR, Moscow, pp 5–9

Yaskawa K, Nakajima T, Kawai N, Torii M, Natsuhara N, Horie S (1973) Palaeomagnetism of a core from Lake Biwa (I). J Geomagn Geoelect 25(4):447–474

Yasonov PG (1981) Methods of magnetic-mineralogic studies with reference to the Upper Permian and Lower Triassic red beds in the eastern part of the Russian plate. Synopsis of thesis. Izd Inst Fisiki Zemli Acad Sci USSR, Moscow, 19 pp

Yeroshkin AF (1980) Paleomagnetic characteristic of Neogene deposits of the north-western Fergana. In: Nauchn trudy Tashkent Univ, 618. Problemy geol, Tashkent, pp 80–86

Yukutake T, Tachinaka H (1968) The non-dipole part of the Earth's magnetic field. Bull Earthqu Res Inst 46:1027–1074

Zakharov SA (1964) The main problem of tectogenesis related with oil-and-gas exploration in the Tajik depression and seismotectonic zonation of the Southern Tajikistan. In: Problemy geologii Tajikistana. Izd Acad Sci Tajik SSR, pp 33–78

Zhilyaeva VA, Kolesnikov LV, Petrova GN (1970) On the partial self-reversal of magnetization in natural ferromagnetics of the Fe_2TiO_4-$MgTiO_4$ series. Izv Acad Sci USSR Ser Fisika Zemli 10:59–71

Zijderveld JDA (1967) A c demagnetization of rocks: analysis of results. In: Collinson DW, Creer KM (eds) Methods in palaeomagnetism. Elsevier, Amsterdam, pp 254–286

Zonenshain LP (1976) Reconstructions of the Paleozoic Oceans. In: Dreif Kontinentov. Nauka, Moscow, pp 28–71

Zotkevich IA (1966) Transformations of coordinates of the NRM vector in paleomagnetic studies. Izv Acad Sci USSR Ser Fisika Zemli 6:104–108

Zubakov VA, Kochegura VV (1976) Magnetochronologic scale of the Recent time (5 millions of years). Geomagnitn issled, 17. Sov Radio, Moscow, pp 37–44

Zubakov VA, Bogatina NV, Pisarevsky SA (1982) The Karangat horizon, its detailed division, stratigraphic range and age, the Black sea area. Dokl Acad Sci USSR 267(2):426–429

Zudin AN (1973) Anomalous magnetization of the Brunhes epoch. In: Materialy IX konf po vopr postoyan geomagn polya, magn gornykh porod i paleomagn III. Izd Acad Sci Azerb SSR, Baku, pp 79–80

Zvegintsev AG, Grankin PI, Bugayev MS (1974) Self-reserval of thermoremanent magnetization in synthetic solid solutions of hemoilmenite under pressure. Izv Acad Sci USSR Ser Fisika Zemli 8:47–54

Subject Index

MIX
Papier aus verantwortungsvollen Quellen
Paper from responsible sources
FSC® C105338

If you have any concerns about our products,
you can contact us on
ProductSafety@springernature.com

In case Publisher is established outside the EU,
the EU authorized representative is:
Springer Nature Customer Service Center GmbH
Europaplatz 3, 69115 Heidelberg, Germany

Printed by Libri Plureos GmbH
in Hamburg, Germany